Heterophase Network Polymers

Heterophase Network Polymers

Synthesis, Characterization and Properties

Edited by

Boris A. Rozenberg & Grigori M. Sigalov

Translated from Russian by Marina Z. Aldoshina and Yurii B. Scheck

London and New York

First published 2002
by Taylor & Francis
11 New Fetter Lane, London EC4P 4EE

Simultaneously published in the USA and Canada
by Taylor & Francis Inc,
29 West 35th Street, New York, NY 10001

Taylor & Francis is an imprint of the Taylor & Francis Group

Publisher's note

This book has been prepared from camera-ready-copy supplied by the authors

Printed and bound in Great Britain by The Cromwell Press, Trowbridge, Wiltshire

British Library Cataloguing in Publication Data
A catalogue record for this book is available from the British Library

Library of Congress Cataloging in Publication Data
A catalog record for this book has been requested

ISBN 0-415-28417-1

CONTENTS

PREFACE

Polymer science is a material- and technology-driven science. Material and technological breakthroughs stimulate fundamental and theoretical comprehension of experimental observations. The latter is a good background for further development. In the recent three decades, one of such breakthroughs has been utilization of cure reaction-induced microphase separation (CRIMPS) of initially homophase multicomponent thermosetting systems containing rubber or thermoplastic additives. This approach combines convenient processing with higher plasticity, fracture toughness, impact strength, resistance to dynamic loads, water resistance, and size stability of resulting heterophase network polymers compared to unmodified rigid and brittle glassy thermosetting polymers. This modification is often not accompanied by any essential decrease in the glass transition temperature, modulus of elasticity, and tensile strength.

To date, heterophase thermosetting polymer blends (in particular, epoxy-based thermosets) obtained by CRIMPS technology have become widely used as polymer matrices in advanced fiber-reinforced composites, structural adhesives, coatings, and high-performance materials. In addition to the effect of toughening, a heterophase structure also provides the basis for creating polymer-dispersed liquid crystals for different electrooptical devices, molding coatings, electronic packaging devices, and memory devices with low internal stresses.

This book is a collection of selected original papers on different aspects of synthesis, characterization, and properties of heterophase network polymers obtained by CRIMPS technology. These papers have been presented at annual workshops (1997–1999) held at the Institute of Problems of Chemical Physics, Russian Academy of Sciences (IPCP), within the framework of a joint project with the International Science and Technology Center (ISTC) on heterophase polymer networks and related materials.

The book is divided into five sections, each containing both novel theoretical and experimental contributions. All papers (except three) were submitted by the ISTC Project participants, so that all sections have much in common and deal with different aspects of CRIMPS.

Synthesis and thermodynamic characterization of some new well-defined modifiers with controlled solubility in epoxies are discussed in

Section I. The CRIMPS theory is further developed in Section II. The kinetics and mechanism of cure reactions and the effect of cure kinetics on CRIMPS are considered in Section III. Different aspects of the structure–properties relation are presented in Section IV. Section V considers some novel approaches to investigating CRIMPS processes by conventional experimental techniques. This section also contains two papers that describe new important techniques for determining (a) MWD of linear and network polymers in the bulk and (b) shrinkage- and heat-induced internal stresses in composites.

The editors are grateful to all of the contributors for concise presentation of their new and intriguing results, techniques, theories, and well-documented review papers. We hope that the multidisciplinary reports on the different aspects of formation and characterization of heterophase networks collected in this book will be interesting and instructive to a wide audience: researchers in the field of polymer chemistry and physics, chemical and mechanical engineering (both at universities and industrial companies), graduate and postgraduate students, etc.) and may be expected to stimulate further progress in this field.

We are also grateful to Marina Z. ALDOSHINA and Yurii B. SCHECK for translating the papers submitted in Russian and to Nikolai F. SURKOV for his help in preparation of the CRC.

On behalf of all Russian authors, participants of the ISTC Project, we are grateful to the Moscow Office of ISTC for fruitful cooperation, to the US government, and to the governments of the European Community for financial support. Without this cooperation and support, the book would never have appeared altogether.

Boris ROZENBERG
Grigori SIGALOV

Chernogolovka, Moscow Region
August, 2000

LIST OF CONTRIBUTORS

Alekseev A.K.
Ph.D., Leading research worker
Scientific and Industrial Corporation "Energy"
Korolev, Moscow Region
Russia

Badamshina E.R.
Ph.D., Head of the Laboratory of Polymer Binders
Institute of Problems of Chemical Physics
(IPCP), Russian Academy of Sciences
14 Institutskii Prospect
Chernogolovka, Moscow Region 142432
Russia

†Baturin S.M.
Professor, Director of the Institute of Problems of Chemical Physics
(IPCP), Russian Academy of Sciences
14 Institutskii Prospect
Chernogolovka, Moscow Region 142432
Russia

Bogdanova L.M.
Ph.D., Senior research worker
Laboratory of Physical Chemistry of Polymer Matrices
Institute of Problems of Chemical Physics
(IPCP), Russian Academy of Sciences
14 Institutskii Prospect
Chernogolovka, Moscow Region 142432
Russia

Brichkin S.B.
Ph.D., Senior research worker
Laboratory of Supramolecular Photochemistry
Institute of Problems of Chemical Physics
(IPCP), Russian Academy of Sciences
14 Institutskii Prospect
Chernogolovka, Moscow Region 142432
Russia

Chalykh A.E.
Professor, Head of the Laboratory of Electron Microscopy

Institute of Physical Chemistry
(IPC), Russian Academy of Sciences
31 Leninskii Prospect
Moscow V-312, 117915 GSP-1
Russia

Chiba T.
Research assistant
Department of Organic and Polymeric Materials
Tokyo Institute of Technology
Ookayama, Meguro-ku,
Tokyo 152-8552
Japan

Chiu H.-W.
Graduate student
Institute of Polymer Engineering
College of Polymer Science and Polymer Engineering
The University of Akron
Akron, OH 44325-0301
USA

Deyun E.V.
Ph.D., Senior research worker
Laboratory of Chemical Mechanics
Institute of Problems of Chemical Physics
(IPCP), Russian Academy of Sciences
14 Institutskii Prospect
Chernogolovka, Moscow Region 142432
Russia

Doronina I.V.
Junior research worker
Laboratory of Polymer Binders
Institute of Problems of Chemical Physics
(IPCP), Russian Academy of Sciences
14 Institutskii Prospect
Chernogolovka, Moscow Region 142432
Russia

Dzhavadyan E.A.
Ph.D., Senior research worker
Laboratory of Physical Chemistry of Polymer Matrices
Institute of Problems of Chemical Physics
(IPCP), Russian Academy of Sciences

14 Institutskii Prospect
Chernogolovka, Moscow Region 142432
Russia

Elizarova T.L.
Research worker
Laboratory of Photodynamic Processes
Institute of Problems of Chemical Physics
(IPCP), Russian Academy of Sciences
14 Institutskii Prospect
Chernogolovka, Moscow Region 142432
Russia

Erofeev L.N.
D.Sc., Head of Laboratory of NMR
Institute of Problems of Chemical Physics
(IPCP), Russian Academy of Sciences
14 Institutskii Prospect
Chernogolovka, Moscow Region 142432
Russia

Estrin Ya.I.
D.Sc., Leading research worker
Laboratory of Polymer Binders
Institute of Problems of Chemical Physics
(IPCP), Russian Academy of Sciences
14 Institutskii Prospect
Chernogolovka, Moscow Region 142432
Russia

Ganina L.V.
Ph.D., Senior research worker
Laboratory of Energetic Polymer Systems
Institute of Problems of Chemical Physics
(IPCP), Russian Academy of Sciences
14 Institutskii Prospect
Chernogolovka, Moscow Region 142432
Russia

Gerasimov V.K.
Ph.D., Senior research worker
Laboratory of Electron Microscopy
Institute of Physical Chemistry
(IPC), Russian Academy of Sciences
31 Leninskii Prospect

Moscow V-312, 117915 GSP-1
Russia

Gorbushina G.A.
Research worker
Laboratory of Polymer Binders
Institute of Problems of Chemical Physics
(IPCP), Russian Academy of Sciences
14 Institutskii Prospect
Chernogolovka, Moscow Region 142432
Russia

Grigorieva V.A.
Ph.D., Senior research worker
Laboratory of Polymer Binders
Institute of Problems of Chemical Physics
(IPCP), Russian Academy of Sciences
14 Institutskii Prospect
Chernogolovka, Moscow Region 142432
Russia

Grishchuk A.A.
Research worker
Laboratory of Polymer Binders
Institute of Problems of Chemical Physics
(IPCP), Russian Academy of Sciences
14 Institutskii Prospect
Chernogolovka, Moscow Region 142432
Russia

Inoue T.
Professor
Department of Organic and Polymeric Materials
Tokyo Institute of Technology
Ookayama, Meguro-ku,
Tokyo 152-8552
Japan

Irzhak V.I.
Professor, Head of the Laboratory of Structural Relaxation of Polymer Matrices
Institute of Problems of Chemical Physics
(IPCP), Russian Academy of Sciences
14 Institutskii Prospect
Chernogolovka, Moscow Region 142432
Russia

Ivanova A.N.
D.Sc., Leading research worker
Laboratory of Mathematical Physics
Institute of Problems of Chemical Physics
(IPCP), Russian Academy of Sciences
14 Institutskii Prospect
Chernogolovka, Moscow Region 142432
Russia

Jacob K.I.
Ph.D., Assistant professor
School of Textile and Fiber Engineering
801 First Drive
Georgia Institute of Technology
Atlanta GA 30332-0295
USA

Jenkins S.
Ph.D., Research scientist
Kimberly-Clark Corp.
1400 Holcomb Bridge Rd.
Roswell GA 30076
USA

Kasumova L.T.
Research worker
Laboratory of Polymer Binders
Institute of Problems of Chemical Physics
(IPCP), Russian Academy of Sciences
14 Institutskii Prospect
Chernogolovka, Moscow Region 142432
Russia

Khitrin A.K.
Ph.D., Head of the Group of Chemical Physics of Polymer Systems
Institute of Problems of Chemical Physics
(IPCP), Russian Academy of Sciences
14 Institutskii Prospect
Chernogolovka, Moscow Region 142432
Russia

Komratova V.V.
Research worker
Laboratory of Polymer Binders
Institute of Problems of Chemical Physics

(IPCP), Russian Academy of Sciences
14 Institutskii Prospect
Chernogolovka, Moscow Region 142432
Russia

Korotkov V.N.
D.Sc., Head of the Laboratory of Macrokinetics of the Processes of Polymer and Composite Formation
Institute of Problems of Chemical Physics
(IPCP), Russian Academy of Sciences
14 Institutskii Prospect
Chernogolovka, Moscow Region 142432
Russia

Krysyuk B.E.
Professor, Leading research worker
Group of Chemical Physics of Polymer Systems
Institute of Problems of Chemical Physics
(IPCP), Russian Academy of Sciences
14 Institutskii Prospect
Chernogolovka, Moscow Region 142432
Russia

Kumar S.
Professor
School of Textile and Fiber Engineering
801 First Drive
Georgia Institute of Technology
Atlanta GA 30332-0295
USA

Kuzaev A.I.
D.Sc., Chief research worker
Laboratory of Polymer Binders
Institute of Problems of Chemical Physics
(IPCP), Russian Academy of Sciences
14 Institutskii Prospect
Chernogolovka, Moscow Region 142432
Russia

Kyu T.
Professor
Institute of Polymer Engineering
College of Polymer Science and Polymer Engineering
The University of Akron

Akron, OH 44325-0301
USA

Lee J.-H.
Graduate student
Institute of Polymer Engineering
College of Polymer Science and Polymer Engineering
The University of Akron
Akron, OH 44325-0301
USA

Lodygina V.P.
Research worker
Laboratory of Polymer Binders
Institute of Problems of Chemical Physics
(IPCP), Russian Academy of Sciences
14 Institutskii Prospect
Chernogolovka, Moscow Region 142432
Russia

Lesnichaya V.A.
Ph.D., Senior research worker
Laboratory of Physical Chemistry of Polymer Matrices
Institute of Problems of Chemical Physics
(IPCP), Russian Academy of Sciences
14 Institutskii Prospect
Chernogolovka, Moscow Region 142432
Russia

Manevich L.I.
Professor, Head of the Division of Polymer Physics and Mechanics
N.N. Semjonov Institute of Chemical Physics
(ICP), Russian Academy of Sciences
4 Kosygin Street
Moscow V-334, 117977 GSP-1
Russia

Manevitch E.L.
Ph.D., Senior research worker
Laboratory of Physical Chemistry of Polymer Matrices
Institute of Problems of Chemical Physics
(IPCP), Russian Academy of Sciences
14 Institutskii Prospect
Chernogolovka, Moscow Region 142432
Russia

Mikhailov Yu.M.
D.Sc., Head of Laboratory of Energetic Polymer Systems
Institute of Problems of Chemical Physics
(IPCP), Russian Academy of Sciences
14 Institutskii Prospect
Chernogolovka, Moscow Region 142432
Russia

Nosov G.B.
Ph.D., Senior research worker
Laboratory of Physical Chemistry of Polymers
Nesmeyanov Institute of Organoelement Compounds
(INEOS), Russian Academy of Sciences
28 Vavilov Street
Moscow V-334, 117813 GSP-1
Russia

Novikov G.F.
D.Sc., Head of Laboratory of Photodynamic Processes
Institute of Problems of Chemical Physics
(IPCP), Russian Academy of Sciences
14 Institutskii Prospect
Chernogolovka, Moscow Region 142432
Russia

Pavlov G.A.
D.Sc., Senior research worker
Laboratory of Polymer Binders
Institute of Problems of Chemical Physics
(IPCP), Russian Academy of Sciences
14 Institutskii Prospect
Chernogolovka, Moscow Region 142432
Russia

Razumov V.F.
Professor, Head of Laboratory of Supramolecular Photochemistry
Institute of Problems of Chemical Physics
(IPCP), Russian Academy of Sciences
14 Institutskii Prospect
Chernogolovka, Moscow Region 142432
Russia

Rozenberg B.A.
Professor, Head of the Department of Polymer and Composite Materials
Institute of Problems of Chemical Physics

(IPCP), Russian Academy of Sciences
14 Institutskii Prospect
Chernogolovka, Moscow Region 142432
Russia

Shaginyan S.A.
Ph.D., Research worker
N.N. Semjonov Institute of Chemical Physics
(ICP), Russian Academy of Sciences
4 Kosygin Street
Moscow V-334, 117977 GSP-1
Russia

Shiryaev A.A.
Ph.D., Senior research worker
Laboratory of the Dynamics of Microheterogeneous Processes
Institute of Structural Macrokinetics and Problems of Materials
(ISMAN), Russian Academy of Sciences
Chernogolovka, Moscow Region 142432
Russia

Sigalov G.M.
Ph.D., Senior research worker
Laboratory of Physical Chemistry of Polymer Matrices
Institute of Problems of Chemical Physics
(IPCP), Russian Academy of Sciences
14 Institutskii Prospect
Chernogolovka, Moscow Region 142432
Russia

Smirnov L.P.
Professor, Head of Laboratory of Chemical Mechanics
Institute of Problems of Chemical Physics
(IPCP), Russian Academy of Sciences
14 Institutskii Prospect
Chernogolovka, Moscow Region 142432
Russia

Smirnov V.S.
Research worker
Laboratory of Energetic Polymer Systems
Institute of Problems of Chemical Physics
(IPCP), Russian Academy of Sciences
14 Institutskii Prospect
Chernogolovka, Moscow Region 142432
Russia

Smirnov Yu.N.
Ph.D., Senior research worker
Laboratory of Physical Chemistry of Polymer Matrices
Institute of Problems of Chemical Physics
(IPCP), Russian Academy of Sciences
14 Institutskii Prospect
Chernogolovka, Moscow Region 142432
Russia

Sonin A.S.
Professor, Leader of the Group of Liquid Crystals
Laboratory of Physical Chemistry of Polymers
Nesmeyanov Institute of Organoelement Compounds
(INEOS), Russian Academy of Sciences
28 Vavilov Street
Moscow V-334, 117813 GSP-1
Russia

Stovbun E.V.
Ph.D., Senior research worker
Laboratory of Polymer Binders
Institute of Problems of Chemical Physics
(IPCP), Russian Academy of Sciences
14 Institutskii Prospect
Chernogolovka, Moscow Region 142432
Russia

Tarasov V.P.
Ph.D., Senior research worker
Laboratory of NMR
Institute of Problems of Chemical Physics
(IPCP), Russian Academy of Sciences
14 Institutskii Prospect
Chernogolovka, Moscow Region 142432
Russia

Veretennikov A.V.
Graduate student
Laboratory of Supramolecular Photochemistry
Institute of Problems of Chemical Physics
(IPCP), Russian Academy of Sciences
14 Institutskii Prospect
Chernogolovka, Moscow Region 142432
Russia

Yang Y.
Ph.D., Research worker
Department of Organic and Polymeric Materials
Tokyo Institute of Technology
Ookayama, Meguro-ku,
Tokyo 152-8552
Japan

Section I

SYNTHESIS AND CHARACTERIZATION OF MODIFIERS

Chapter 1

Synthesis of Well-Defined Bifunctional Oligobutadienes as Initiated by Dilithium Alkanes Soluble in Hydrocarbons

Aleksandr A. GRISHCHUK, Larisa T. KASUMOVA, and Yakov I. ESTRIN*

Institute of Problems of Chemical Physics, Russian Academy of Sciences, Chernogolovka, Moscow Region, 142432 Russia

*e-mail: estrin@icp.ac.ru

ABSTRACT

Disecondary dilithium alkanes of general formula $CH_3CHLi(CH_2)_nCHLiCH_3$ (DSDLAs) are efficient initiators (soluble in hydrocarbons) for 'living' polymerization of conjugated dienes. These compounds are prepared by reaction between lithium and appropriate dichloroalkanes in hydrocarbons under mild conditions. Polymerization of butadiene in the presence of initiators leads (under some conditions) to formation of oligomers with narrow molecular weight distribution (polydispersity indices below 1.2) over a wide range of molecular weights (3000–12 000). Upon treatment with functionalizing agents, the functionality of synthesized oligomers may attain a value about 1.9. The amount of 1,4-units (both *cis* and *trans*) in the oligomers was found to be above 80%.

INTRODUCTION

Living polymerization [1, 2] in the presence of lithiumorganic compounds was found [3, 4] to yield (in hydrocarbons) polyisoprene and polybutadiene with a high content of 1,4-units. It could also be expected that using dilithium alkanes as initiators would allow preparing oligodienes with two functional groups, with any desired molecular weight (MW), and narrow molecular weight distribution (MWD) typical of living polymers with largely a 1,4-microstructure. Upon processing a monomer containing the C–Li end groups with appropriate reagents, we may in principle synthesize polymers with different functional end groups.

According to Eberly [5], α,ω-dilithium alkanes, $Li(CH_2)_nLi$ ($n > 2$), synthesized through reaction of Li with dihalides are insoluble in hydrocarbons [5], which makes them inapplicable to synthesis of bifunctional oligodienes with a desired MW and narrow MWD because of slow initiation (see also [6]). Further efforts aimed at synthesizing efficient dilithium initiators soluble in hydrocarbons (e.g., see review [7]). But the synthesis of dilithium initiators in hydrocarbon media has not been performed so far. The most of suggested initiators are either synthesized in the presence of polar molecules (largely electron-donating) or exhibit low (about 10^{-3} M) solubility in hydrocarbons. Moreover, the precursors are inaccessible and can be prepared only by using multistage synthetic procedures.

Previously, we have found [8] that disecondary dichloroalkanes readily react with Li in hydrocarbon solvents and yield concentrated solutions of respective dilithium alkanes. In contrast to similar reaction of Li with diprimary dihalides that require mechanical replenishment of the surface (ball mill, ultrasound), the reaction of disecondary derivatives proceeds in a standard stirred reactor under mild conditions.

In this paper, we will discuss the synthesis, characteristics, and applications of soluble dilithium alkanes. These can be used as initiators of polymerization for conjugated dienes (in particular, butadiene) that yield bifunctional oligomers with a high degree of molecular homogeneity.

EXPERIMENTAL

Reagents

All solvents (except gasoline) were treated as described in [9], purified in a 1-m packed column (to 0.01% purity, by GLC), and stored over CaH_2. Before use, solvents were placed in an Ar flow over CaH_2, repeatedly degassed, frozen out into a vessel containing small amount of Li-alkane and then into a degassed vessel for storage. In 3-l and 50-l reactors, we used dearomatized gasoline or benzene as solvents. These were subjected to azeotropic drying in fractionating columns (to a moisture content below 0.002%), purged with Ar, and pumped into batchers (preliminary purged with Ar).

Butadiene was rectified over triisobutylaluminum, the fraction containing the C_3 hydrocarbons being eliminated. Rectified butadiene was condensed into a vessel with Li-alkane. After thawing, monomer was stirred until coloration typical of living polymer. Then it was rapidly frozen out into a degassed batcher.

Isoprene was rectified over CaH_2 and then handled just as butadiene.

Synthesis of Dichlorides

Disecondary chloroalkanes—2,5-dichlorohexane (DCHx), 2,6-di-chloroheptane (DCHp), and 2,7-dichlooctane (DCO)—were prepared by hydrochlorination of hexadiene-1,5 [10], heptadiene-1,6, and octadiene-1,7*, respectively. Since we failed to find in the literature any prescription for synthesis and properties of DCHp and DCO, these dichlorides (after rectification in vacuum) are characterized in Table 1.

Table 1. Some physical parameters of disecondary dichloroalkanes

Dichloroalkane	T_b, K/mm Hg	n_D^{25}	d_4^{20}
2,5-dichlorohexane	330–331/10	1.4465	1.0440
2,6-dichloroheptane	307.5–308/1	1.4500	1.0317
2,7-dichloroctane	340/1	1.4529	1.0081

*Heptadiene-1,6 and octadiene-1,7 were synthesized by V.Sh. Feldblum and coworkers at the Yaroslavl' Institute of Monomers for Synthetic Rubber.

Synthesis and Analysis of Dilithium Alkanes

Dilithium alkanes were synthesized (in Ar) in an evacuated glass reactor with a magnetic stirrer or in a metallic reactor with a mechanical stirrer. The volume of metallic reactors ranged between 1.0 and 10 dm^3. Li used in our experiments contained up to 2.5% Na. Reaction was carried out at 281–285 K for at least 8 h upon intense stirring, up to complete consumption of chlorides. After short sedimentation, lithium chloride and excess metal were filtered out in Ar.

Hydrolyzate formed in dilithium alkane solutions was analyzed chromatographically (343 K, katharometer, He as a carrier gas) by using a Chrom 4 or Chrom 5 apparatus (columns 3 mm in diameter and 3700 mm long filled with Celite impregnated with a 3% $AgNO_3$ solution in polyethylene oxide). In similar way, we determined the amount of alkanes and alkenes in a hydrolyzate of dilithium alkanes formed during storage of the solutions at room temperature (calibrations alkene/heptane and alkene/alkane/heptane).

Synthesized dilithium alkanes had a structure of starting dichlorides because a product isolated by carbonylation of 2,5-dilithium hexane was found to contain two forms of 1,4-dimethyladipic acid, *meso*-form (m.p. 417 K) and *d,d-l,l*-racemate (m.p. 344–345 K). The published values are 415 K and 344–345 K, respectively [11].

Synthesis of Oligomers

Dienes were polymerized in an evacuated glass reactor with a magnetic stirrer or in a metallic reactor with a mechanical stirrer. The volume of metallic reactors (water-cooled) was 1.0, 3.0 or 50 dm^3. The reactors were equipped with pressure/temperature gauges and glass window for visual monitoring of the system behavior. The metallic reactors were preheated (263 K) in vacuum, cooled, and flushed with Ar. After filling with solvent, a solution of dilithium alkane was added, the mixture was heated to 303–305 K, and an aliquot amount of butadiene was admitted. The initial concentration of monomer was 20–25 wt %. Polymerization was carried out at 303–305 K. When the pressure ceased falling down ethylene oxide (2:1 with respect to Li) was added, which resulted in formation of dense gel.

Oligomer solution was neutralized by treatment with water (up to 100 ml/mol Li) and CO_2. Insoluble lithium carbonate was filtered out [12].

Characterization of Oligomers

MWD was determined by GPL with a Waters chromatograph [13] (set of styrogel columns with a pore size of $3\cdot10^2$, $3\cdot10^3$, and $3\cdot10^4$ nm, THF as eluent, refractometric detection) or Milikhrom chromatograph [14] (Lichrospher Si 100 sorbent, dioxane as eluent, UV detection at 210 nm).

The content of OH groups was determined by isocyanate technique [15]. The microstructure was assessed from the intensity of IR bands (UR-20 spectrophotometer) at 970 (1,4-*trans*), 3010 (1,4-*cis*), and 910 cm^{-1} (1,2-units) in CCl_4 [16]. The distribution of functionality type (DFT) was determined as described elsewhere [17].

Kinetic Measurements

In case of 3-l reactors, sampling (into a flask with ethanol flushed with Ar) was performed through the lower faucet (at elevated pressure in reactor). After neutralization, oligomers were dried in vacuum to constant weight. Calorimetric analysis was performed as described elsewhere [18, 19].

RESULTS AND DISCUSSION

Compared to primary ones, secondary lithium alkanes are more efficient initiators for polymerization of conjugated dienes and styrene in hydrocarbons [20, 21] probably owing to lower degree of their association [22]. In this context, we expected that disecondary dilithium alkanes might turn out more reactive than the above mentioned α,ω-dilithium alkanes. We found that 2,5-dichlorohexane vigorously reacts (under normal conditions and without mechanical replenishment of metal surface) with Li suspension both in benzene and aliphatic hydrocarbons and yields a solution of 2,5-dilithium hexane (DLHx):

$$CH_3CHCl(CH_2)_2CHClCH_3 + 4Li \rightarrow CH_3CHLi(CH_2)_2CHLiCH_3 + 2LiCl. \quad (1)$$

Detection of hexane in hydrolyzed solutions and isomers of 1,4-dimethyladipic acid upon carboxylation confirms formation of a dilithium derivative. Alkalinity of solutions is in agreement (within 1–2 %) with a measured (by GLC) hexane content of hydrolyzate. This indicates the absence of other lithium-containing compounds in solution. Meanwhile, the yield of DLHx strongly depends on process conditions and ranges between 30 and 90 % theoretical. This indicates the occurrence of side reactions that do not yield lithiumorganic products (e.g., elimination of LiCl from intermediate monolithium chloride yielding substituted cycloalkanes).

The DLHx concentration may attain a value above 0.5 M, solutions are clear and virtually colorless. Evaporation of solvent in vacuum (without heating) leaves a white solid that decomposes without melting around 353 K (heating rate 1 deg/min). Dried DLHx is insoluble in hydrocarbons but becomes readily soluble upon addition of monomers (butadiene or isoprene). During storage of DLHx solutions (at 258 K) for 7–10 days, half amount of a compound undergoes irreversible crystallization. But again, the precipitate (apparently the *meso*-form) becomes soluble in the presence of monomers. Prolonged (several months) storage at 258 K did not result in other visible changes.

At room temperature, DLHx solutions gradually decompose which is accompanied by color change (up to reddish-brown) and precipitation. Hydrolysis of partially decomposed solution yields (besides hexane) hexenes and hydrogen. This implies that the decomposition involves largely elimination of LiH, while coloration may be attributed to formation of polylithium compounds upon metallation of lithium hydrocarbons.

Behavior of 2,6-dichloheptane is similar to that of 2,6-dichlorohexane. Its reaction with Li in hexane yields 2,6-dilthium heptane (DLHp). The reaction yield is 80–90 %. In case of 2,7-dichlorooctane, the situation is somewhat different. High-purity 2,7-dichlorooctane is difficult to prepare (its rectification in vacuum is accompanied by marked decomposition and isomerization). Our first experiments on the synthesis of dilithium octane (DLO) were carried out with a product that contained about 20 % of structural isomer (perhaps 2,6-dichloroctane). The yield of dilithium derivative was found to be 85 %. Meanwhile, chromatographically pure 2,7-dichlorooctane does not react with Li in heptane. In benzene, the reaction proceeds slowly at a yield below 40 % theoretical. Lower solubility of pure DLO may be associated with higher symmetry of its molecules. In any case, dissolution of product replenishes the metal surface, thus favoring reaction of dichloride with lithium. Contrary, poor solubility leads to mechanical screening of the metal surface. It may be assumed that DSDLA solutions in hydrocarbons are metastable. The lower the molecular symmetry and higher the number of structural stereoisomers, the lower crystallization rate can be expected for these solutions.

Since the parameters of all three DSDLAs as initiators are essentially the same, our further discussion will concern only the more accessible DLHx.

The early experiments with DLHx showed that this initiator ensures preparation of oligomers with a MW close to theoretical, narrow MWD ($\overline{M}_w / \overline{M}_n \approx 1.1$), and functionality close to 2.0 (with respect to OH groups after treatment of living polymer with ethylene oxide). For oligomers with MW below 3000, the content of 1.4-units is below 80 %. For oligomers with higher MW, this parameter attains a value of 85–90 % (Table 2).

The synthesis of bifunctional high-uniformity oligodienes assisted by accessible initiators might be considered as a resolved problem. At least, similar results are normally regarded as satisfactory. However, more detailed analysis of the synthetic procedure and products shows [23] that the measured overall parameters of products do not reflect their real structure.

Table 2. Characterization of some hydroxyoligobutadienes prepared in the presence of 2,5-dilithium hexane

$\overline{M}_n$, calc.	MWD $\overline{M}_n$	MWD $\overline{M}_w/\overline{M}_n$	[OH], %	$\bar{f}_n$	Microstructure, % 1,2-	Microstructure, % 1,4-*cis*	Microstructure, % 1,4-*trans*
3000	2440	1.09	1.37	1.96	27.2	40.2	32.6
3200	3010	1.14	1.15	2.04	21.2	43.1	35.7
5200	5350	1.11	0.63	1.98	13.6	48.8	37.6
5200	6190	1.06	0.55	2.00	10.2	50.3	39.5
7200	8200	1.06	0.41	1.97	12.6	48.6	38.8
7200	11400	1.12	0.30	2.02	11.5	49.9	38.6

As is known, polymerization of dienes initiated by lithium alkanes in hydrocarbons is homogeneous, and reactive mixtures exhibit no special behavior. But in the presence of DSDLAs, the process yields a precipitate of insoluble bifunctional living oligomer in the form of viscous mass sticking to reactor walls and stirrer. The rest of solution remains transparent with a viscosity close to the initial one. The amount of precipitate gradually grows, it becomes more and more mobile, and then dissolves (upon stirring). The solution viscosity sharply increases and then keeps rapidly growing. At the initial monomer concentration 10–20 wt %, the resultant solution reminds stiff dough that exhibits the pronounced Weissenberg effect. Upon addition of ethylene oxide, the thick gel disintegrates into a loose mass (provided stirring is sufficiently intense).

When reaction is carried out in a 1-l reactor, the reactive mixture was found to exhibit (in 2.5–3 h) a temperature rise that could be suppressed by water cooling; at constant water temperature, this rise may be well above 10 K. In a 50-l reactor, this temperature rise cannot be suppressed merely by cooling, it requires some additional amount of solvent. The effect (observed at a conversion degree above 50 %) cannot be attributed to reaction acceleration because the kinetics of monomer consumption is not S-shaped and the reaction rate is largest at the initial stage [8]. We explain this heating by a sharp drop in the heat transfer coefficient caused by increasing viscosity. This phenomenon was theoretically predicted by Zhirkov and Estrin [24].

Heterogeneity of reactive mixture cannot help affecting the molecular structure of resultant oligomers. First of all, we found the bimodal behavior of MWD during polymerization. This behavior markedly depends on the initial concentration of monomer $[M]_0$. For $[M]_0 > 20$ wt %, the intensity of the low-molecular peak (initially much weaker than the high-molecular one) in gel chromatograms increases, the peak positions get closer each to other and finally merge into one peak. For $[M]_0 < 20$ wt %, the bimodality of gel chromatograms is retained until the end of reaction.

Table 3. Characteristics of molecular inhomogeneity of OH-terminated oligobutadienes synthesized in the presence of 2,5-dilithium hexane

Solvent	$\overline{M}_n$ (GPC)	$\bar{f}_n$ (chem. anal.)	Mole fraction MF and MW of macromolecules with functionality f							$\bar{f}_n$ (DFT)
			$f=0$	$f=1$		$f=2$		$f=3$		
			MF	MF	MW	MF	MW	MF	MW	
One-step process										
Gasoline	5580	1.67	0.01	0.25	8600	0.64	6360	0.11	9640	1.84
—	3630	1.97	0.01	0.29	4400	0.56	3940	0.14	5010	1.83
—	5520	2.20	0.01	0.23	8010	0.66	5200	0.10	7920	1.86
—	7560	1.80	0.01	0.23	3610	0.75	7670	—	—	1.74
—	9060	1.65	0.05	0.29	6890	0.62	7360	0.04	10850	1.64
Benzene	6100	1.47	0.02	0.23	8470	0.71	6900	0.03	14500	1.77
—	6350	1.66	0.01	0.19	12100	0.78	6700	0.03	13000	1.83
Two-step process										
Heptane	5420	1.82	0.001	0.04	17960	0.93	6320	0.03	9540	1.98
Benzene	7800	2.34	0.004	0.10	11720	0.90	7990	—	—	1.89
—	9200	1.73	0.015	0.10	10300	0.89	8860	—	—	1.87
—	11300	1.93	0.008	0.08	13380	0.89	9860	—	—	1.87

The bimodality of MWD during diene polymerization in the presence of dilithium initiators was also investigated in [25–27]. However, the effect was related to reduced reactivity of intermolecular cyclic associates of bifunctional living chains. Inconsistency of this explanation as well as distinct relation between the bimodality of MWD and heterogeneity of reactive mixture was substantiated in [28, 29].

Earlier, we analyzed the oligomers formed to the moment of their phase separation from solution and found that the amount of reactive species in precipitate is above $0.66[M]_0$, while MW is below 1000. In this case, the chain microstructure is other than that of polybutadiene prepared by lithium polymerization: the amount of the 1,4-chains is always below 55 % (Table 2). This could be expected because the concentration of reactive species in precipitate must be fairly high (above 1 M). As this concentration increases (above 0.1 M), the amount of 1,4-units decreases [30].

The oligomers separated from solution after precipitation exhibit a normal microstructure (about 90 % 1,4-units). Apparently, dissolution of precipitate is accompanied by averaging the concentration of reactive species, so that further growth of macromolecules yields normal microstructure. This implies that the macromolecules that underwent the stage of growth in precipitate can be expected to have a block microstructure.

The above data show that heterogeneity of reactive mixture leads to different kinds of molecular inhomogeneity in synthesized oligomers.

One of the most important parameters of functional oligomers is the so-called distribution of functionality types (DFT). The DFT of hydroxyoligobutadienes prepared in the presence of DSDLAs was found [31] to exhibit marked amount of mono- and trifunctional fractions.

We managed to determine a cause for formation of macromolecules with other functionality [32, 33]: at high concentration of active species in the precipitate, the following reaction between still unreacted *sec*-lithiumalkyl groups and lithiumallyl living chain ends becomes possible:

$$\sim CHLiCH_3 + \sim CH_2CH{=}CHCH_2Li \rightarrow \sim CH_2CH_3 + \sim C_4H_5Li_2. \qquad (2)$$

This reaction competes with the reaction of initiation:

$$\sim CHLiCH_3 + \sim CH{=}CH{-}CH{=}CH_2 \rightarrow \sim CH(CH_3){-}CH_2CH{=}CHCH_2Li. \qquad (3)$$

Besides, reaction (2) is accompanied by elimination of LiH. All this leads to a decrease in functionality.

On the one hand, we demonstrated feasibility of preparation of well-soluble (in hydrocarbons) efficient dilithium initiators from readily accessible compounds. On the other hand, oligodienes prepared in the presence of these initiators strongly differ in their molecular uniformity from similar oligodienes synthesized in the presence of monolithium initiators. It may be taken for granted that the polymodality of MWD and block microstructure are the implication of heterogeneity of reactive mixture arising owing to insolubility of bifunctional living macromolecules.

Performing two-stage polymerization can diminish the amount of functionality defects. In this case, a precipitate of low-molecular bifunctional oligomer formed at the first stage is separated and used as a stable dilithium initiator at the second stage [34, 35].

Molecular inhomogeneity of oligomers synthesized in a one- and two-step process in the presence of DLHx is characterized in Table 3.

CONCLUSIONS

Disecondary dilithium alkanes can be readily prepared in the form of hydrocarbon solutions from respective dichlorides. The latter can be synthesized in a one-step process from available reagents. Synthesis of dilithium alkanes is essentially the same as that of commercially produced monolithium alkanes (*n*-butyllithium, *sec*-butyllithium, etc.). The synthesized dilithium alkanes are efficient initiators for polymerization of hydrocarbon monomers yielding bifunctional oligomers. They can also be used for preparation of different ABA copolymers, including those that cannot be obtained with multifunctional initiators.

Polymerization started by bifunctional initiators was found to be much more complicated than that started by monolithium initiators: it is accompanied by side reactions giving rise to several types of molecular inhomogeneity in polymerization products. The side reactions take place at the initial stages. This can be related to high local concentration of reactive species caused by poor solubility of low-molecular living dilithium oligomers in hydrocarbons. Nevertheless, using some technological know-how may minimize relative contribution from the side reactions.

Acknowledgments. This work was supported by the International Science and Technology Center (grant no. 358-96).

REFERENCES

1. Szwarc M., *Nature*, **178**, 1168 (1956).
2. Szwarc M., Carbanions, Living Polymers, and Electron Transfer Processes, Wiley, New York, 1968.
3. Stavely F.W. *et al.*, *Ind. Eng. Chem.*, **48**, 778 (1956).
4. Foster F.C. and Binder J.L., *Adv. Chem. Ser.*, **17**, 7 (1957).
5. Eberly K.C., US Patent 2 947 793, 1961.
6. Kuzaev A.I., Linde V.A., Estrin Ya.I., Afanas'ev N.A., Baturin S.M., Entelis S.G., *Vysokomol. Soedin.*, *Ser. A*, **18**, 585 (1976).
7. Estrin Ya.I., *Polym. Sci.*, *Ser. B*, **39**, 64 (1997).
8. Estrin Ya.I., Kasumova L.T., Baturin S.M., and Radugin V.S., *Polym. Sci.*, *Ser. A*, **38**, 813 (1996).
9. Weisberger A., Proskauer E.S., Riddick J.A., and Toops E.E., *Organic Solvents: Physical Properties and Methods of Purification*, Wiley, New York, 1955.
10. Cortese F., *J. Am. Chem. Soc.*, **52**, 1519 (1930).
11. Lean B., *J. Chem. Soc.*, **65**, 1007 (1894).
12. USSR Inventors's Certificate 829 635, *Byull. Izobr.*, no. 18, 109 (1981).
13. Atovmyan E.G. and Kuzaev A.I., *Dokl. Akad. Nauk SSSR*, **248**, 104 (1979).
14. Estrin Ya.I., *Vysokomol.. Soedin.*, *Ser. A*, **30**, 1560 (1988).
15. Gafurova M.P., Lodygina V.P., Grigor'eva V.A., Tchernyi G.I., Komratova V.V., Baturin S.M., *Vysokomol. Soedin.*, *Ser. A*, **24**, 858 (1982).
16. Simak P., and Fahrbach G., *Angew. Makromolek. Chem.*, **12**, 73 (1970).
17. Estrin Ya.I. and Kasumova L.T., *Russ. J. Phys. Chem.*, **68**, 1620 (1994).
18. Radugin V.S., Barzykina R.A., and Estrin Ya.I., *Vysokomol. Soedin.*, *Ser. A*, **26**, 350 (1984).
19. Estrin Ya.I., *Kinet. Katal.*, **26**, 373 (1985).
20. Shatalov V.P., Kirchevskaya I.Yu., Samotsvetov A.R., and Proskurina N.P., *Vysokomol. Soedin.*, *Ser. A*, **15**, 2042 (1973).
21. Bywater S. and Worsfold D.J., *J. Organomet. Chem.*, **10**, 1 (1967).

22. Talalaeva T.V. and Kocheshkov K.A., *Metody elementorganicheskikh soedinenii: Litii, natrii, kalii, rubidii, tsezii* (Methods of Elementorganic Chemistry: Lithium, Sodium, Potassium, Rubidium, and Cesium), Vol. 1, Nauka, Moscow, 1971.
23. Estrin Ya.I., Kasumova L.T., and Kol'tover V.K., *Polym. Sci.*, **33**, 2283 (1991).
24. Zhirkov P.V. and Estrin Ya.I., *Polym. Process. Eng.*, **2**, 219 (1984).
25. Quirk R.and Ma J.-J., *Polym. Int.*, **24**, 197 (1991).
26. Lo G.Y.-S., Otterbacher E.W., Gatzke A.L., and Tung L.H., *Macromolecules*, **27**, 2233 (1994).
27. Bredeweg C.J., Gatzke A.L., Lo G.Y.-S., and Tung L.H., *Macromolecules*, **27**, 2225 (1994).
28. Estrin Ya.I., *Polym. Sci., Ser. A*, **38**, 450 (1996).
29. Estrin Ya.I., *Polym. Sci., Ser. A*, **40**, 651 (1998).
30. Yudin, V.P., *Vysokomol. Soedin., Ser. A*, **20**, 1001 (1978).
31. Gorshkov A.V., Verenich S.S., Estrin Ya.I., Evreinov V.V., and Entelis S.G., *Vysokomol. Soedin., Ser. A.*, 30, 945 (1988).
32. Barzykina R.A., Kasumova L.T., and Estrin Ya.I., *Vysokomol. Soedin., Ser. B*, **23**, 643 (1981).
33. Barzykina R.A., Kasumova L.T., Lodygina V.P., and Estrin Ya.I., *Vysokomol. Soedin., Ser. B*, **26**, 1442 (1984).
34. Estrin Ya.I., Kasumova L.T., and Grishchuk A.A., *Polym. Sci., Ser. A*, **39**, 128 (1997).
35. Russ. Patent 2 114 123, 1998.

Chapter 2

Synthesis of Block Copolymers with Controlled Solubility in Epoxy Resins

Elmira R. BADAMSHINA*, Yakov I. ESTRIN, Valentina A. GRIGOR'EVA, Aleksandr A. GRISHCHUK, Galina A. GORBUSHINA, Larisa T. KASUMOVA, Viktoriya V. KOMRATOVA, Anatolii I. KUZAEV, and Vera P. LODYGINA

Institute of Problems of Chemical Physics, Russian Academy of Sciences, Chernogolovka, Moscow Region, 142432 Russia

---

*e-mail: badamsh@icp.ac.ru

ABSTRACT

To produce polymers with controlled solubility in epoxy resins, block copolymers (BCPs) of two types were synthesized: (a) A–B, B–A–B of the polymerization type produced by step-by-step 'living' polymerization of non-polar and polar monomers, butadiene and ε-caprolactone; (b) $(AX)_n$–$(BX)_m$ of the polyaddition type produced from bifunctional non-polar hydroxyoligobutadienes, polar polyoxypropyleneglycols and diisocyanates. It has been shown that BCPs solubility depends on the ratio of polar to non-polar fractions in the macromolecule that ensures uniform composition of resulting products. This ratio can be controlled by dosing monomers (for BCPs of the first type) or by variation in blend composition and/or in MW of oligomers (for BCPs of the second type).

INTRODUCTION

Modification of rigid glassy polymer matrices by elastomeric inclusions is the most promising way for production of composites with high fracture toughness. A modifying agent must be soluble in the monomer or oligomer and undergo microphase separation during thermoset cure. For composites based on epoxy resins, only a few elastomers meet these requirements. Hydrocarbon elastomers are poorly soluble in the initial resin, whereas more polar polymers (including polyesters and polyethers, polyetherurethanes) that are soluble in the initial resin and do not form a separate phase upon cure.

One of possible approaches to solving the problem is synthesis of chemically bonded blends with macromolecules containing both polar and non-polar fractions, that is, block copolymers (BCPs). BCPs of two types may be used: (a) polymerization type A–B, B–A–B block copolymers produced by step-by-step polymerization of non-polar and polar monomers; (b) polyaddition type $(AX)_n$–$(BX)_m$ produced from bifunctional nonpolar, polar oligomers and diisocyanates. Controlled amount of polar and non-polar blocks in such copolymers provides a certain solubility of these polymers in resins and allow controlling their ability to phase separation during cure. In this work, we explored the feasibility of synthesizing the above BCPs with controlled solubility in epoxy resins.

EXPERIMENTAL

Chemicals

ε-Caprolactone (CL) (Aldrich) was distilled over KOH in vacuum, stored under argon, and placed in an Ar flow over CaH_2 before using.

Table 1. Characterization of starting oligomers

No.	Oligomer grade	% OH	$\overline{M}_n$	$\overline{M}_w/\overline{M}_n$	$\bar{f}_n$ [a]
		Bifunctional hydroxyl terminated oligobutadienes			
1	PB-79	1.04	3050	1.17	1.92
2	PB-69	0.41	6400	1.06	1.76
3	PB-100	0.52	6300	1.19	1.88
4	PB-78	1.31	2600	1.23	1.98
5	PB-72	0.91	3500	1.10	1.87
6	PBS-6	1.22	3070	1.07	2.20
		Polypropylene glycols			
7	PPG-500	6.23	530	1.06	1.95
8	PPG-2000	1.79	1750	1.06	1.82
9	PPG-3000	1.31	2120	1.23	1.68

[a] Averaged functionality

2,4-Toluylenediisocyanate (TDI) was distilled at 323–325 K/12 Pa (d_4^{20} = = 1.2178, n_D^{20} = 1.5669) and stored in sealed ampoules.

Synthesis and analysis of oligobutadienes with terminal hydroxyl groups (HOBD) are described in [1].

Polyoxypropyleneglycols (PPG) obtained by anionic polymerization were dried in a column in a thin layer at 353 K/13.3 Pa. Parameters for the starting HOBD and PPG samples are presented in Table 1.

Synthesis of Block Copolymers

For synthesis of BCPs by step-by-step polymerization of butadiene (BD) and CL in benzene, either sec-butyllithium (for A–B type) or 2,5-dilithiumhexane (for B–A–B type) was used as initiator, where A is a polybutadiene block (PBD) and B is a polycaprolactone one (PCL). Preparation of chemicals and synthesis of polybutadiene block were similar to HOBD synthesis [1]. With the only difference that, after butadiene polymerization was completed and the living polymer was treated with ethylene oxide, CL was added to the reactor upon intense stirring.

BCPs of polyaddition type were synthesized in two steps. At first prepolymer based on HOBD and TDI was prepared at different values of [NCO]/[OH] ≥ 2, at 298–313 K. Then PPG was added to the prepolymer, so that [OH] was equal to residual [NCO], and the reaction was performed at 313 K until all diisocyanate groups (as detected from IR spectra) disappeared.

Characterization of Starting Oligomers and Block Copolymers

Molecular weight distribution (MWD) was evaluated by size exclusion chromatography by using either a Milikhrom chromatograph in double-wave recording mode, 210 and 270 nm [2], or a Waters chromatograph equipped with a refractometric detector.

Composition of PBD–PCL block copolymers was estimated by the intensity of IR bands of carbonyl groups at 1773 cm^{-1} (extinction coefficient in benzene 630 l/(mol·cm), poly-ε-caprolactone as a reference sample).

RESULTS AND DISCUSSION

Solubility of BCPs in epoxy resins depends on chemical composition and structure of macromolecules, all other factors being identical. Meanwhile, there always exists some distribution of macromolecules over their composition. This may be a factor that defines the uniformity of copolymer solubility. Indeed, the polymer fraction having the structure almost similar to that of non-polar starting oligomer is minimal, whereas 'over-modified' fragments may not participate in phase separation during the monomer cure.

Qualitative considerations suggest that, to achieve the most uniform composition of BCPs, the starting non-polar polymer should have as narrow MWD as possible. Living polymerization, in particular polymerization of conjugated dienes (butadiene, isoprene) under the action of organolithium initiators, is the most promising way for synthesis of polymers with controlled MW [3].

A–B and B–A–B Block Copolymers

These BCPs were produced by step-by-step living polymerization of non-polar and polar monomers, butadiene and CL.

CL undergoes polymerization under moderate conditions with anionic initiators via a mechanism similar to living polymerization. Poly-ε-caprolactone is compatible with many polymers, including PVC, nitrocellulose, styrene, and acrylonitrile copolymers, polyepichlorohydrin, as well as with epoxy resins [4]. One may expect that BCPs of BD and CL synthesized directly by living polymerization of the latter on the functional group of the former will have high MWD and good uniformity. The length of each block as well as their ratio is easily set by the monomer ratio.

Synthesis of BCPs based on both polystyryllithium and poly-butadienyllithium with CL is described in [5]. To avoid side reactions upon interaction of living polymers with CL, chain ends were modified with

1,1-diphenylethylene. However, this procedure is time-consuming, especially for polybutadienyllithium.

CL undergoes polymerization easily and quickly at ambient temperature under the action of lithium alcoholates in aromatic solvents. This property makes feasible the synthesis of BCPs by the following scheme:

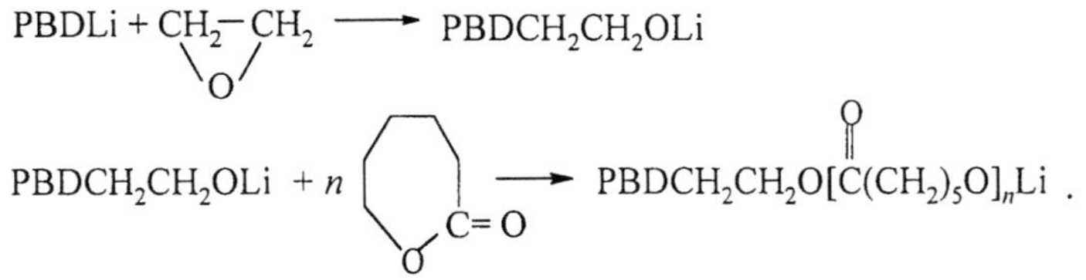

(1)

(2)

Balsamo *et al.* [5] note that the alcoxide form of the growing PCL chains may participate in intermolecular exchange with the ester bonds, thus resulting in broadening the MWD in blocks. In addition, intramolecular exchange is responsible for depolymerization yielding cyclic oligomers. Therefore, the authors recommend that the duration of lactone polymerization be a few minutes.

When the reaction blend is exposed for a long time after the addition of CL to OBD–OLi, products with a broad MWD are formed, although starting oligobutadiene completely converts to the BCP (there is no fraction soluble in heptane). Meanwhile, the product contains a low-molecular fraction which most probably is a blend of cyclic oligomers of CL. If the reaction did not occur for a long time (below 1 hour), the product MWD was relatively narrow (only slightly broader than that of starting OBD–OLi). In this case, the oligomer fraction was also present.

In all cases, chromatograms exhibit a peak of a low-molecular product whose intensity is higher than that of other peaks. At the same time, a crystal product is sublimed at 373 K upon drying of BCPs in vacuum. According to masspectral data (ESI TOF method), the crystal product is a cyclic dimer, and its retention time (in chromatograms) coincides with the retention time for of the strongest peak. Therefore, the anionic polymerization of CL seems to be accompanied by intramolecular chain transfer reaction [5–7], with cyclic oligomers as its products (dimer as a main product):

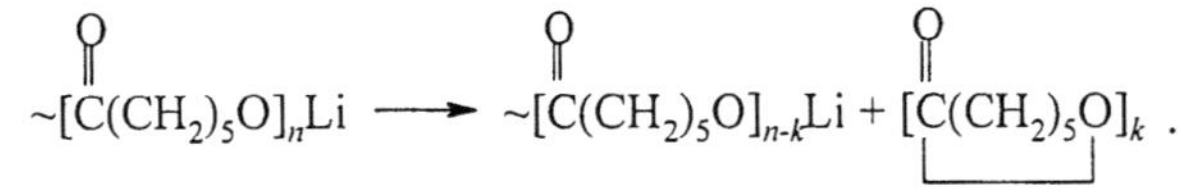

(3)

In [6] the results of investigation of the chain initiation, transfer and termination mechanism upon CL polymerization in this system are presented.

Table 2. Conditions for synthesis of PBD-PCL (1-8) and PCL-PBD-PCL (9) block copolymers, their composition and molecular-weight parameters

Nos.	T, K/t, h	$\overline{M}_n$ / ($\overline{M}_w/\overline{M}_n$) of initial PB	PB/PCL weight fraction ratio Calc.	PB/PCL weight fraction ratio Meas.	$\overline{M}_w/\overline{M}_n$ of BCP without oligomer fraction	Oligomer weight fraction in BCP, %
1	343/2	3000/1.03[a]	44/56	—	1.69	8.6
2	313/7	3000/1.03[a]	40/60	37/63	1.61	9.6
3	323/1	3000/1.03[a]	67/33	63/37	1.42	6.4
4	323/1	3000/1.03[a]	80/20	78/22	1.29	< 1
5	303/0.5	3000 / 1.03[a]	65/35	61/39	1.39	8.1
6	298/1	9000/1.1[b]	60/40	65/35	1.33	8.6
7	298/0.25	5000[c]	50/50	45/55	1.45	13.4
8	298/0.25	3000/1.03[a]	30/70	<40/60	1.90	21.6
9	298/0.33	5000/1.17[d]	60/40	60/40	1.19	13.6

[a]Sample no. 1.
[b]Sample no. 9 in Table 1.
[c]Initial PB was not recovered. MW is calculated.
[d]Sample no. 7 in Table 1.

PCL–PBD–PCL type block copolymer obtained using 2,5-dilithium-hexane as initiator is the most interesting, because it can be used for studying the influence of total block length and block lengths ratio, as well as of macromolecule structures on the products properties. Table 2 presents some parameters of the BCPs obtained.

It can be concluded from the data of Table 2 that the lesser the duration of lactone polymerization and the lower its temperature, the more narrow the BCP's MWD. However, it should be noted that oligomer fraction does not depend essentially on synthesis conditions (except sample 4) and is maximal for the moderate conditions (samples 7–9). Apparently, at proceeding of intrachain exchange reactions, the equilibrium is achieved rather quickly.

As follows from comparison of samples 6 and 7 obtained under similar conditions but with a different exposure time, BCPs with uniform composition and MWD similar to that of PBD-block can be synthesized (see Fig. 1). Comparison of samples 8 and 9 shows (Fig. 2) that, under similar synthetic conditions, the longer the length of PCL block chain, the broader MWD of BCP compared to PB-block precursor. Therefore, composition and molecular parameters of BCP of the polymerization type may be varied within a wide range.

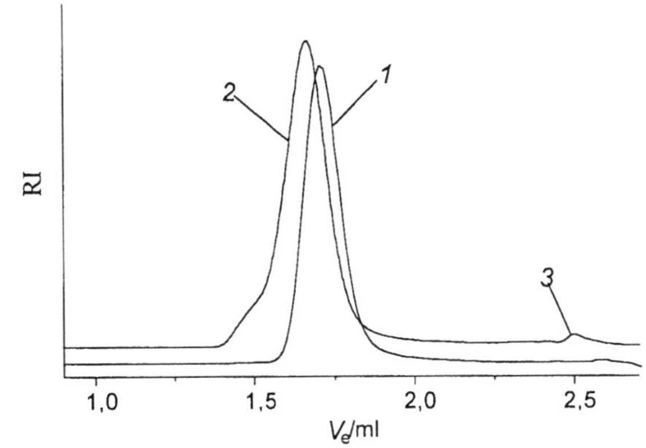

Figure 1. SEC curves: (*1*) PBD-block, (*2, 3*) Sample 4 (Table 2). Peak *3* refers to the oligomer. Waters chromatograph, Column 250×4 mm, Silasorb 600, THF eluent, 0.3 ml/min, 398 K, refractometric detection.

$(AX)_n$–$(BX)_m$ Block Copolymers

The following scheme illustrates the synthesis of random multiblock polyurethane copolymers:

HOBD + TDI → X–A–X   prepolymer at [NCO]/[OH] ≥ 2

–NCO + –OH → –NHCOO–

X–A–X + B + X → $(AX)_n(BX)_m$ random multiblock copolyurethanes,

where A, B and X are the bifunctional residual fragments of HOBD, PPG and TDI, respectively.

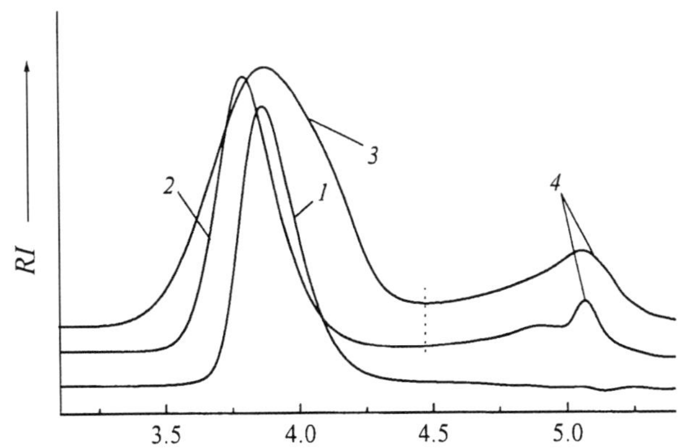

Figure 2. SEC curves: (*1*) PB precursor, (*2, 4*) sample 9, (*3*) sample 8, (*4*) oligomer. The conditions were similar to those in Fig. 1.

Table 3. MW parameters of prepolymers

Initial HOBD		[NCO]/[OH]	Prepolymer	
$\overline{M}_n$	$\overline{M}_w/\overline{M}_n$		$\overline{M}_n$	$\overline{M}_w/\overline{M}_n$
6900	1.16	1.7	10100	1.32
6900	1.16	2.0	8700	1.25
3070	1.07	8.5	3900	1.09
3500	1.12	11.2	4300	1.15

BCP production involves the step of synthesis of prepolymer based on HOBD and TDI at 298–313 K. It was shown [8] that prepolymer synthesis is accompanied by the chain growth reactions and, at temperatures above 323 K, by side reactions. The ratio of the rates of the prepolymer formation to the chain growth and, hence, the composition of the final product depends on the initial [NCO]/[OH] ratio.

As follows from Table 3, the higher the [NCO]/[OH] ratio, the lower is the probability of chain growth, and polydispersity of the prepolymer for these cases does not differ much from that of initial HOBD. SEC chromatograms of the prepolymer and initial HOBD presented in Fig. 3 confirm this conclusion. This figure also suggests that for high [NCO]/[OH] ratios, the final product contains considerable amount of the free TDI, together with the prepolymer.

BCP samples were prepared by interaction between prepolymer and PPG. The content of the polar fragment (TDI and PPG) was varied by changing [NCO]/[OH] ratio upon the prepolymer synthesis, and, accordingly, PPG amount, because equimolar ratio of functional groups was maintained. To obtain BCPs with different content of the polar fragment, prepolymers were synthesized at [NCO]/[OH] = 2–36.

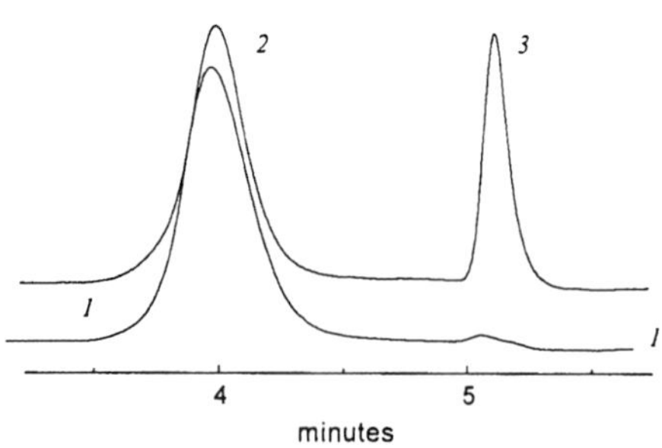

Figure 3. SEC curves: (*1*) oligobutadienediol PB-72, (*2*) prepolymer based on PB-72, (*3*) TDI at the initial value of [NCO]/[OH] = 11.2. The SEC conditions were similar to those in Fig. 1.

Table 4. MW parameters of block copolymers

BCP[a]	Initial components: $\overline{M}_n$ of HOBD	$\overline{M}_n$ of PPG	$\frac{[NCO]}{[OH]}$ [b]	P [c]	BCP: $\overline{M}_n$	$\overline{M}_w/\overline{M}_n$
1	6400 (2[a])	530 (7[a])	3.0	0.27	16600	4.98
2	6400 (2)	530 (7)	12.5	0.50	9700	6.02
3	6400 (2)	530 (7)	35.6	0.75	8400	4.77
4	3050 (1)	530 (7)	15.1	0.76	7200	3.03
5	6300 (3)	530 (7)	28.6	0.75	9100	5.57
6	3050 (1)	1750 (8)	9.8	0.85	11600	2.16
7	6300 (3)	1750 (8)	10.5	0.75	19800	2.81
8	2600 (4)	1750 (8)	8.0	0.85	17200	2.17
9	3500 (5)	1750 (8)	11.2	0.85	16900	2.23
10	3070 (6)	1750 (8)	8.5	0.85	18800	2.48
11	3070 (6)	2100 (9)	6.7	0.85	20200	1.89
12	3500 (5)	2100 (9)	8.6	0.85	21100	1.91

[a]Numbers in parentheses correspond to the numbers of HOBD or PPG in Table 1.
[b]The ratio of functional groups upon prepolymer synthesis.
[c]Weight fraction of polar components, $P = ([PPG] + [TDI])/([PPG] + [TDI] + [HOBD])$.

Table 4 presents the parameters of synthesized BCPs. All BCPs based on PPG with a MW of 530 are characterized by a wide MWD, apparently because of composition uniformity of the product. We assume that in this case, together with interaction between the prepolymer and PPG yielding the desired BCP, the reaction of PPG and the free TDI available in the system yields polyurethane containing both linear molecules and macrocycles [9]. A decrease in the initial MW of HOBD, with unchanged content of the polar fragment (samples 4 and 5) results in MWD narrowing (Fig. 4). A decrease in BCP polydispersity is also caused by an increase in the initial PPG's MW (samples 5, 7) (Fig. 5). If MW for initial oligomers HOBD and PPG are comparable, the polydispersity is close to 2 (samples 11, 12). Apparently, in this case macromolecules of similar type are formed, this testifying the composition uniformity of the BCP.

Our data show that, upon synthesis of BCPs of the polyaddition type, the polar to nnpolar component ratio may be controlled over a wide range by varying the initial blend composition and oligomer MW. The closer the MW of initial oligomers, the more uniform is the BCP composition.

The results of the study of BCP solubility in DGEBA have been reported in [10].

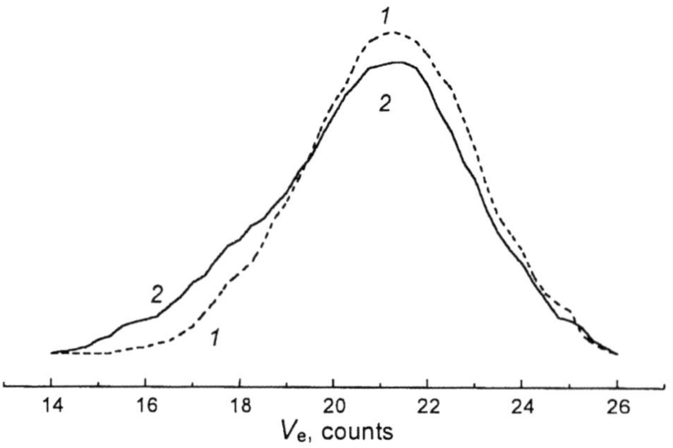

Figure 4. SEC curves of block copolyurethanes with a polar fraction content of P = 0.75: (*1*) block copolymer sample 4 and (*2*) block copolymer sample 5. Waters-2000 chromatograph, THF eluent, 1.2 ml/min, 398 K, refractometric detection.

CONCLUSIONS

BCPs with a controlled ratio of polar to nonpolar fractions may be obtained in two ways.

One is the consecutive living polymerization of nonpolar (butadiene) and polar (CL) monomers. BCPs of this kind have uniform composition and

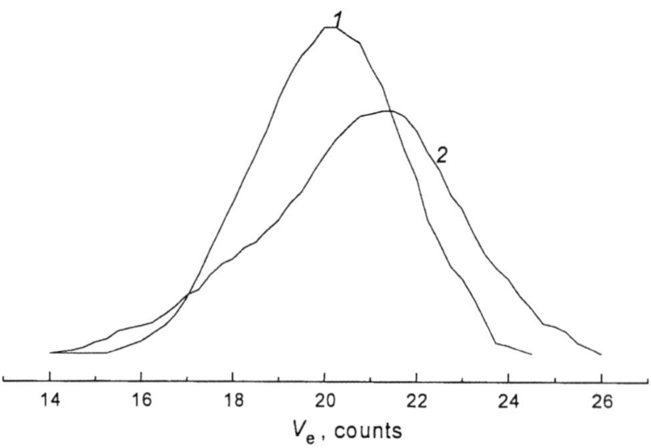

Figure 5. SEC curves of block-co-polyurethanes with a polar fraction content of P = 0.75: (*1*) block-copolymer sample 7, (*2*) block-copolymer sample 5. The chromatography conditions were similar to those in Fig. 4.

the parameters—MW, block lengths ratio, narrow MWD—that are controllable over a wide range of parameters.

The second way is polyaddition of polar (PPG) and nonpolar (HOBD) oligomer diols with bifunctional low-molecular binding agent (TDI). In this case, the closer are the MWs of starting oligomers, the more uniform is the resulting BCP composition.

Acknowledgments. This work was supported by the International Science and Technology Center (grant no. 358-96).

REFERENCES

1. Grishchuk A.A., Kasumova L.T., and Estrin, Ya.I., *this book*. p. 3.
2. Estrin Ya.I., *Vysokomolek. Soedin.*, **30**, 1560 (1988).
3. Swarc M., Carbanions, Living Polymers and Electron Transfer Processes, Wiley, New York, 1968.
4. Slomkowski S., *Macromol. Symp.*, **103**, 23 (1996).
5. Balsamo V., von Gildenfeld F., and Stadler R., *Macromol. Chem. Phys.*, **197**, 1159 (1996).
6. Rozenberg B.A., Estrin Ya.I., and Estrina G.A., *Macromol. Symp.*, **153**, 197 (2000).
7. Rozenberg B.A., Irzhak V.I., and Enikolopyan N.S., *Mezhtsepnoi obmen v polimerakh* (Interchain Exchange in Polymers), Khimiya, Moscow, 1975.
8. Stovbun E.V., Lodygina V.P., Kuzaev A.I., Romanov A.K., and Baturin S.M., *Vysokomolek. Soedin., Ser. A*, **26**, 1449 (1984).
9. Grigor'eva V.A., Komratova V.V., Kuzaev A.I., and Baturin S.M., *Vysokomolek. Soedin., Ser. A*, **25**, 2182 (1983).
10. Mikhailov Yu.M., Ganina L.V., and Smirnov V.S., this book, p. 33.

Chapter 3

Universal Calibration in Gel Permeation Chromatography of Oligomers

Anatolii I. KUZAEV*

Institute of Problems of Chemical Physics, Russian Academy of Sciences, Chernogolovka, Moscow Region, 142432 Russia

ABSTRACT

The universal calibration based on the dependence of the retention volume on the van der Waals volume of macromolecules was found applicable to gel permeation chromatography (GPC) of oligomers. For THF as eluent, the size of analyzed molecules containing functional groups should be corrected for the size of eluent

*e-mail: kuzaev@icp.ac.ru

molecule. The suggested technique was used to investigate complex formation of metal halides with 2,4-toluylene-diisocyanate (TDI) in acetone solutions. In this way, complex composition and their molecular weight (MW) were determined. The technique was also used to investigate formation of urethanes (and their MW parameters) in the metallocomplex–polyester system.

INTRODUCTION

In gel permeation chromatography (GPC) macromolecules are separated by their size, so that there is some universal calibration dependence between the hydrodynamic molecular volume and retention volume. As is known [1], this calibration is inapplicable to oligomers. Further studies are needed in order to determine a universal parameter that characterizes the size of oligomer molecules.

This work aims at finding a universal and absolute calibration curve that would allow direct determination of the MW of oligomers and polymers.

The suggested method was checked experimentally on determining the composition and MW of iron halide complexes with 2,4-toluylene-diisocyanate (TDI). These complexes are being used in synthesis of polyurethane-based metallocomplex compositions [2, 3].

EXPERIMENTAL

Gel Chromatography

Composition of products and complex formation kinetics were investigated with a Waters GPC instrument (set of three styrogel columns with pore size of 200, 500 and 1000 Å, THF as eluent, flow rate of 1.2 ml/min, 298 K). Samples in the form of 0.2–0.3% solutions were injected during one count (one count = 5 ml eluent).

MWD distribution was determined by using a universal calibration curve that gives the relation between the retention volume V_r and van der Waals (VDW) volume of polymer molecules ΔV [4]. The molecular volume was calculated from the volume increment arising upon chemical bonding [5]. The atomic volume was calculated as the volume of a sphere with the VDW radius R. The intermolecular distance d_i between two chemically bonded atoms is always greater than the sum of their atomic radii.

The increment ΔV_i in the volume of a given atom can be found from the following expression:

$$\Delta V_i = 4/3\pi R^3 - \sum_i \ 1/3\pi\, h_i^2(3R - h_i),$$

where ΔV_i is the increment of the volume of atoms or groups of atoms, h_i is the segment height as determined from the expression

$$h_i = R - (R^2 + d_i^2 - R_i^2)/2d_i,$$

R is the VDW radius of a given atom, d_i is the interatomic distance between chemically bound atoms, and R_i is the radius of a chemically bound atom. Molecular volume of inorganic compounds was calculated with regard to their ionic radius [6]. A universal calibration curve was plotted by using the available GPC data for oligomers of styrene [7], propylene oxide, and ethylene oxide with various functional groups. The parameters of MWD were calculated as described in [1].

Reagents

Acetone (Ac), tetrahydrofuran (THF), and 2,4-toluylene-diisocyanate (TDI) were purified and rectified by standard techniques. Ethanol, anhydrous $FeBr_2$, and anhydrous $FeCl_3$ were used as purchased, with no further purification.

Synthesis of Metal Complexes

A weighed amount of iron salt was dissolved in an aliquot amount of acetone and then an aliquot amount of TDI was added to the solution. The mixture was held at a given temperature (normally 293 K). The obtained mixture containing metal complexes was analyzed by GPC. Prior to analysis, the mixture was processed with ethanol and then sampled as 0.3% solutions in THF. The solutions were stored in an icebox.

RESULTS AND DISCUSSION

Despite a number of advantages, GPC ensures only relative measurements. In order to determine MW and parameters of MWD, we have to obtain the calibration dependence between MW and retention volume by using some reference compounds with a well-known MW or very narrow MWD. For this reason, high precision and reproducibility of GPC measurements cannot often be realized because of insufficiently correct parameters of reference sample. In view of this, we suggested a procedure for plotting universal calibration curve for oligomers that requires no reference sample. Essentially, the procedure is based on separation of polymer homologues by their degree of polymerization, with a well-defined retention volume for each homologue [4]. The calibration procedure developed initially for oligomers of styrene and ethylene oxide was then successfully applied to

Table 1. Molecular characteristics of the complexes under investigation: V_r is the retention volume; ΔV_c and ΔV_m are the calculated and measured VDW volume of complexes, respectively; and MW is their molecular weight

Complex	V_r	ΔV_c	ΔV_m	MW
TDI + 2Ac	22.80	288	263	290
$FeBr_2$ + TDI + 3Ac	21.47	418	407	564
$FeBr_2$+ TDI +2Ac	20.59	511	537	680
$FeBr_2$ + 3 TDI + Ac	20.50	604	562	796
$FeBr_2$ + 4 TDI	20.00	676	696	912
2$FeBr_2$ + 2 TDI + 4Ac	19.68	706	741	1012
2$FeBr_2$ + 3 TDI +3Ac	19.30	801	851	1128
2$FeBr_2$ + 5 TDI + Ac	18.95	985	955	1360
2$FeBr_2$ + 6 TDI	18.48	1076	1096	1476
3$FeBr_2$ + 8TDI	17.60	1457	1479	2040

oligomers of epichlorohydrin, butadiene, methyl methacrylate, propylene oxide, epoxy resins, and oligohydroquinones [1].

For the above oligomers, the universal dependence between the retention volume and size of macromolecules also holds true, provided that the latter is expressed in terms of chain length or VDW volume and with regard to solvation of chains with eluent molecules [7].

Since calculation of chain length faces some problems, the VDW volume seems to be a parameter that adequately characterizes the size of macromolecules during GPC. In this work, we applied the method to individual hydrocarbons, oligosulfones, oligoisocyanurates, and oligomers of octene oxide.

As follows from Fig. 1, the data obtained comply well with the universal dependence. Keeping in mind the different nature of polymers used for calibration, we can state that the VDW volume (just as the hydrodynamic volume in case of polymers) properly defines the size of oligomer molecules. As already mentioned, this calibration is not only universal but also absolute. This is because the curve is based not on average molecular weights, $\overline{M}_n$ or $\overline{M}_w$, but on the values calculated from molecular structure.

This calibration curve was used to identify the complexes formed in the TDI–$FeBr_2$–Ac system The TDI/$FeBr_2$ ratio ranged between 5 and 40.

The composition and molecular characteristics of the complexes under investigation as detected by using the universal calibration plot are presented in Table 1. The VDW volume was calculated by assuming the *n*-nuclear bridging structure of iron halides in Ac [8–10] and octahedral environment of iron ions in its complexes with TDI [11]. The calculated and measured VDW radii are in good agreement, which confirms the validity of the suggested approach.

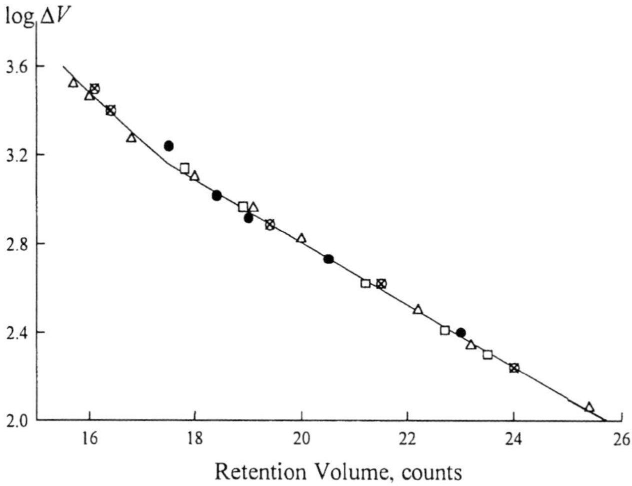

Figure 1. Universal calibration curve for oligostyrene and oligo(ethylene oxides) as taken from [4] (solid line). The data points correspond to (•) oligosulphones, (⊗) oligo(octene oxides), (Δ) hydrocarbons, and (□) oligoisocyanurates.

The kinetics of complex formation as studied by GPC sheds light on the processes taking place in the system. The amount of the polynuclear form increases with time while that of monomers and dimers diminishes. The overall composition of $FeBr_2{\cdot}TDI$ complexes shows the presence of free NCO groups (between 2 and 8 in their number) that may take part in reactions yielding urethane.

Inclusion of polyesters with the end OH groups into metallocomplexes leads (upon elimination of solvent) to formation of cross-linked polyurethanes. The presence of polynuclear $FeBr_2{\cdot}TDI$ complexes favours formation (at room temperature) of a polymer insoluble in THF but soluble in dimethylformamide. In terms of donor–acceptor interactions, this implies the formation of intermolecular coordination bonds between iron ions and oxygen atoms from the carboxyl groups of the polyester or urethane groups of polymer.

CONCLUSIONS

The use of the VDW molecular volume as a universal size parameter of oligomer molecules was found not only to simplify the calibration procedure in GPC but also to provide the absolute method for determination the MW of oligomers and polymers. Application of this technique shed light on complicated overall view of complex formation in the $FeBr_2{\cdot}TDI$ system. The functionality (with respect to NCO) of the complexes formed in this system was found to range between 2 and 8, which ensures synthesis of cross-linked polyurethanes.

Acknowledgments. This work was supported by the International Science and Technology Center (grant no. 358-96).

REFERENCES

1. Entelis S.G., Evreinov V.V., and Kuzaev A.I., *Reaktsionnosposobnye polimery* (Reactive Polymers), Khimiya, Moscow, 1985.
2. Kirpichnikov P.A., Davletbaeva I.M., Dorozhkin V.P., and Parfenov V.V., *Dokl. Akad. Nauk SSSR*, **227**, 1430 (1984).
3. Davletbaeva I.M., Parfenov V.V., Dorozhkin V.P., and Kirpichnikov P.A., *Vysokomolek. Soedin., Ser. A*, **31**, 1215 (1989).
4. Kuzaev A.I., *Vysokomolek. Soedin., Ser. A*, **22**, 1146 (1980).
5. Askadskii A.A., *Deformatsiya polymerov* (Deformation of Polymers), Khimiya, Moscow, 1973.
6. Cotton F.A. and Wilkinson G., *Advanced Inorganic Chemistry: A Comprehensive Text*, Interscience, New York, 1966.
7. Kuzaev A.I., *Vysokomolek. Soedin., Ser. B*, **22**, 202 (1980).
8. Vertes A., Konec Z.L., and Burger K., in *Mössbauer Spectroscopy*, Academiai Kiado, Budapest, 1979, p. 138.
9. Vertes A., Nagy-Czako J., and Burger K., *J. Phys. Chem.*, **82**, 1469 (1978).
10. Beck M., *Chemistry of Complex Equilibria*, Van Nostrand, London, 1970.
11. Davletbaeva I.M., *Cand. Sci. (Chem.) Dissertation*, Kazan', Inst. Chem. Technology, 1984.

Chapter 4

Phase Equilibrium in Binary Polymer Systems Based on Diglycidyl Ether of Bisphenol A

Yurii M. MIKHAILOV*, Lyudmila V. GANINA, and Vladimir S. SMIRNOV

Institute of Problems of Chemical Physics, Russian Academy of Sciences, Chernogolovka, Moscow Region, 142432 Russia

ABSTRACT

Using light scattering and microinterference methods, the phase separation was studied in the systems based on diglycidyl ether of bisphenol A with hydroxyl- and bromo-containing derivatives of oligobutadiene, random copolymers of nonyl acrylate with acrylamide and acrylic acid, and polyurethane block copolymers of various MW and composition with limited solubility. Obtained phase diagrams were treated in the framework of Flory–Huggins (FH) theory. Thermodynamic behavior

*e-mail: yumm@icp.ac.ru

of polymer solutions was shown to depend on the degree of intermolecular interactions. The phase diagrams for copolymers of nonyl acrylate with 30 % and 40 % of acrylic acid were shown to be similar to the diagrams of chemically crosslinked systems and not to obey FH theory. The upper critical solution temperature was plotted against solubility parameters for a few systems.

INTRODUCTION

Investigation of phase equilibrium is one of the most important challenges for physical chemistry of multicomponent polymer systems. Phase diagrams (PD) for many binary systems based on polymers and oligomers were reported in [1–6]. The multicomponent system based on diglycidyl ether of bisphenol A (DGEBA) and epoxy resins + modifying additives undergo phase separation during thermoset cure is one of the most promising from the technical point of view. Nevertheless, information in the literature about their thermodynamic characteristics (determining factor for practical application of these systems) is still limited [1–3, 7–10]. Information about the influence of the network of physical bonds in systems with a different degree of components interaction on a thermodynamic behavior of polymer solutions, in particular, on phase boundary concentrations and position of critical points is scarce. This work aims at filling this gap using (as a second component in the binary systems DGEBA + modifying agent) some new well-defined oligomers, polymers, random and block copolymers of various molecular weight (MW), composition, and with a controlled ratio of polar to nonpolar components in the compounds synthesized in this Institute.

EXPERIMENTAL

Investigation was focused on diglycidyl ether of bisphenol A with an epoxide number of 180 and MW of 340. Table 1 presents the characteristics of oligomers and polymers used as a second component in the DGEBA + additive system. Three types of additives were studied:

- oligobutadienes (OB), hydroxyl-terminated oligobutadienes (OBM, OBD), and brominated derivatives of OBD-1 [8, 9];
- polynonylacrylate (PNA) and copolymers of nonyl acrylate with acrylamide (CNAA) and acrylic acid (CNAAA) with $\overline{M}_n \approx 10^5$ and an acrylic acid (AA) fraction of 10, 20, 30, and 40% [7, 10];
- urethane block copolymers based on oligobutadiene, toluylenediisocyanate, and polypropyleneglycol (BCP).

Table 1. Molecular characteristics of the objects under study

Objects	Grade	$\overline{M}_w$	$\overline{M}_w/\overline{M}_n$	Polar groups content, %
Oligobutadienes	OB-1	1960	1.13	—
	OB-2	3460	1.10	—
	OB-3	10500	1.13	—
Oligobutadiene monools	OBM-1	1830	1.15	0.74
	OBM-2	3520	1.09	0.45
	OBM-3	10100	1.17	0.18
Oligobutadiene diols	OBD-1	3300	1.10	0.91
	OBD-2	5560	1.15	0.51
	OBD-3	9200	1.34	0.32
Copolymers of nonyl acrylate with acrylamide	CNAA-1	1900	1.83	20*
	CNAA-2	3000	1.50	20*
	CNAA-3	9700	2.23	12.5*
	CNAA-4	7400	2.16	7.5*
Urethane block copolymers	BCP-1**	18800	2.48	0.85
	BCP-2**	20000	2.15	0.905
	BCP-3**	21100	1.91	0.85
	BCP-4	16500	2.35	0.7
	BCP-5	25750	1.89	0.47

*Acrylamide concentration in the copolymer.

**These block copolymers correspond to BCP-10, 11, 12, see [13].

The polymers were obtained by radical polymerization as described in [11–13]. The molecular weight distribution was determined by GPC [14]. All additives were kindly provided by the authors cited in Refs.

To study the phase equilibrium, we used the techniques based on the analysis of concentration distribution (within the area of mutual dissolution of components) by microinterference [15–17] and by automatic recording of the optical transmission of samples heated according to some program [18]. Both techniques are widely used for studying phase equilibrium in polymer solutions and are known to yield reliable results confirmed by other experimental techniques [17].

RESULTS AND DISCUSSION

Figures 1–8 show the PD for the systems studied. The most of PD are asymmetrical, this become apparent in prior DGEBA solubility in the

Table 2. Comparison of the data obtained by microinterference and light scattering and used in plotting the phase diagrams

System based on	T, °C	DGEBA solubility in the polymer, %		Polymer solubility in DGEBA, %	
		Light scattering	Micro-interference	Light scattering	Micro-interference
BCP-5	50	40	38	1.0	1.5
	60	—	41	2.0	2.0
	120	60	58	—	—
	135	65	62	—	—
BCP-2	50	65	62	1.0	1.0
	60	70	68	1.5	1.0
	70	80	78	2.0	2.0
OBD-1	64	13	12	0.3	0.5
	85	16	15	0.5	0.6
	93	18	17	—	—

second component and in shifting of the upper critical solution temperature (UCST) to diepoxide region. Table 2 presents the data from both methods, which are quite consistent for various composition regions. This compatibility confirms reliability of the results obtained.

The data indicate that solubility grows with temperature increasing for all considered systems. However, UCSTs were not experimentally reached in many systems, owing to decomposition of the components at high temperatures. For such cases, UCST values were estimated from Flory–Huggins (FH) theory [19]. Relationship (1) was used, and linear dependence of χ_{12} on inverse temperature was extrapolated to χ_{cr} derived from (2).

$$\chi_{12} = \frac{\left[\ln\varphi_1' + \left(1 - \frac{\overline{V}_1}{\overline{V}_2}\right)\varphi_2'\right] - \left[\ln\varphi_1'' + \left(1 - \frac{\overline{V}_1}{\overline{V}_2}\right)\varphi_2''\right]}{(\varphi_2'')^2 - (\varphi_2')^2} \qquad (1)$$

$$\chi_{cr} = \frac{1}{2} + \sqrt{\frac{\overline{V}_1}{\overline{V}_2}} + \frac{\overline{V}_1}{2\overline{V}_2}, \qquad (2)$$

where φ_1, φ_2 are components volume fractions, $\overline{V}_1$ и $\overline{V}_2$ are their molar volumes, χ is the Flory–Huggins interaction parameter; the symbols related to the equilibrium phases are primed.

The values for critical points show satisfactory coincidence with the majority of experimentally determined UCSTs (Table 3).

Table 3. Comparison of experimental and calculated UCST values

Copolymer	χ_c	UCST, °C	
		Calculation	Experiment
CNAA-1	1.20	125	125
BCP-2	0.61	74	75
BCP-3	0.62	90	95
BCP-4	0.64	134	140
BCP-5	0.62	198	—

These results show that such binary systems may be described by the FH theory. Meanwhile, some systems were found not to obey to this theory, in particular, copolymers of NA with AA.

The following regularities were derived from analysis of experimental and calculated data. For oligobutadiene derivatives (Figs. 1–3), oligomer MW, hydroxyl group concentration, and functionality influence the degree of thermodynamic compatibility. For oligomers with terminal groups, MW correlates with concentration, so that a decrease in $\overline{M}_n$ results in the increase in hydroxyl group concentration. Thus, both factors favor the increase in thermodynamic affinity. As follows from Figs. 1–3, for hydroxyl-containing oligomers with similar functionality, $\overline{M}_n$ influence is more pronounced than for non-functional oligomers. Meanwhile, for mono- and bifunctional oligomers with almost similar hydroxyl groups content and $\overline{M}_n$ values differing by a factor of two, solubility is almost equal. Generally, for systems based on DGEBA and hydroxyl-terminated

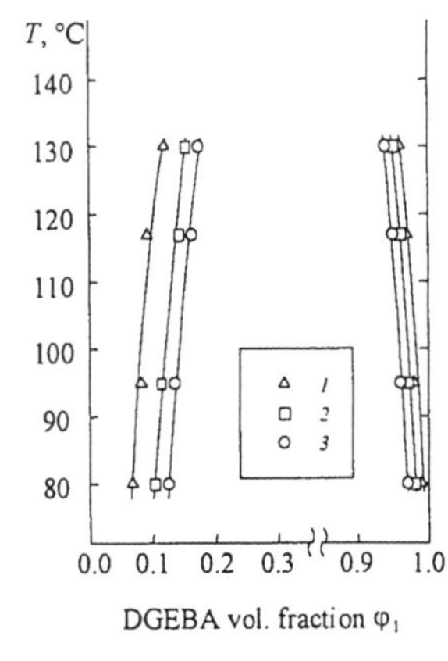

Figure 1. Phase diagrams for the DGEBA–oligobutadiene system with no terminal hydroxyl groups: (*1*) OB-1, (*2*) OB-2, (*3*) OB-3.

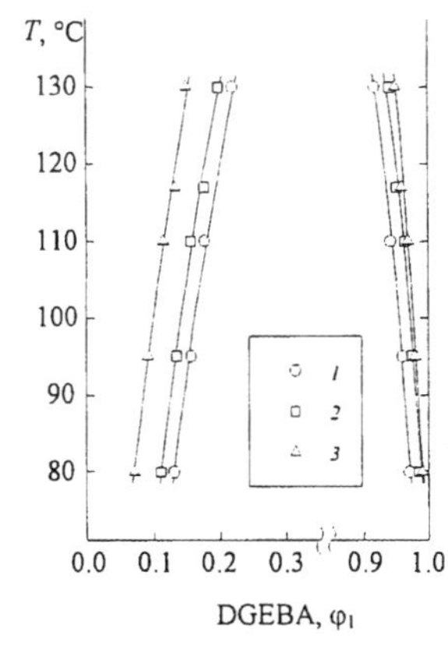

Figure 2. Phase diagrams for the DGEBA–oligobutadiene monool system: (*1*) OBM-1, (*2*) OBM-2, (*3*) OBM-3.

oligobutadiene derivatives, MW and corresponding OH-groups content essentially affect the phase equilibrium.

If the phase boundary concentrations at T = const are plotted as a function of hydroxyl concentration (Fig. 9), the data for mono- and bifunctional oligomers form a common dependence. Bromination of oligomers leads to the growth of the components thermodynamic compatibility. In this case, the increase in the bromine content from 0 to

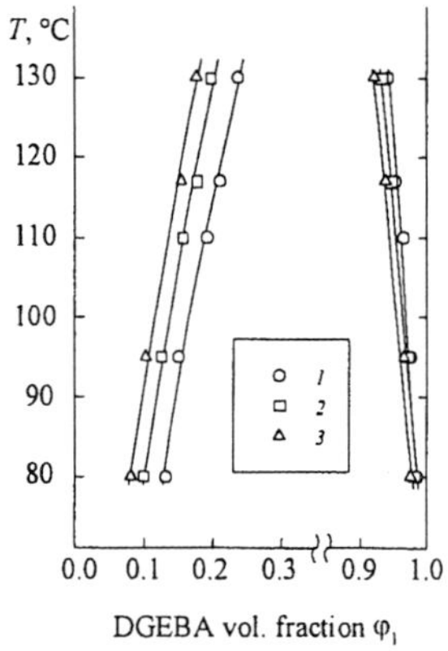

Figure 3. Phase diagrams for the DGEBA–oligobutadienediol system: (*1*) OBD-1, (*2*) OBD-2, and (*3*) OBD-3.

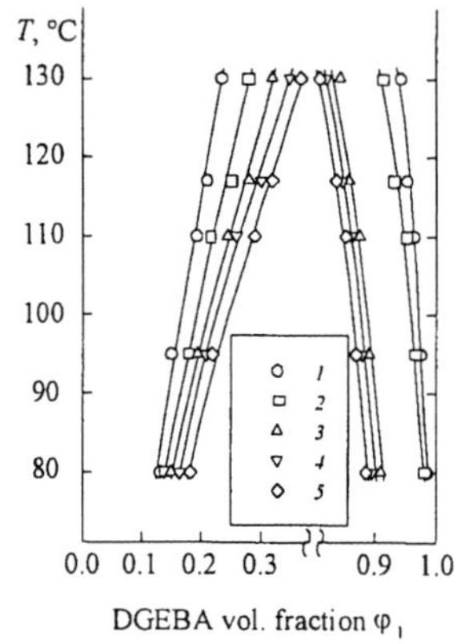

Figure 4. Phase diagrams for the DGEBA–brominated oligobutadienediol system: (*1*) OBD-1, (*2*) 12% Br, (*3*) 20% Br, (*4*) 40% Br, (*5*) 50% Br.

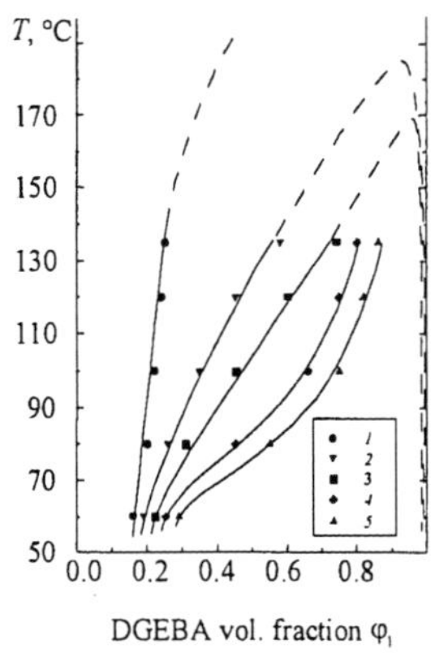

Figure 5. Phase diagrams for the binary system NA copolymer and AA–DGEBA: (*1*) Pure PNA, (*2*) 10% AA, (*3*) 20% AA, (*4*) 30% AA, (*5*) 40% AA.

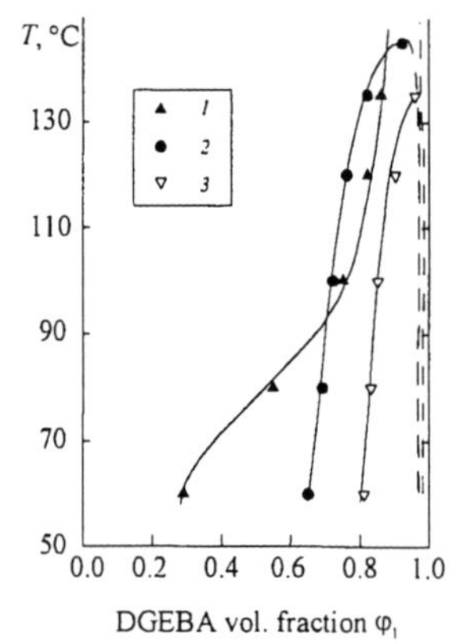

Figure 6. Phase diagrams for the binary systems of NA copolymer and (*1*) 40% AA–DGEBA, (*2*) diisopropyladipinate, (*3*) diethyladipinate.

50 wt % leads to reduction of the UCST value by about 40°C and to the increase in the DGEBA solubility (Fig. 4).

Interesting interaction features were found for random copolymers of NA and AA. Solubility of the components grows, and the region of two-phase state of solutions decreases with growing content of polar acrylic acid (Fig. 5). However, UCSTs were not reached in the studied temperature range for any systems. CNAAA solubility in DGEBA is low, a few percent,

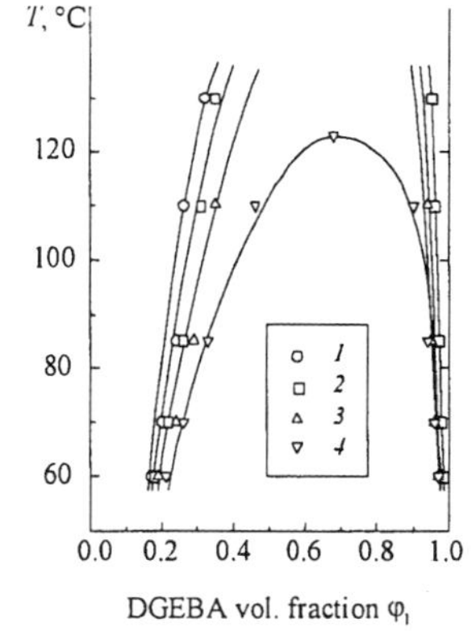

Figure 7. Phase diagrams for binary systems of NA copolymers with AA–DGEBA: (*1*) CNAA-4, (*2*) CNAA-3, (*3*) CNAA-2, (*4*) CNAA-1.

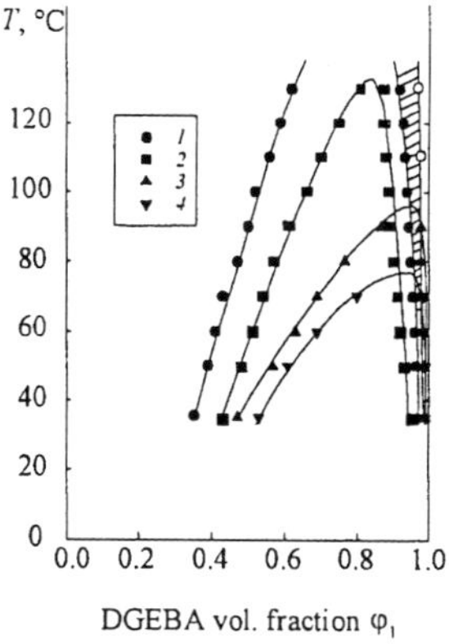

Figure 8. Phase diagrams for binary systems of block copolymers–DGEBA: (*1*) BCP-5, (*2*) BCP-4, (*3*) BCP-3, (*4*) BCP-2.

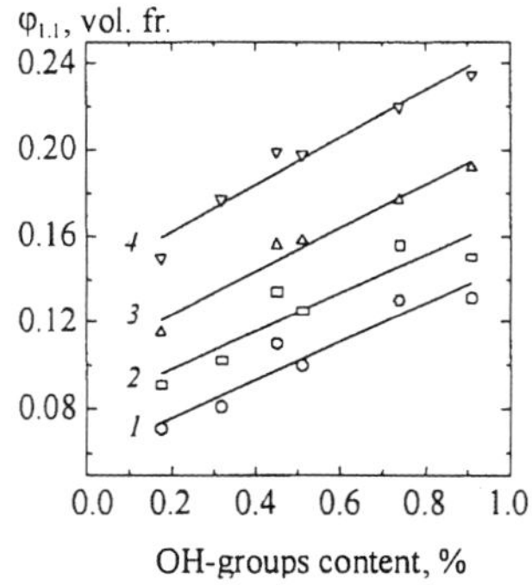

Figure 9. Phase boundary concentrations *vs.* OH-groups concentrations in oligomers based on butadiene (*1*) 80°C, (*2*) 95°C, (*3*) 110°C, (*4*) 130°C.

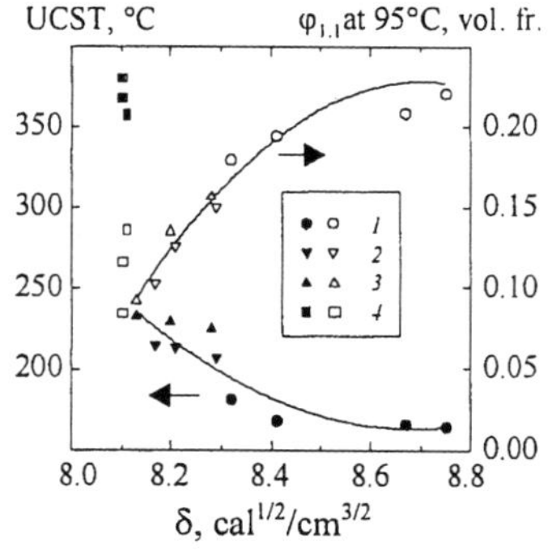

Figure 10. Calculated UCTS and DGEBA solubility in butadiene-based oligomers *vs.* their solubility parameter. (*1*) oligobutadienes, (*2*) oligobutadienemonools, (*3*) oligobutadienediols, (*4*) brominated oligobutadienediols.

therefore the right branches of binodals are shown by dashed lines in Fig. 5. It should be noted that DGEBA solubility increases essentially in copolymers containing more than 20% of acrylic acid and at temperatures above 100°C. Apparently, the reduction of the hydrogen bond influence, due to the drastic decrease of their stability in this temperature range, is responsible for this fact. For a rather small content of acrylic acid (10 and 20 wt %) PDs are satisfactorily described by FH theory (Fig. 5). However, the transition to the copolymer with AA content of 30 and 40 wt % results in a drastic change of the binodal shape: it becomes similar to the diagram for chemically cross-linked systems, and it is not governed by FH theory. Apparently, this behavior is due to rather weak interaction of copolymers with DGEBA, and to strong network of hydrogen bonds, which cannot be destroyed by diepoxide even at elevated temperatures. If DGEBA having a large, low-polar and stiff molecule is replaced by more polar adipinates having a few orders higher coefficients of self-diffusion, PDs have a normal shape and obey the FH theory (Fig. 6).

Phase diagrams for copolymers of nonyl acrylate with acrylamide are described adequately by the FH theory (Fig. 7).

As follows from the phase diagrams of systems of DGEBA and polyurethane block copolymers based on polypropyleneglycol (PPG), oligobutadiendiol and toluylenediisocyanate (Fig. 8), the increase of PPG MW and of its content in the block copolymer results in the growth of UCST and of the region of two-phase state at corresponding temperatures.

It was found that the phase separation of block copolymers in the region of diluted DGEBA solutions was characterized by the signs of unfinished phase separation in the field of concentration gradient of the interference picture. This appears as a region with isolated phases in a diffusion zone, which can be distinctly seen as dark points in the visual field of the microscope. For this composition range, the signs of the phase separation are also registered by light scattering technique. For all systems studied, the phase boundary and a region of monophase solutions are observed, in addition to the above microphase separation. In this case, the region of microphase separation adjoins the phase boundary from the side of diluted solutions. In particular, in BCP-1–DGEBA the zone of microphase separation (without the phase boundary that is typical for separation region) lies between $\varphi_2' > 0.5$ wt % and $\varphi_2'' < 8$ wt % for T=50°C, and between $\varphi_2' > 0.5$ wt % and $\varphi_2'' < 5$ wt % for $T = 60$°C. Typically for this system, heating above 115°C leads to disappearance of the microphase particles, and cooling resumes the phase separation, i.e., it is reversible. For separation in this solution region it is common that the phases grow up to a certain size and do not coalesce.

Such a behavior is evidently due to selective dissolving ability of DGEBA towards polypropyleneglycol blocks. It can be assumed that in this

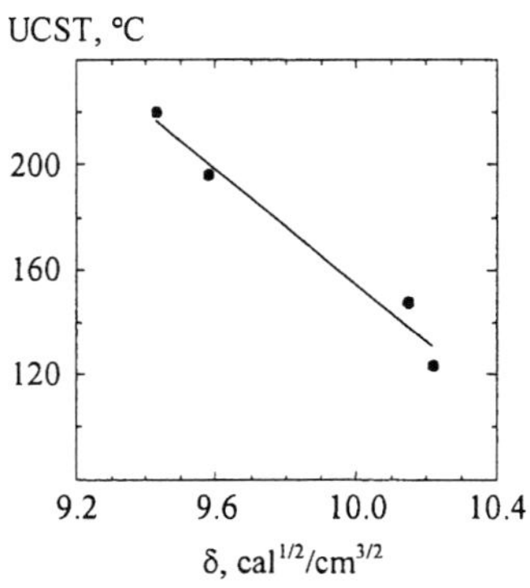

Figure 11. Calculated UCST vs. solubility parameter for copolymers of nonylacrylate and acrylamide.

region oligobutadiene blocks insoluble in DGEBA form microphases surrounded by swollen polypropyleneglycol. Blocks of similar MW can be present in different phases simultaneously. These data suggest that separation in these systems was unfinished, and PDs were non-equilibrium.

To derive a general dependence of the parameter of thermodynamic affinity degree, solubility parameters δ were calculated for all studied oligomers and polymers, using the method of group contributions [20]. However, we failed to derive a general dependence of UCST on δ. Figures 10 and 11 show the data for some investigated systems. The data for hydroxyl-containing and brominated hydroxyl-containing oligomers fit the graph of solubility parameter dependence of UCST and phase boundary concentrations (Fig. 10). For oligobutadienes without polar groups, and for copolymers of nonylacrylates with acrylamide (Fig. 11) UCST and boundary concentration values cannot be described by such dependence.

CONCLUSIONS

We failed to establish a regularity that could summarize the results obtained for all of the systems. At the same time, we found such regularities for some systems of similar nature. It should be noted that the relationship between the degree of intermolecular interaction in the polymer (copolymer, oligomer) and the solvent activity affects the shape of PD. When functional groups of a certain type are present in the polymer in a sufficient concentration, so that a network of physical bonds has some critical strength (e.g., for random copolymers of NA with 20% AA), their quasi-chemical influence on the PD form may be considered. For hydroxyl-containing oligomers and polymers having amide and carboxyl groups (< 20%), which form a system of hydrogen bonds, no influence on the PD shape is observed. In addition, our data show that the solvent should have a certain thermodynamic activity, because the effect was observed for an inert DGEBA but disappeared upon its replacement with adipinates.

Acknowledgments. This work was supported by the International Science and Technology Center (grant no. 358-96).

REFERENCES

1. Williams R.J.J., Rozenberg B.A., and Pascault J.-P., *Adv. Polym. Sci.*, **128**, 95 (1997).
2. Chalykh A.E., Volkov V.P., Roginskaya G.F., Avdeev N.N., Matveev V.V., and Rozenberg B.A., *Plast. Massy*, 25 (1981).
3. Volkov V.P., Roginskaya G.F., Chalykh A.E., and Rozenberg B.A., *Usp. Khim.*, **51**, 1733 (1982).
4. Mikhailov Yu.M., Ganina L.V., and Shapaeva N.V., *Polymer Sci., Ser. A*, **37**, 642 (1995).
5. Mikhailov Yu.M., Ganina L.V., and Chalykh A.E., *J. Polym. Sci., Part B: Polym. Phys.*, **32**, 1799 (1994).
6. Mikhailov Yu.M., Ganina L.V., and Baturin S.M., *Vysokomol. Soedin., Ser. A*, **33**, 51 (1992).
7. Mikhailov Yu.M., Ganina L.V., Smirnov V.S., and Makhonina L.I., *Proc. VI Int. Conf. on Chemistry and Physical Chemistry of Oligomers*, Kazan', Russia, 1997, p. 78.
8. Mikhailov Yu.M., Ganina L.V., Smirnov V.S., and Estrin Ya.I., *Proc. VI Int. Conf. on Chemistry and Physical Chemistry of Oligomers*, Kazan', Russia, 1997, p. 79.
9. Mikhailov Yu.M., Ganina L.V., Smirnov V.S., Estrin Ya.I., and Rozenberg B.A., *J. Appl. Polym. Sci.*, **71**, 953 (1999).
10. Mikhailov Yu.M., Ganina L.V., Makhonina L.I., Smirnov V.S., Shapayeva N.V., *J. Appl. Polym. Sci.*, in press.
11. Estrin Ya.I., Kasumova L.T., Baturin S.M., and Radugin V.S., *Polymer Sci., Ser. A.*, **38**, 813 (1996).
12. Korolev G.V., Makhonina L.I., Grachev V.P., and Baturina A.A., *Izv. Vyssh. Uchebn. Zaved., Khim. Tekhnol.*, **36**, 70 (1993).
13. Badamshina E.R., Estrin Ya.I., Grigor'eva V.A., Grishchuk A.A., Gorbushina G.A., Kasumova L.T., *et al.*, *this book.* p. 15.
14. Entelis S.G., Evreinov V.V, and Kuzaev A.I., *Reaktsionnosposobnye oligomery* (Reactive Oligomers), Khimiya, Moscow, 1985.
15. Malkin A.Ya., Askadskii A.A., Chalykh A.E., and Kovriga V.V., *Experimental Methods of Polymer Physics*, Prentice Hall, New York , 1983.
16. Mikhailov Yu.M., Ganina L.V., and Chalykh A.E., *J. Polym. Sci., Part B: Polym. Phys.*, **32**, 1799 (1994).
17. Chalykh A.E., Gerasimov V.K., and Mikhailov Yu.M., *Diagrammy sostoyaniya polimernykh system* (Phase Diagrams of Polymer Systems), Yanus-K, Moscow, 1998.
18. Mikhailov Yu.M., Ganina L.V., and Smirnov V.S., in *Struktura i dinamika molekulyarnykh sistem* (Structure and Dynamics of Molecular Systems), Unipress, Kazan', vol. 6, 90 (1999).
19. Flory P.J., *Principles of Polymer Chemistry*, Cornell University Press, Ithaca NY, 1953.
20. Van Krevelen D.W., Properties of Polymers: Correlation with Chemical Structure, Elsevier, Amsterdam, 1972.

Section II

THEORY OF MICROPHASE SEPARATION IN CURING SYSTEMS

Chapter 5

Cure Reaction-Induced Microphase Separation in Multicomponent Blends: Theory of Nonequilibrium Nucleation and Growth

Grigori M. SIGALOV* and Boris A. ROZENBERG

Institute of Problems of Chemical Physics, Russian Academy of Sciences, Chernogolovka, Moscow Region, 142432 Russia

*e-mail: sigalov@icp.ac.ru

ABSTRACT

The model of cure-reaction induced microphase separation proceeding via nucleation and growth mechanism with a moderate reaction rate and far enough from the critical point is proposed. The classical frequency of elementary acts of nucleation was adopted. The model of particle growth with equilibrium boundary conditions was developed to describe the changes in concentration profiles both inside and outside the particle and particle growth caused by diffusion. With the help of numerical modeling, single particle nucleation and growth at different reaction rates and nucleation delays was studied. Scaling laws for short and long times were found; in particular, the final particle size, r_{max}, depends on the overall reaction time, t_{proc}, as $r_{max} \propto t_{proc}^{1/2}$. The particle size distribution was qualitatively characterized. Ways to deal with particle ensembles and take into account effective interaction of particles were indicated.

INTRODUCTION

It is widely recognized that smartly designed composite materials are often more effective from different viewpoints than the original simple materials which have been combined to achieve new or improved properties. However, it is sometimes desirable but difficult or even impossible to bring together a couple of materials due to their incompatibility. Mechanical mixing of incompatible materials usually gives very rough dispersions, that is why this method is often ineffective. Therefore, heterophase polymer systems with a small size of inhomogeneities are generally obtained starting from systems compatible on the molecular level.

As the cure reaction proceeds and the average molecular weight (MW) of the thermoset increases, the solubility of the additive decreases, and the latter is demixing to form a heterophase structure. The increase in MW and the crosslinks density finally leads to gelation or vitrification of the matrix phase. This results in exhausting of the system mobility and fixation of the phase structure made up at a certain cure reaction conversion. This is a sketch description of the cure reaction-induced microphase separation (CRIMPS).

The operating properties of the heterophase material are mainly determined by its structure and composition of phases. Therefore, special care should be taken to control them by optimizing the initial blend composition and reaction conditions. General principles of such optimization and correlation between numerous factors that may affect the final structure became mainly known on a qualitative level after much experimental and theoretical work during last four decades [1–4]. Meantime, up to date no quantitative conclusions may be made as far as predicting the material structure and properties is concerned.

It is well known that the phase separation may proceed, depending on the initial conditions and the driving force parameters, through two mechanisms: nucleation-and-growth (NG) and spinodal decomposition (SD). In this paper we consider CRIMPS proceeding via NG mechanism in a quasibinary blend of components A and B, A being a thermosetting system (e.g., diepoxide + diamine and products of their reactions) while B, a non-reactive modifying oligomeric additive (e.g., liquid rubber). The components are compatible at the very beginning of the process. In the course of isothermal cure, MW of A increases, which leads to a decrease in the compatibility of A and B and finally to liquid–liquid phase separation of this system. In contrast to other systems exhibiting phase separation, in the case of CRIMPS the process is virtually stopped due to vanishing of the diffusion coefficient at gelation or vitrification of the thermoset-rich phase [1–4].

Among various factors that influence the final material morphology and properties, the cure reaction kinetics is still underestimated. The properties of the final material, in particular, its elastic moduli and fracture toughness, at the same volume fraction of the dispersed phase are mainly determined by the particle size distribution (PSD) and interparticle distances [5]. The consideration of the PSD function in such systems was first tried by Williams, Pascault, *et al.* [6]. They silently supposed that phase separation is much faster than chemical reaction. This means that phase separation is an equilibrium process at any extent of chemical reaction. In this case, the branches of the phase diagram directly give the composition of phases, and the volume fractions of phases are easily obtained from the initial and current compositions. The concentrations of components within each phase were considered to be uniform. In fact, if the diffusion is fast in comparison with the chemical reaction then there is enough time for the composition of phases to be equilibrated. This approach allowed us to qualitatively analyze how the morphology parameters depend on the initial composition and cure conditions. The applicability of this theory is restricted to the case of very slow cure reaction, which is generally not the case in the real systems.

In this paper, we have allowed for the real cure rate and its influence on the processes of nucleation and growth of the particles of new phase. In this case, the spatial distribution of the blend components within each phase is essentially non-uniform and is also taken into account. We were dealing with the case when the rate of reaction is of the same order of magnitude or moderately greater than the rate of diffusion, and the additive concentration is off-critical (in the limit of very high reaction rate and in the vicinity of the critical point, the phase separation proceeds through SD mechanism, which is not considered here). The nucleation act was described on the basis of Volmer–Frenkel–Zeldovich classical theories [7–9] with proper account of additive conservation at nucleation, the frequency of elementary acts being purely classical. As nucleation is a barrier-overcoming process and the barrier threshold of a critical nucleus formation decreases as the reaction goes on, all particles are born with some certain delay relative to the moment of reaching the binodal. The model of particle growth with equilibrium boundary conditions was developed [10] to describe changes in the concentration profiles both inside and outside the particle and particle growth caused by diffusion. Nucleation and growth of a single particle at different reaction rates and nucleation delays were simulated numerically to reveal new regularities of the process. Scaling laws for short and long times were found; in particular, the final particle size, $r_{\max}$, was found to depend on the overall reaction time, t_{proc}, as $r_{\max} \propto t_{\text{proc}}^{1/2}$. The particle size distribution was qualitatively characterized. Although the numerical data were obtained only for single particles and are valid for ensembles of particles, which are rather far from each other (low additive concentration), it was shown how to deal with denser particle ensembles and to take into account the effective interaction between particles.

RESULTS AND DISCUSSION

Phase Diagrams of Reacting Systems

As a rule, the mutual solubility of various non-reacting substances (in particular, polymers) is studied under equilibrium conditions. This yields the phase diagrams plotted as temperature against composition. In contrast to non-reacting systems, the compatibility of systems that contain a curing thermoset depends, besides temperature, on the changing curing oligomer MWD. The latter, in turn, is determined by reaction conversion. This suggests expressing the phase diagrams for isothermal CRIMPS processes conversion against system's composition [3]. Such diagrams (Fig. 1) are often well described within the framework of the Flory–Huggins theory [11]. The density of the free energy of mixing for the quasibinary system described above may be written down as

$$\Delta g_{\text{m}} = RT\left(\frac{\rho_{\text{A}}}{M_{\text{A}}}(1-\phi)\ln(1-\phi) + \frac{\rho_{\text{B}}}{M_{\text{B}}}\phi\ln\phi + \chi\phi(1-\phi)\right),$$

where M_{A} and M_{B} are weight-averaged MWs of the components, ρ_{A} and ρ_{B} are their densities, ϕ is the volume fraction of B component, and the interaction parameter χ depends on temperature and conversion $q(t)$. The changes in density during reaction are usually neglected. The value M_{B} is constant, while M_{A} increases in line with conversion. The latter dependence may be found from the chemical kinetics of the curing reaction.

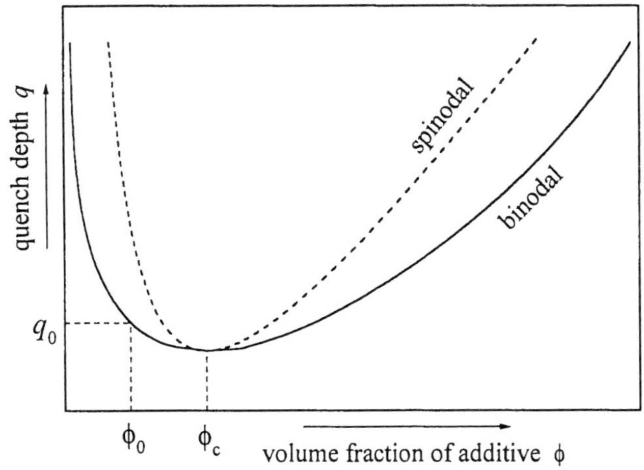

Figure 1. Typical phase diagram of a blend of curing thermoset and modifying additive.

It is worth noting that this consideration is based on the principles of equilibrium thermodynamics and therefore it is most accurate in the limit case of small reaction rate or high diffusion rate. The criterion of equilibrium of CRIMPS [10, 12] allows determining whether the given process may be considered as quasi-equilibrium and, therefore, whether the above approach may be used.

When CRIMPS proceeds under quasi-equilibrium conditions, then the phase separation starts shortly (in terms of conversion) after the moment when the binodal of the phase diagram is achieved. This allows plotting the phase diagram on the basis of cloud point measurements. Under nonequilibrium conditions, the phase separation is essentially retarded, i.e., it starts at a higher reaction conversion.

Model of Nucleation Act

Let $\Delta g_m(\phi, q)$ be the specific (per unit volume of the blend) free energy of mixing for the given pair of components, ϕ being the volume fraction of the second component, or the blend composition, and q the extent of cure reaction. Using the conventional procedure [11], it is possible to calculate the binodal and spinodal curves, i.e., the phase diagram (Fig. 1). Let us represent the branches of the binodal by functions $\phi = \phi_{1e}(q)$ and $\phi = \phi_{2e}(q)$. Let the initial composition of the system be ϕ_0; we suppose that ϕ_0 is not close to the critical composition ϕ_c. For definiteness, let ϕ_0 be less than ϕ_c. While q is increasing, the system remains homogeneous at least until the binodal is reached at q_0 such that $\phi_{1e}(q_0) = \phi_0$; and the metastable region of the phase diagram is entered. If the reaction is slow enough, q_0 corresponds to the cloud point.

It is expected that the phase separation gets started at this moment. We shall describe it on the basis of the classical theory [7–9] applied to the following model of an elementary nucleation act. We suppose that, at every $\Delta q = q - q_0 > 0$, the birth of a nucleus of radius r has a probability depending on r, q_0, and Δq, that the nucleus is a spherical particle, and that every act of nucleation is instantaneous. Here Δq has a sense of the depth of chemical quenching.

We also suppose that nucleation is a local process and leads to concentration change only inside the sphere of an unknown radius $R > r$, its center coinciding with that of the nucleus (see Fig. 2). The concentration profile immediately after nucleation is taken to be a stepwise function of the radial coordinate. Namely, the composition is uniform within every of the three regions, I (nucleus shell), II (nucleus core) and III (undisturbed region). The initial concentrations inside the nucleus core and shell are supposed to be equal to the equilibrium (at given q) concentrations $\phi_2 = \phi_{2e}(q)$ and $\phi_1 = \phi_{1e}(q)$ of the secondary and primary phases, respectively. In the region III $\phi = \phi_0$ at this moment.

Let us denote the volumes of the nucleus core and shell and their total volume as follows:

$$V_2 = \frac{4}{3}\pi r^3, \quad V_0 = \frac{4}{3}\pi R^3, \quad V_1 = V_0 - V_2, \tag{1}$$

respectively. It is then easy to find from the principle of matter conservation that

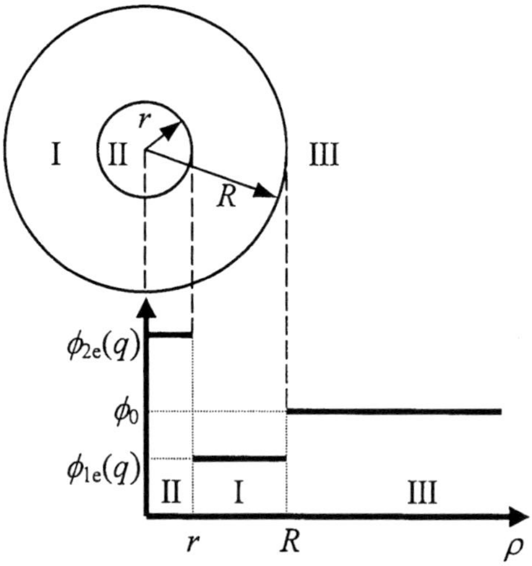

Figure 2. The model of nucleus and the additive concentration profile just after the nucleation.

$$\omega = \frac{V_2}{V_0} = \frac{\phi_0 - \phi_1}{\phi_2 - \phi_1} = \frac{\phi_{1e}(q_0) - \phi_{1e}(q)}{\phi_{2e}(q) - \phi_{1e}(q)}, \quad \frac{V_1}{V_0} = 1 - \omega, \tag{2}$$

where $\omega = \omega(q_0, q)$ is an auxiliary dimensionless parameter. Therefore, the change in the system free energy due to formation of a nucleus is as follows:

$$\Delta G = V_1 \Delta g_m(\phi_1) + V_2 \Delta g_m(\phi_2) - V_0 \Delta g_m(\phi_0) + 4\pi r^2 \sigma = -V_0 \varepsilon + 4\pi r^2 \sigma, \tag{3}$$

where Δg_m is the energy of mixing per unit volume, σ is the surface tension, and $\varepsilon = \varepsilon(q_0, \Delta q)$ is the average density of the free energy of phase separation taken with the opposite sign:

$$\varepsilon = -\Delta g_m(\phi_{1e}(q)) + \Delta g_m(\phi_{1e}(q_0)) - \omega\left[\Delta g_m(\phi_{2e}(q)) - \Delta g_m(\phi_{1e}(q))\right]. \tag{4}$$

The phase separation may occur if $\varepsilon > 0$. Let us express V_0 through V_2 with help of (1) and substitute $V_2 = 4/3\pi r^3$, then the free energy of a particle formation (3) may be rewritten as

$$\Delta G(r, q) = -\frac{4\pi r^3 \varepsilon}{3\omega} + 4\pi r^2 \sigma. \tag{5}$$

The use of the stationary condition $\partial G/\partial r = 0$ yields the critical nucleus radius,

$$r_c = 2\sigma\omega\varepsilon^{-1}. \tag{6}$$

Since $\omega = V_2 / V_0 = r^3/R^3$, the radius of the critical nucleus shell is

$$R_c = r_c \omega^{-1/3} = 2\sigma\omega^{2/3}\varepsilon^{-1}. \tag{7}$$

The free energy of the critical nucleus formation is as follows:

$$\Delta G^*(q) = \Delta G(r_c, q) = \frac{16\pi}{3} \frac{\sigma^3 \omega^2}{\varepsilon^2}. \tag{8}$$

The probability of the birth of a critical nucleus in a unit volume in a unit time is taken, according to the classical theory (see, e.g., [13]), in the form:

$$p = A \exp(-\Delta G^* / RT), \quad A = A' \exp(-\Delta G_\eta / RT), \tag{9}$$

where ΔG_η is the free energy of viscous flow activation and A' is a factor. Substitution of (8) into (9) yields

$$p = A\exp\left(-(16\pi/3RT)\,\sigma^3\omega^2\varepsilon^{-2}\right). \tag{10}$$

If the chemical quench depth Δq is small enough, it is possible to expand the functions $\phi_{ie}(q)$ into series around $q = q_0$ and thus obtain simple estimates for the values ω, and so on. In particular, from equation (2) it follows that

$$\omega \approx -\psi_1\left[\phi_{2e}(q) - \phi_{1e}(q)\right]^{-1}\Delta q\,, \tag{11}$$

where

$$\psi_1 = \left(\frac{d\phi_{1e}}{dq}\right)_{q=q_0} = \left[\left(\frac{dq}{d\phi}\right)_{\phi=\phi_0}\right]^{-1}.$$

To estimate ε which is given exactly by equation (4) note that

$$\begin{aligned}&\Delta g_{\mathrm{m}}(\phi_{1e}(q)) - \Delta g_{\mathrm{m}}(\phi_{1e}(q_0)) \approx \\ &\approx \left(\frac{d\Delta g_{\mathrm{m}}}{d\phi}\right)_{\phi=\phi_0}\left(\frac{d\phi_{1e}}{dq}\right)_{q=q_0}\Delta q = \psi_1\Delta q\left(\frac{d\Delta g_{\mathrm{m}}}{d\phi}\right)_{\phi=\phi_0}.\end{aligned} \tag{12}$$

Let us recall now that the common tangent to two conditional minima of the plot $\Delta g_{\mathrm{m}}(\phi)$ gives the binodal points ϕ_{1e} and ϕ_{2e} (see Fig. 3). Therefore, in these points the derivative $d\Delta g_{\mathrm{m}}/d\phi$ is equal to the slope of the line connecting these two points, $[\Delta g_{\mathrm{m}}(\phi_{2e}) - \Delta g_{\mathrm{m}}(\phi_{1e})]/(\phi_{2e} - \phi_{1e})$, and we find that

$$\begin{aligned}\omega\left[\Delta g_{\mathrm{m}}(\phi_{2e}(q)) - \Delta g_{\mathrm{m}}(\phi_{1e}(q))\right] &\approx -\frac{\psi_1\Delta q}{\phi_{2e}(q) - \phi_{1e}(q)}\left(\frac{d\Delta g_{\mathrm{m}}}{d\phi}\right)_{\phi=\phi_{1e}(q)}\times \\ &\times\left[\phi_{2e}(q) - \phi_{1e}(q)\right] = -\psi_1\Delta q\left(\frac{d\Delta g_{\mathrm{m}}}{d\phi}\right)_{\phi=\phi_{1e}(q)}\end{aligned} \tag{13}$$

Substituting (12) and (13) into equation (5), we obtain

$$\varepsilon \approx -\psi_1\Delta q\left(\frac{d\Delta g_{\mathrm{m}}}{d\phi}\right)_{\phi=\phi_{1e}(q_0)} + \psi_1\Delta q\left(\frac{d\Delta g_{\mathrm{m}}}{d\phi}\right)_{\phi=\phi_{1e}(q)} \approx \kappa\psi_1^2\Delta q^2\,, \tag{14}$$

where $\kappa = \left(d^2\Delta g_{\mathrm{m}}/d\phi^2\right)_{\phi=\phi_0}$. Let us further simplify equation (11):

$$\omega \approx -\psi_1\Phi^{-1}\Delta q, \quad \Phi = \phi_{2e}(q) - \phi_{1e}(q) \approx \phi_{2e}(q_0) - \phi_{1e}(q_0)\,.$$

Therefore, ω and ε are found to be proportional to Δq and $(\Delta q)^2$, respectively, the proportionality factors being constant (within the approximation involved) and composed only of some values and derivatives (Φ, ψ_1, κ) taken at the moment when the system reaches the boundary of the metastable region ($q = q_0$).

Finally, we can substitute (11) and (14) into equations (6) and (7) to find that (see Fig. 4a)

$$r_c \approx \frac{2\sigma}{\kappa\Phi\,|\,\psi_1\,|\,\Delta q}, \quad R_c \approx \frac{2\sigma}{\kappa(\Phi\psi_1^2\Delta q^2)^{2/3}}. \tag{15}$$

The free energy (8) and probability (10) of a critical nucleus formation take the form (Fig. 4b)

$$\Delta G^* \approx \frac{16\pi}{3}\frac{\sigma^3}{(\kappa\Phi\psi_1\Delta q)^2}, \quad p \approx A\exp\left(-\frac{16\pi}{3RT}\frac{\sigma^3}{(\kappa\Phi\psi_1\Delta q)^2}\right). \tag{10'}$$

Model of Particle Growth

When the phase separation is essentially nonequilibrium, the additive distribution in the bulk of each phase is not uniform. In this case the thermodynamic theory (in particular, the Flory–Huggins approach) cannot be applied without additional assumptions, and a more adequate

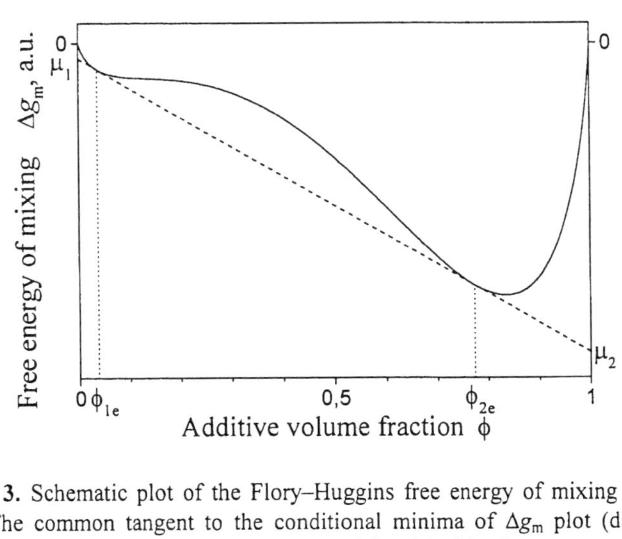

Figure 3. Schematic plot of the Flory–Huggins free energy of mixing Δg_m (solid line). The common tangent to the conditional minima of Δg_m plot (dashed line) yields the binodal points and chemical potentials of the blend components.

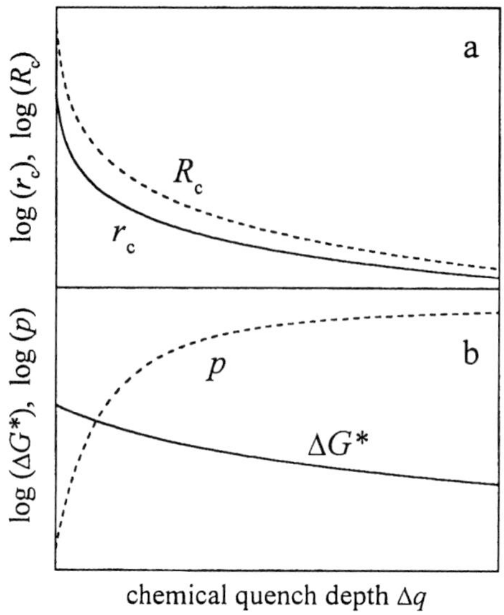

Figure 4. Schematic representation of the behavior of critical nucleus characteristics as functions of the quench depth.

nonequilibrium theory, e.g., the Cahn–Hilliard–de Gennes (CHdG) approximation [14] or equation-of-state theory, are to be used. Unfortunately, the CHdG equation may only be solved numerically, and, despite recent progress in this field [15, 16], it takes too much computer time to simulate realistic processes using CHdG.

This is a reason to apply the model of particle growth with equilibrium boundary conditions [10] that combines the simplicity of the Flory–Huggins theory and the generality of the Cahn–Hilliard approach. This model allows describing the evolution of the initial concentration profile given in Fig. 2 due to continuing reaction. Consider a particle born at the moment t_1, the corresponding quench depth being $\Delta q \approx \upsilon\tau_1$, where υ is the reaction rate and $\tau_1 = t_1 - t_0$. The concentrations of the additive in the particle shell and core at the moment of the nucleus birth are $\phi_{1e}(q(t_1)) \approx [\phi_{1e}(q(t_0)) + \psi_1\upsilon\tau_1]$ and $\phi_{1e}(q(t_1)) \approx [\phi_{2e}(q(t_0)) + \psi_2\upsilon\tau_1]$, respectively, where $\psi_2 = d\phi_{2e}/dq$. The radii of the particle shell and core, r_c and R_c, may be found with help of equations (15). If the quench were infinitely slow, the concentration profile across the particle center would be kept constant; in particular, at a moment $t_2 = t_1 + \tau_2$ the concentrations in the particle shell and core would be $\phi_{1e}(q(t_2)) \approx [\phi_{1e}(q(t_1)) + \psi_1\upsilon\tau_2]$ and $\phi_{2e}(q(t_2)) \approx [\phi_{2e}(q(t_1)) + \psi_2\upsilon\tau_2]$, respectively. However, due to finite rate of diffusion the real profile will be different, leading to nonuniformity of the additive distribution inside both phases. To estimate this distribution at the moment t_2 let us solve the diffusion equation in the spherical coordinates:

$$\frac{\partial \phi}{\partial t} - D\nabla^2 \phi = 0 \tag{16}$$

with the interdiffusion coefficient D depending on reaction conversion. D is inversely proportional to the blend viscosity, which is increasing due to an increase in M_A. According to [17], the following equation may be adopted:

$$D(q) = D_0[M_A(q)/M_A(0)]^{-3.4},$$

where D_0 and $M_A(0)$ are the initial values of D and M_A, respectively. Note that the dependence of M_A on conversion q must be specific of the chemical kinetics of cure reaction under consideration.

The initial and boundary conditions for equation (16) are as follows: (*i*) inside the particle core:

$$\phi = \phi_{2e}(q(t_1)), \quad 0 < \rho < r_c, \quad t = t_1; \tag{17}$$

$$\phi = \phi_{2e}(q(t)), \quad \rho = r, \quad t > t_1; \tag{18}$$

and (*ii*) outside the particle core:

$$\begin{aligned} &\phi = \phi_{1e}(q(t_1)), \quad r_c < \rho < R_c, \quad t = t_1; \\ &\phi = \phi_0, \quad r_c > R_c, \quad t = t_1, \end{aligned} \tag{19}$$

$$\phi = \phi_{1e}(q(t)), \quad \rho = r, \quad t > t_1. \tag{20}$$

The boundary equations take into account that the concentrations inside each region tend to the equilibrium value but only on the boundary $r = r_c$ it may be reached even at finite diffusion and reaction rates.

The solution of the diffusion problems stated above may be found by using conventional methods [18]. In particular, the quantity of the additive, which passed through the boundary, may be found. In general case, the quantity Q_1 that left from the particle shell does not equal the quantity Q_2 that entered the particle core. Within approximation of small enough τ_1 and τ_2, one may find:

$$Q_1 \approx |\psi_1| v \left\{ \tau_2 \left(\frac{2}{\sqrt{\pi}} \sqrt{D\tau_2} + \frac{D\tau_2}{r_c} \right) + \frac{\tau_1}{\sqrt{\pi}} \frac{D\tau_2}{r_c} \left(\frac{(R_c - r_c)^2}{4D\tau_2} \right)^{-3/2} \exp\left(-\frac{(R_c - r_c)^2}{4D\tau_2} \right) \right\}, \tag{21}$$

$$Q_2 \approx \psi_2 \nu \tau_2 \left\{ \frac{2}{\sqrt{\pi}} \sqrt{D\tau_2} \left[1 + \left(\frac{r_c^{\,2}}{D\tau_2} \right)^{-1} \exp\left(-\frac{r_c^{\,2}}{D\tau_2} \right) \right] - \frac{D\tau_2}{r_c} \right\}. \quad (22)$$

Due to inequality of material flows Q_1 and Q_2, the particle boundary will be shifted by

$$\Delta r = \frac{4\pi r^2 (Q_1 - Q_2)}{\phi_{2e}(q(t)) - \phi_{1e}(q(t))} \approx 4\pi r^2 (Q_1 - Q_2)/\Phi . \quad (23)$$

Thus, at every moment we can calculate the radius $\rho = r_c + \Delta r$ for every particle born at the moment t_1. Substituting the expression for r_c from equation (15) and keeping only major terms in equations (21) and (22), one obtains:

$$\rho(\tau_1, \tau_2) = \frac{2\sigma}{\Phi k |\psi_1| \nu \tau_1} + \frac{\nu \tau_2}{\Phi} \left[\frac{2}{\sqrt{\pi}} \left(|\psi_1| - \psi_2 \right) \sqrt{D\tau_2} + \left(|\psi_1| + \psi_2 \right) D\tau_2 \frac{\Phi k |\psi_1| \nu \tau_1}{2\sigma} \right].$$

Since the number of particles born at the moment t_1 is proportional to the probability p given by equation (10'), numerical integration over time period (t_0, t_2) will yield the particle size distribution (PSD) function at arbitrary $t_2 > t_0$.

In any case of practical importance, the above equations do not seem to be solvable analytically except for very short times. Meanwhile, it is the final structure attained on process completion that is most interesting from the practical viewpoint. Thus, we have resorted to numerical modeling to study the nucleation and growth of dispersed phase particles in a diepoxy/diamine/nonreactive rubber system. Whenever possible, real values of parameters involved have been used, so that the results are physically meaningful, though direct fitting of experimental data was beyond our goals. So, the method presented above may be characterized as semi-quantitative.

Numerical Results

We have simulated the CRIMPS process in the quasibinary system described above with different bimolecular reaction rate constants, k. Every program run yielded a growth plot, $r(t)$, t being varied up to gel point, for a single particle. In the first numerical experiment the nucleation delay, Δt_n, was varied within a range such that the nucleation probability was reasonable (not too small or too large). The data obtained for a set of particles born at different moments for the same k are shown on Fig. 5. It can be easily seen that PSD is very narrow for this ensemble, while the birth probability changes considerably within given range of Δt_n. It may be argued that bigger particles ($\Delta t_n < 200$) will virtually not appear because their birth probability is too small. Moreover, smaller particles ($\Delta t_n > 300$) will not appear since all the additive will be consumed by particles born earlier. Therefore, the statistical ensemble involved is representative, and the final PSD would be very narrow if the neighboring particles would not interact with each other.

The model allows taking into account such interaction approximately, in particular to estimate the concentration profiles between the centers of neighboring particles. In the profile evolution two stages may be distinguished. At first, the 'concentration fronts' are propagating from the particle boundaries independently, so that the additive concentration in the center of the segment connecting the particles centers, ϕ_{mid}, virtually does not change (Fig. 6). Then ϕ_{mid} and, consequently, the supersaturation start to decrease. This must result in hindering of further growth of particles due to depletion in their vicinity. The latter effect, however, cannot be quantitatively described within the framework of the given model.

At higher reaction conversions the interdiffusion coefficient may decrease so that the diffusion radius may become less than the half-distance between particles. Then ϕ_{mid} will be virtually fixed, while ϕ_{1e} will be still decreasing due to continuing reaction. Thus, because the supersaturation ($\phi_{mid} - \phi_{1e}$) is increasing this may lead to new wave of nucleation. The final size of such particles will be much less than the mean size of the first generation particles because of smaller supersaturation, diffusion coefficient and time left to grow. In fact, bimodal and even trimodal PSD's with

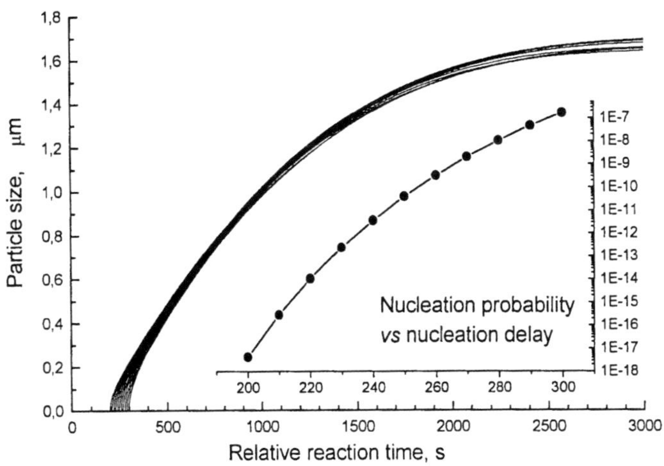

Figure 5. Particle growth plots for different nucleation delay times. Times are counted from the moment when the binodal was reached. The inlay plot shows corresponding nucleation probabilities. $\phi_{20} = 0.05$, $k = 10^{-6}$ mol^{-1} s^{-1}.

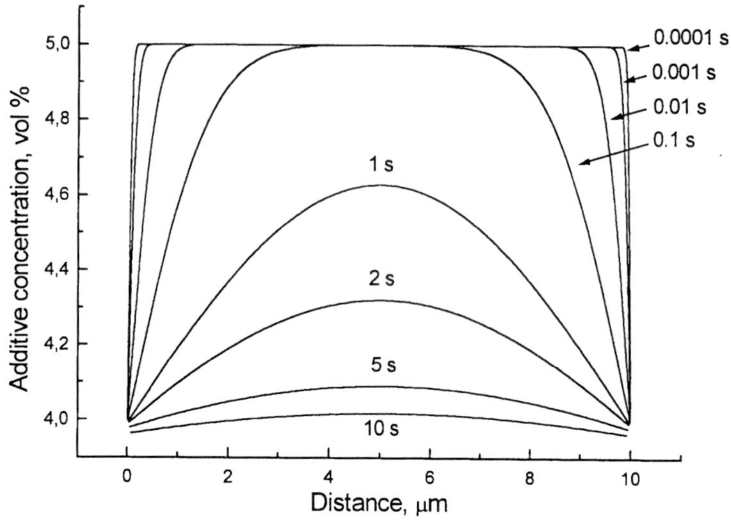

Figure 6. Concentration profiles between two particles with centers at 0 and 10 μm at different times indicated on the graph. Particles start to interact at $t \approx 0.1$ s. At larger times the curves are rather approximate and give estimates from below the real profiles.

particle sizes differing by an order of magnitude for successive generations have been observed earlier [3,19,20].

When the reaction rate constant was varied and the particles were chosen as born at the same reaction conversion (though at different times), growth plots were found to be very similar but differing from one other by a factor depending on k (Fig. 7). As expected, the faster the reaction, the smaller the particle. The final particle sizes, $r_{\max}$, plotted against k exhibit obvious power law $r_{\max} \propto k^x$, where $x = -0.4994 \pm 0.0005$ as found by the least-square method. In the case of bimolecular reaction, the total time of the process $t_{\text{proc}} \propto 1/k$, therefore, $r_{\max} \propto t_{\text{proc}}^{1/2}$ with high accuracy. This equation looks like the well-known law $r \propto t^{1/2}$ found earlier for diffusion-controlled growth [13]. However, note an important difference: in our case *neither of individual particles* obey the latter law as they grow, whereas the *ultimate radii* of the particles formed with different k's depend on t_{proc} in a misleadingly similar way. The reason for such dependence is still not absolutely clear.

Mutual Influence of Growing Particles

We have considered above the model describing the birth and growth of a single particle. It is still valid for particle ensembles if the interparticle distances are at least several times larger than the particle sizes. If this is not

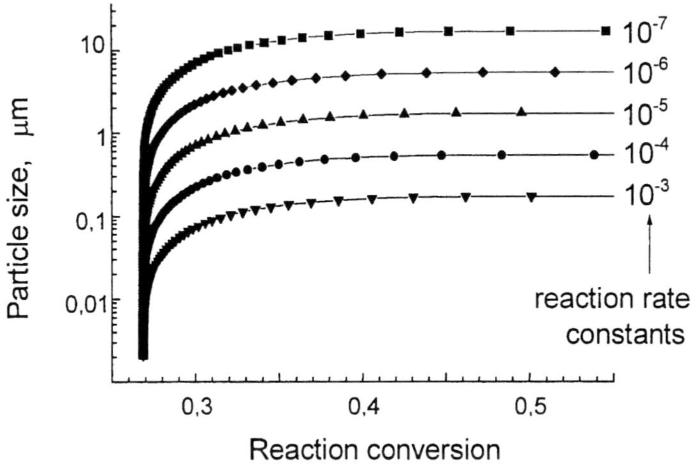

Figure 7. Particle growth plots for $\phi_{20} = 0.05$ and different reaction rate constants.

true, the influence of other particles on evolution of the given particle should be taken into account. This may be done within the framework of the proposed model.

Compare the above single-particle model to an infinite chain of particles of equal sizes placed along a line at a distance $2a$ from each other (Fig. 8). In the former case one of the boundary condition for the diffusion equation (16) is implicitly set at infinity since $\phi(\rho) = \phi_0$ for $\rho \to \infty$. In the latter case, the additive concentration profile between the centers of every two neighboring particles is symmetric relative to the middle point of this segment. Hence, such a one-dimensional multiparticle problem may be

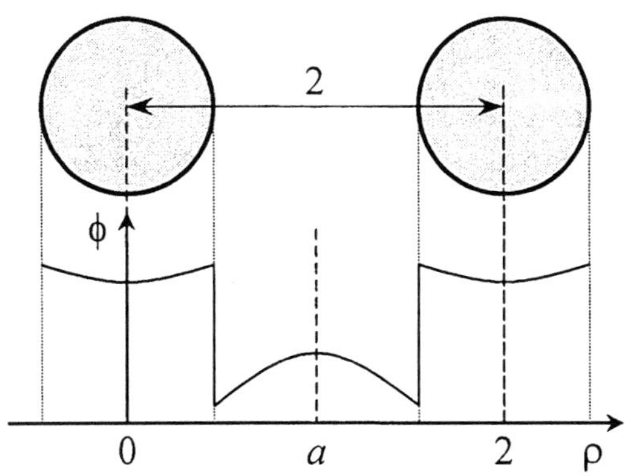

Figure 8. One-dimensional model of equidistant growing particles of α-phase (only two neighbors are shown) and a schematic plot of the additive concentration profile. The plot is symmetric relative each of the points 0, $\pm a$, $\pm 2a$, … .

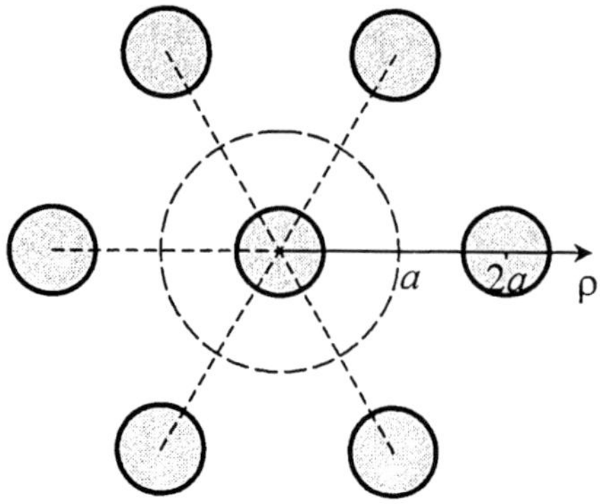

Figure 9. Three-dimensional version of the model shown in Fig. 8. The dashed circle approximately corresponds to the isoconcentrational curve for the additive.

reduced to the one-particle problem by modification of the boundary conditions. Namely, equations (18), (20) should be complemented by the condition of impermeable wall at $\rho = a$,

$$\partial\phi/\partial\rho = 0; \quad \text{at } \rho = a, \ t > t_1. \tag{24}$$

In the three-dimensional case the influence of the neighboring particles can be taken into account in a similar way. Let us consider particles of equal size located in the nodes of the hexagonal lattice with period $2a$ (Fig. 9). It is clear that the concentration profile along any edge of the lattice is symmetric relative to the center of the edge. Plot the surface of equal concentration through middle points of the lattice edges around a given particle. The analysis shows that this surface is close to spherical. Hence, the angular dependence of the additive concentration in the vicinity of the given particle may be neglected in spherical coordinates with center in the center of this particle. For the radial dependence of concentration, the boundary condition (24) should be used.

Such a consideration gives even better approximation if the particle size distribution is narrow and their spatial distribution is close to hexagonal lattice. These conditions are best met at high rates of nucleation as compared to the rate of particle growth.

Let us note that, as was shown above, the existing particles also affect the birth of new particles by formation of the area with lowered supersaturation of the additive in the matrix phase around themselves (see Fig. 6). As a result, new particles are usually born as far as possible from existing particles. Moreover, both the particle nucleation and growth rates in the second and further generations are always less than in the previous generation of particles due to fast decrease in the molecular mobility at cure [19,20].

CONCLUSIONS

Particle nucleation and growth at cure reaction-induced microphase separation (CRIMPS) is a peculiar process, which requires adequate methods of description. If this process is governed by sufficiently fast reaction, the existing CRIMPS theories seem to be inadequate. In this work, we have developed a new theory of CRIMPS as a combination of improved classical nucleation theory and the model of particle growth with equilibrium boundary conditions introduced recently by the authors. The new theory provides the possibility for the accurate description of the birth and evolution of a particle of new phase in the course of liquid-liquid phase separation in solution. We have first taken into account: (*i*) the influence of the real cure kinetics on the phase separation process and (*ii*) the spatial inhomogeneity of the modifying additive concentration both at the moment of nucleation and in the course of the particle growth. The latter means that the above model allows taking into account the influence of already existing particles on the birth of new particles in their vicinity as well as the mutual interaction of neighboring particles. Numerical modeling yields results that are valid for an ensemble of particles with interparticle distances much larger than the particle size. However, the evolution of an arbitrary ensemble of the dispersed phase particles may also be described with help of the above theory completed with proper statistical consideration.

The particle size distribution (PSD), which is one of the most important morphological characteristics of heterophase polymers, was qualitatively characterized. It was shown that it would be very narrow in absence of mutual influence of growing particles. It is possible to conclude that PSD must be generally wider for larger additive concentrations. More accurate prediction of final PSD requires explicit incorporation of particle interactions.

The results obtained demonstrate the strong dependence of the morphology of materials formed via CRIMPS on the reaction kinetics and encourage further research, which should be directed, first of all, to explicit consideration of interacting particle ensembles. Better understanding of the origin of the above scaling law is also desirable.

Acknowledgments. This work was supported by the International Science and Technology Center (grant no. 358-96) and the Russian Foundation for Basic Research (project no. 96-03-32027).

REFERENCES

1. Bucknall C.K., *Toughened Plastics*, Applied Science Publishers, London, 1977.
2. Paul D.R and Newman S., *Polymer Blends*, Academic Press, New York, 1978.
3. Williams R.J.J., Rozenberg B.A., and Pascault J.-P., *Adv. Polym. Sci.*, **128**, 95 (1997).
4. Inoue T., *Progr. Polym. Sci.*, **20**, 129 (1995).
5. Wu S., *Polymer*, **26**, 1855 (1985).
6. Moschiar S.M., Riccardi C.C., Williams R.J.J., Verchere D., Sautereau H., and Pascault J.-P., *J. Appl. Polym. Sci.*, **42**, 717 (1991).
7. Volmer M., *Kinetik der Phasenbildung*, Steinkopff, Dresden, 1939.
8. Frenkel Ya.I., *Teoriya zhidkostei* (Theory of Liquids), Nauka, Moscow, 1944.
9. Zeldovich Ya.B., *Zhurn. Eksp. Teor. Fiz.*, **12**, 525 (1942).
10. Rozenberg B.A. and Sigalov G.M., *Polym. Adv. Technol.*, **7**, 356 (1996).
11. Olabisi O., Robertson L.M., and Shaw M.T., *Polymer–Polymer Miscibility*, Academic Press, New York, 1979.
12. Rozenberg B.A. and Sigalov G.M., *Polymer Science, Ser. A*, **37**, 1049 (1995).
13. Christian J.W., The Theory of Transformations in Metals and Alloys: Part I. Equilibrium and General Kinetic Theory, Pergamon Press, Oxford, 1975.
14. de Gennes P.G., *J. Chem. Phys.*, **72**, 4756 (1980).
15. Shaginyan Sh.A. and Manevich L.I., *Polymer Science, Ser. A*, **39**, 908 (1997).
16. Shaginyan Sh.A., Manevich L.I., and Rozenberg B.A., *Vysokomol. Soedin. A*, 40, 2011 (1998).
17. Irzhak V.I., Rozenberg B.A., and Enikolopyan N.S., *Setchatye polimery: Sintez, struktura, svoistva* (Network Polymers: Synthesis, Structure, Properties), Nauka, Moscow, 1979.
18. Crank J., *The Mathematics of Diffusion*, Clarendon Press, Oxford, 1957.
19. Roginskaya G.F., Volkov V.P., Dzhavadyan E.A., Zaspinok G.S., Rozenberg B.A., and Enikolopyan N.S., *Dokl. Akad. Nauk SSSR*, **290**, 630 (1986).
20. Rozenberg B.A., *Makromol. Chem., Macromol. Symp.*, **41**, 165 (1991).

Chapter 6

Analytical Description of Microphase Formation in Curing Polymer Blends

Leonid I. MANEVICH*[1], Boris A. ROZENBERG[2], and Shagen A. SHAGINYAN[1]

[1]*Semenov Institute of Chemical Physics, Russian Academy of Sciences, 4 Kosygin Street, Moscow, 117977 Russia*

[2]*Institute of Problems of Chemical Physics, Russian Academy of Sciences, Chernogolovka, Moscow Region, 142432 Russia*

ABSTRACT

A previously proposed model is applied to analytical description of the microphase structures formation during cure-reaction-induced microphase separation (CRIMPS). The wave number of the mode of maximum growth of concentration

*e-mail: lmanev@chph.ras.ru

heterogeneities is determined. The behavior of this wave number in the regions of instability and stability of the initial blend homogeneous states is explored. Regularities of the spatial concentration structures formation at the linear stage of CRIMPS are investigated. The characteristic times for evolution of these structures as well as a change in the conversion degree and molecular weight of cured product are estimated.

INTRODUCTION

Formation of microphase heterogeneities during cure of multicomponent blends of reactive oligomers with modifiers has been experimentally studied for a long time [1–7] but only few papers have been devoted to theoretical explanation of this phenomenon [8–11]. The purpose of this study was the development of a theoretical model of the process.

The process of cure-reaction-induced microphase separation (CRIMPS) can involve the mechanism of nucleation followed by growth (NG) or spinodal decomposition (SD). The conditions for realization of both the mechanisms are discussed in detail in [1]. In this work, we investigated the process that involves SD.

It is supposed that the starting blend of thermosets is initially present in its stable homogeneous state. As cure proceeds, the initial homogeneous state becomes unstable with respect to composition fluctuations in the region between the binodal and spinodal curves of the phase diagram. At the intersection with the spinodal, this state becomes unstable relative to small long-wavelength fluctuations.

Available theoretical approaches to description of the formation of phase heterogeneities upon curing reactive oligomers are mainly based on the modified Flory–Huggins–de Gennes theory for binary polymer blends [7–11]. There exist the following three approaches to theoretical description of the spontaneous CRIMPS process.

- Process simulation is performed by successive increase in the quenching degree (by stepwise changing the system temperature) using the nonlinear Cahn–Hilliard equation of diffusion and new Flory theory [8]. The formation of microphase structures is described by a numerical solution to the nonlinear equation of diffusion. Under these assumptions, the authors have not found any qualitative difference between CRIMPS and usual SD induced by temperature change. As could be expected, the final system morphology is determined by the quenching rate.
- Self-consistent model of microphase separation induced by polymerization [9]. This takes into account only the composition corresponding to a critical point in the phase diagram of starting blend.

- Numerical simulation of the early stage of CRIMPS by using the linear Cahn–Hilliard equation of diffusion and the data on cure kinetics [10]. An important conclusion is made that the space scale of fluctuations is not constant in time as at the early stage of the conventional SD but steadily decreases. Experimental data on CRIMPS in an epoxy system obtained by light scattering technique have confirmed this theoretical conclusion [10].

Therefore, the results obtained up-to-date give an evidence of specific character of the SD separation in curing systems. Hence, an analytical study initiated in [11] becomes important for a deeper insight into the mechanism of this process. This work aims at a more detailed analysis of the early stage of the microheterogeneities formation during cure of a multicomponent blend of thermosets. The mutual influence of the curing and microphase separation is studied by taking an epoxy–amine system modified with a chemically inert linear polymer as an example. The equation describing the diffusion of modifier in the curing system may be solved analytically in the linear approximation. Using this equation, the system behavior is analyzed in the regions above and below the spinodal of the starting system.

RESULTS AND DISCUSSION

Description of the Model

The system under consideration is simulated as a pseudo-binary blend of chemically inert additive and polymer with increasing average degree of polymerization $N_2(t)$.

An additive with the polymerization degree N_1 is characterized by the average (over the bulk) invariant molar fraction φ_0. The chemical reaction is described by the conversion of functional groups, $\alpha(t)$, which determines the average degree of polymerization of the product. The function $\alpha(t)$ is dependent on the distribution of diffusible additive after beginning of phase separation. The reactive blend is characterized by the initial state (in the absence of chemical reaction), current state (at arbitrary moment of time, $t > 0$), and final state (after completion of chemical reaction, $t \to \infty$), where the initial and final states may be present in thermodynamic equilibrium.

The current state of reactive blend is characterized by spatial distribution of the additive concentration $\varphi(r,t)$ at a given time moment t (or Fourier transforms $u_k(t)$ of the deviation of this distribution from the average concentration φ_0) and by conversion of functional groups $\alpha(t)$. A blend of the additive and reaction product having the average chain length $N_2[\alpha(t)]$ corresponding to the current state.

The system in its initial state represents a blend of bifunctional diepoxy and tetrafunctional diamine with chemically inert linear polymer enclosed in volume V. Chemical reactions include the growth of linear chains by polycondensation mechanism. This growth leads to formation of a three-dimensional (3D) network. The curing kinetics for the epoxy–amine system in the absence of linear polymer is well studied in [1, 12, 13]. The curing of this system is clearly defined by the autocatalytic mechanism because the hydroxyl groups formed catalyze reaction between the amine and epoxy groups. For the time dependence of the conversion $\alpha(t)$ of the amine groups (at a stoichiometric ratio of amine to epoxy groups and upon neglect by noncatalytic reactions), we may write:

$$\alpha(t) = 4A_0 k_x t/(1 + 4A_0 k_x t), \tag{1}$$

In this case, the average polymerization degree of the reaction product may be presented in the form:

$$N_2(\alpha) = 3/[3 - 4\alpha(t)]. \tag{2}$$

Here $\alpha(t) = 1 - A(t)/A_0$; A_0 and $A(t)$ are initial and current molar concentrations of the amine monomer, and k_χ is the rate constant of their bimolecular reaction at a given temperature.

At the stage when $N_2(\alpha)$ depends weakly on time (small conversions) a current state of the reactive blend may be described by the modified Flory–Huggins–de Gennes model for binary blend [14–16]. This model is based on the diffusion equation for linear polymer:

$$\frac{\partial\varphi(r,t)}{\partial t} = div\left(\frac{\Lambda}{k_B T}\,\text{grad}\,\frac{\delta F}{\delta\varphi}\right). \tag{3}$$

The free energy of mixing per a lattice node is given by:

$$\frac{F}{k_B T} = \frac{F_0}{k_B T} + \frac{1}{2}L_s^2(\text{grad}\varphi)^2, \tag{4}$$

where

$$\frac{F_0}{k_B T} = \frac{\varphi\ln\varphi}{N_1} + \frac{(1-\varphi)\ln(1-\varphi)}{N_2(\alpha)} + \chi\varphi(1-\varphi).$$

The influence of the reaction in the second component on the modifier diffusion is taken into account via the time dependence $N_2(\alpha(t))$.

Expression for the free energy (4) involves also the additional contribution from the spatial inhomogeneity of the modifier concentration. This contribution is determined both by monomers linear connectivity in the chain and by potential energy of monomer-monomer interaction. As a result, the expression for the coefficient L_s includes both the entropy- and enthalpy-related contributions [15]:

$$L_s^2 = a^2\left(2\chi + \frac{1}{18\varphi(1-\varphi)}\right).$$

The model is completed by the expression for the Onsager coefficient in the diffusion equation (3):

$$\Lambda(\varphi(r,t)) = d_1\varphi. \tag{5}$$

This approximation [15,17] is applicable to rather short additive chains whose behavior corresponds to the Rouse model. Here χ is the Flory parameter of interaction, k_B is Boltzmann constant, T is temperature, d_1 is the self-diffusion coefficient, a is characteristic size of the blend components segments, and $\delta F/\delta\varphi = \mu$ is the exchange chemical potential of the blend.

The description of the concentration heterogeneity relaxation of the additive in terms of the Flory–Huggins–de Gennes (FHdG) model with the unique (upper) critical point of phase separation allows us to consider the system properties depending on the diffusion only. In this approximation, the blend can be simulated as an incompressible lattice, which is tightly filled with identical segments of the additive and curing oligomer.†

The Flory parameter χ is the temperature-reduced microexchange energy in the blend. In the majority of cases, $\chi > 0$. This implies that aggregation of identical monomers is thermodynamically favored. However, the entropy-related trend to the homogeneous distribution of different monomers competes with this process [18].

The mutual influence of phase separation and chemical reaction is conditioned by the fact that the curing reaction leads to the appearance and increase in supersaturation in the system, on the one hand, and to the decrease in interdiffusion coefficient, on the other hand.

---

† Rejection of the latter hypothesis may lead to appearance of a lower critical point, and the compressibility near this point should be also taken into account. However, this is far beyond the applicability limits for the FHdG model.

Solution of Linearized Diffusion Equation

Linearized diffusion equation for small concentration fluctuations, $u(r,t) = \varphi(r,t) - \varphi_0$, has the form:

$$\frac{\partial u(r,t)}{\partial t} = \Lambda_0 \nabla^2 \left(\frac{1}{N_1 \varphi_0} + \frac{1}{(1-\varphi_0) N_2(\alpha)} - 2\chi - L_{s0}^2 \nabla^2 \right) u(r,t) . \quad (6)$$

To solve equation (6), let us expand the concentration fluctuations into the Fourier series

$$u(r,t) = \sum_k u_k(t) \exp(ikr) ,$$

where $u_k(t)$ is the amplitude of the Fourier components of concentration fluctuation with the wave vector k. The diffusion equation for the Fourier transformants may be presented as

$$\frac{\partial u_k(t)}{\partial t} = R_f(k,t) u_k(t) . \quad (7)$$

Here

$$R_f(k,t) = \Lambda_0 L_{s0}^2 k^2 (k_f^2(t) - k^2) ,$$

$$k_f^2(t) = \frac{2}{L_{s0}^2} \left(\chi - \frac{1}{2N_1 \varphi_0} - \frac{1}{2(1-\varphi_0) N_2(\alpha)} \right) .$$

Equation (7) has an analytical solution for the stoichiometric ratio of the amine to epoxy groups when the expression for $N_2(\alpha)$ is reduced to (2). This solution has the form:

$$u_k(t) = c_k (1 + 4 A_0 k_x t)^{-\beta k^2} \exp(R_* t) . \quad (8)$$

Here c_k is the Fourier component of the initial concentration heterogeneity in the blend,

$$\beta = \frac{\Lambda_0}{3(1-\varphi_0) A_0 k_x} ,$$

and

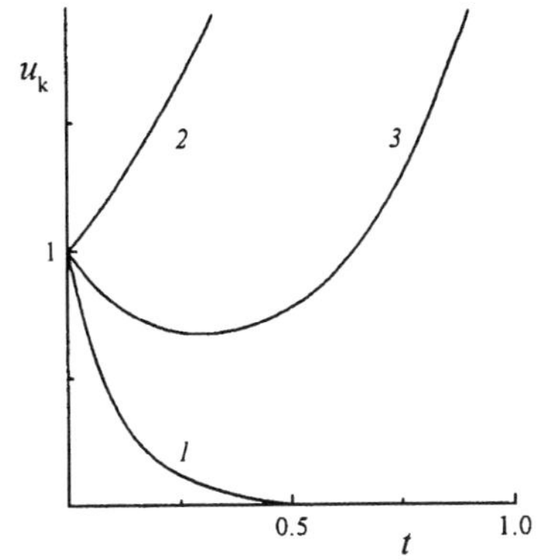

Figure 1. Time dependence of the k-th mode of concentration heterogeneity for the modes with (1) $k < k_c$, (2) $k_c < k < k_s$, and (3) $k > k_s$.

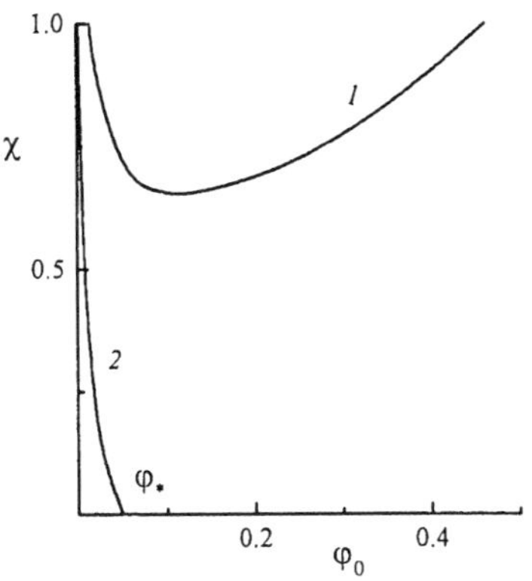

Figure 2. The spinodal curves for the (1) starting and (2) cured blend.

$$R_*(k) = \Lambda_0 L_{s0}^2 k^2 (k_*^2 - k^2), \tag{9}$$

where k_* is the threshold of structural stability of the curing blend given by the following expression:

$$k_*^2 = \frac{2}{L_{s0}^2}\left(\chi - \frac{1}{2N_1\varphi_0} + \frac{1}{6(1-\varphi_0)}\right).$$

As follows from (8), there are three types of the k-th mode of the time dependence of concentration heterogeneity (Fig. 1).

For $R_*(k) < 0$, solution (8) exhibits a fully relaxation behavior and corresponds to approximation to a stable homogeneous state of the reactive blend. In this case, microphase separation is impossible (Fig. 1, curve 1). On the thermodynamic plane of the phase states $(\chi - \varphi_0)$ for fully cured blend, the stability region of spatially homogeneous states is separated by the boundary curve (spinodal of curing system) (Fig. 2, curve 2):

$$\chi_*(\varphi_0) = \frac{1}{2N_1\varphi_0} - \frac{1}{6(1-\varphi_0)}.$$

This curve exists for the concentrations lower than the threshold, $\varphi_*(N_1) = 3/(3+N_1)$. In this concentration range, $\varphi < \varphi_*$ for $\chi < \chi_*(\varphi_0)$ for all k, such that $R_*(k)<0$. In this case, the reacting blend remains always homogeneous up to the end of cure. Any heterogeneity relax according to equation (8). This is the stability region for spatially homogeneous states of the fully cured system. It should be noted that both the existence of this region and corresponding peculiarities of the blend properties were unknown and were not studied earlier[‡] (the region restricted by the coordinate axes and curve *2* in Fig. 2).

In the concentration range $\varphi<\varphi_*$ for $\chi>\chi_*(\varphi_0)$ as well as in the range $\varphi_*<\varphi<1$ at all temperatures, $\chi>0$ and the value of k_*^2 is positive. Accordingly, $R_*(k)<0$ for $k>k_*$, while $R_*(k)>0$ for $k<k_*$. This means that the small-scale concentration heterogeneities in the curing blend relax within the given region of system thermodynamic states, while the blend becomes unstable relative to large-scale heterogeneities with wave numbers $k<k_*$, so that phase separation takes place. Note that this region covers the whole plane of thermodynamic states, except for a small part restricted by curve *2* in Fig. 2 (this is the region of instability for the homogeneous states of curing blend). However, the development of instability in this region has significant distinctions from conventional SD.

For $R_*(k)>0$, a strongly growing factor at $\exp(R_*(k(t)))$ opposes the decreasing power index ($\beta>0$ for any thermodynamic state of reactive blend). Consequently, the initial relaxation of concentration fluctuations is inevitably changed [see Eq. (11)] by their growth. The behavior of the blend at the earlier stage is determined by the rate of initial relaxation:

$$\left.\frac{d\ln u_k(t)}{dt}\right|_{t=0}=R(k)\,. \tag{10}$$

Here $R(k)$ is determined by the expression

$$R(k)=\Lambda_0 L_{s0}^2 k^2(k_c^2-k^2)\,,$$

$$k_c^2=\frac{2}{L_{s0}^2}\left(\chi-\frac{1}{2N_1\varphi_0}-\frac{1}{2(1-\varphi_0)}\right).$$

The dimensionless coefficient βk^2 may be presented in the following form:

$$\beta k^2=(R_*-R)/4A_0k_x\,.$$

[‡] In reality, we have here a non-realizable situation whenever the extension of the spinodal concept on the fully cured system may be useful for general description of the process in the non-cured system.

For the system under consideration, the diffusion relaxation rate is significantly higher than the reaction rate, so that parameter β is very large. This leads to the smallness of power index in solution (8). In the limit $k_x \to 0$ (in the absence of chemical reaction), solution (8) becomes a solution to the Cahn–Hilliard problem for SD in the initial blend with the amplification coefficient $R(k)$. In this case, the curve

$$\chi_s(\varphi_0) = \frac{1}{2N_1\varphi_0} + \frac{1}{2(1-\varphi_0)}$$

is the spinodal for the initial blend (Fig. 2, curve *1*) on the plane of phase states $(\chi - \varphi_0)$. In the region $\chi > \chi_s(\varphi_0)$, the concentration heterogeneities with wave numbers $k < k_c$ grow spontaneously while in the region $\chi < \chi_s(\varphi_0)$ we have $R(k) > 0$ for all k.

Finally, two types of behavior may be distinguished relative to the sign of the initial relaxation rate $R(k)$ in the instability region of the curing blend homogeneous states and for $R_*(k) > 0$ (Fig. 1, curves *2* and *3*).

Development of Initial Inhomogeneity at Low Temperatures

For $R_*(k) > 0$ and $R(k) > 0$ (i.e., when the system is initially below the spinodal), the growth of k-th mode of concentration fluctuation $u_k(t)$ occurs at the linear stage for all $k < k_c$ (Fig. 1, curve *2*). The process described by equation (8) differs from usual temperature-induced SD both by the range of space scale of instabilities and by the rate of amplification for these heterogeneities. Moreover, the dimension of these heterogeneities depends on time.

If the initial (t = 0) concentration distribution $u(r,0)$ with the Fourier expansion

$$u(r,0) = \sum_k c_k \exp(ikr)$$

appears in the blend, then the evolution of the initial distribution of heterogeneities in the linear (by perturbation) approximation $u(r,t)$ has the form:

$$u(r,t) = \sum_k c_k (1 + 4A_0 k_x)^{-\beta k^2} \exp(ikr + R_* t)\,.$$

In the spectrum of concentration heterogeneities corresponding to this solution, there is some most rapidly growing mode with the wave number k_{max}, which is time-dependent unlike usual SD. In the case of $R_*(k) > 0$ and $R(k) > 0$, k_{max} is given by:

$$2k_{\max}(t)^2 = k_c^2 + \frac{4}{3(1-\varphi_0)L_{s0}^2}\left\{1 - \frac{\ln(1+4A_0k_xt)}{4A_0k_xt}\right\} \tag{11}$$

in the region of spontaneously growing wave lengths ($k < k_c$). However, the process gradually leads to diminution of these heterogeneities (as suggested by equation (11)), and the limit heterogeneity dimensions are determined by the value of $k_* / \sqrt{2}$.

Development of Initial Inhomogeneity at High Temperatures

The additional peculiarity of the reactive blend behavior appears is observed for $R_*(k) > 0$ and $R(k) < 0$. In this case, $u_k(t)$ relaxes initially to its minimum value $u_{\min}(k)$ which is realized at

$$t = \tau_{\exp}(k) = \frac{1}{4A_0k_x}\frac{k^2 - k_c^2}{k_*^2 - k^2} \tag{12}$$

and then, for $t > \tau_{\exp}(k)$, $u_k(t)$ grows infinitely [20] (see Fig. 1, curve *3*). Such a situation arises when the reaction runs in the homogeneous blend (out of the spinodal) or even in the heterogeneous blend (below the spinodal) for such part of concentration fluctuations, which should not grow in nonreactive blend. The unrestricted growth of the Fourier amplitude $u_k(t)$ in both cases (for $R_*(k) > 0$ and $R(k) > 0$ as well as for $R_*(k) > 0$ and $R(k) < 0$) for $t > \tau_{\exp}(k)$ could be evidently interrupted only by nonlinear effects. These effects become significant in the subsequent stages and are not considered here.

In case of $R_*(k) > 0$ and $R(k) < 0$, the time dependence of the wave number for the most rapidly growing mode can be presented in the form:

$$2k_{\max}(t)^2 = k_*^2 - \frac{1}{3(1-\varphi_0)L_{s0}^2}\frac{\ln(1+4A_0k_xt)}{A_0k_xt}.$$

This mode can be separated after some certain characteristic delay time $\tau_*(\chi,\varphi_0)$:

$$\tau_* \approx \frac{3(1-\varphi_0)}{4A_0\,k_x}(\chi_s - \chi) + \frac{(1-\varphi_0)}{2A_0\,k_x}(\chi_s - \chi)^2.$$

The reactive blend remains in the spatially homogeneous state during the entire delay time $\tau_*(\chi,\varphi_0)$. After the delay time $\tau_*(\chi,\varphi_0)$, the mode with a maximal growth rate and decreasing with time wavelength appears.

As mentioned above, for $R_*(k) > 0$ and $R(k) < 0$ after the characteristic exposure time $\tau_{\exp}(k)$ (Fig. 3 and 4, curves *1*) determined by equation (11), the amplitude of the k-th mode of concentration heterogeneity $u_k(t)$ has a minimum at $t = \tau_{\exp}(k)$:

$$u_{\min}(k) = c_k \left(\frac{k_*^2 - k^2}{k_*^2 - k_c^2} \right)^{\beta} \exp\left(-\frac{R(k)}{4A_0 k_x} \right). \tag{13}$$

Both the existence of the exposure time $\tau_{\exp}(k)$ and the formation of structures with the amplitude $u_{\min}(k)$ are the result of simultaneous diffusion and chemical reaction in the blend. The function $u_{\min}(k)$ represents the distribution of concentration heterogeneities over the wave numbers (or heterogeneity dimensions) in the reactive blend.

Let us consider also the effect of the diffusion relaxation (and chemical reaction) on the conversion $\alpha(t)$ and average degree of polymerization for resultant polymer, $N_2(\alpha)$. At $t = \tau_{\exp}(k)$, the conversion corresponding to the mode of concentration heterogeneity with the wave number k has the form:

$$\alpha_{\exp}(k) = \frac{3}{4}(1 - \varphi_0) L_{s0}^2 (k^2 - k_c^2) .$$

This conversion is determined for $\chi > \chi_s(\varphi_0)$ in the wave number range $k_c < k < k_*$ and takes values from the range (0; 1). For $\chi < \chi_s(\varphi_0)$ (the range of homogeneous states of initial blend), the conversion $\alpha_{\exp}(k)$ after the exposure time is determined over a wider range of wave numbers $(0; k_*)$ and has the values from the interval (ε; 1).

The chain length after the exposure time is

$$N_2^{\exp}(k)^{-1} = (1 - \varphi_0) L_{s0}^2 (k_\infty^2 - k^2) ,$$

where $k_\infty^2 = 2L_{s0}^{-2}\left(\chi - (2N_1\,\varphi_0)^{-1}\right)$.

The condition of reactive blend at the instant $t = \tau_{\exp}(k)$ may be called the ‘starting’ one. During the exposure time, some distributions of concentration heterogeneities, values of conversion and product chain length are formed. During exposure time the reactive blend is in its spatially homogeneous states and remains optically transparent. With time, these heterogeneity distributions are upset. The following expansion for the

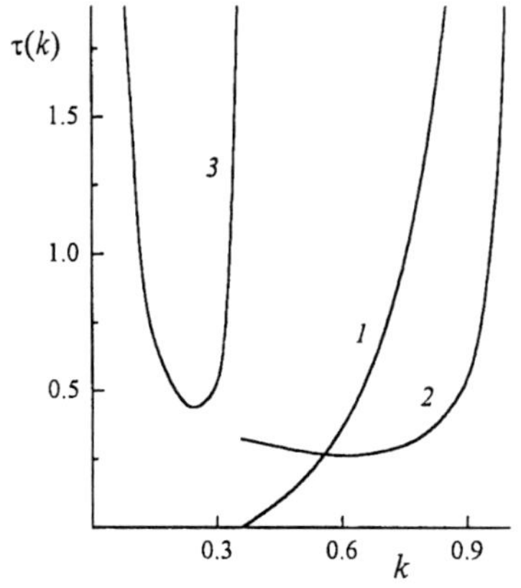

Figure 3. Characteristic times of the system: (*1*) $\tau_{\rm exp}(k)$, exposure times for unstable modes with wave numbers $k_{\rm c} < k < k_{\rm s}$; (*2*) $\omega_0^{-1}(k)$, growth times for those modes; (*3*) $R^{-1}(k)$, growth times for unstable modes in the initial blend.

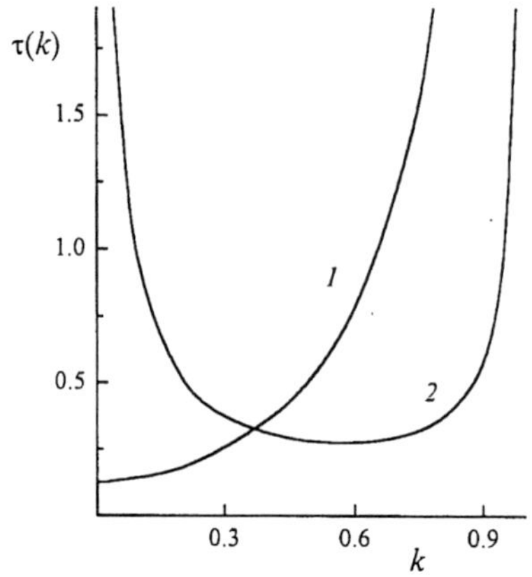

Figure 4. Characteristic relaxation times in the region of the initial blend stability.

amplitude of concentration heterogeneity may be obtained from equation (8):

$$u_k(t') = u_{\min}\,(1 + \frac{1}{2}\omega_0^2\, t'^2)\,. \tag{14}$$

near $\tau_{\rm exp}(k)$ at $t = \tau_{\rm exp}(k) + t'$. This expansion describes the unlimited growth of $u_k(t)$ after passing the minimum of the process. Here the characteristic growth rate for concentration heterogeneity (Figs. 3 and 4, curves *2*) is determined as

$$\omega_0(k) = \left(3\,(1-\varphi_0)\Lambda_0\, A_0\, k_x\right)^{1/2} L_{s0}^2\, k\,(k_*^2 - k^2)\,.$$

This rate has a maximum at $k = k_*/\sqrt{3}$. Note that the modes with the minimum values of $\tau_{\rm exp}(k)$ (i.e., the modes reaching the starting state earlier than the other modes) have a considerably larger scale than the mode corresponding to a minimum of $\omega_0^{-1}(k)$ that grows faster than the other modes.

After the characteristic reaction time, the conversion of the amine groups and product chain lengths are given by the following expressions:

$$\alpha(t') = \alpha_{\exp}(k)\left\{1+\Omega_\alpha(k)t'\right\},$$

$$N_2(t') = N_2^{\exp}(k)\left\{1+\Omega_2(k)t'\right\}.$$

Their growth rates depend on the heterogeneity dimension:

$$\Omega_\alpha(k) = 3(1-\varphi_0)A_0 k_x L_{s0}^2 \frac{(k_*^2 - k^2)^2}{k^2 - k_c^2},$$

$$\Omega_2(k) = \frac{4}{3} N_2^{\exp}(k)\,\alpha_{\exp}(k)\,\Omega_\alpha(k).$$

In case of $R_*(k) > 0$ and $R(k) < 0$ during the characteristic exposure time $\tau_{\exp}(k)$, the starting state is formed in the reactive blend with the specific distribution of concentration, $u_{\min}(k)$, conversion $\alpha_{\exp}(k)$, and the polymerization degree for a growing chain $N_2^{\exp}(k)$. Further blend evolution from the starting state is characterized by the growth times for concentration heterogeneities $\omega_0^{-1}(k)$, conversion $\Omega_\alpha(k)^{-1}$, and chain length $\Omega_2(k)^{-1}$. As a result, at the linear stage of evolution of concentration fluctuations $u_k(t)$ the characteristic time of their formation τ_{ev} can be determined as a sum of characteristic times of relaxation and growth

$$\tau_{ev}(k) = \tau_{\exp}(k) + \omega_0^{-1}(k).$$

In this case, we can obtain the expression for the time of heterogeneities development:

$$\tau_{ev}(k) = \frac{1}{4A_0 k_x}\frac{k^3 - k_c^2 k + B_0}{k\left(k_*^2 - k^2\right)}, \tag{15}$$

where $B_0 = 4L_{s0}^{-2}[A_0 k_x / 3(1-\varphi_0)\Lambda_0]^{1/2}$.

In the region of unstable homogeneous states of initial blend for $\chi > \chi_s(\varphi_0)$, a narrow range of wave numbers near k_c with τ smaller than those for the mode of maximal growth can be distinguished. In this case, the characteristic time of development is minimal for the mode with wave number $\approx B_0^{1/3}$ in the region below the spinodal of the initial blend at $\chi < \chi_s(\varphi_0)$ but located in the range of unstable homogeneous states of the curing blend (Fig. 4). The possibility of formation of localized microphase heterogeneities in the reactive blend will be considered elsewhere.

It must be noted that the starting state of the reactive blend does not correspond to the thermodynamic equilibrium. Both on the stage of formation of this state and its further transformation, irreversible processes that are far from equilibrium take place in the system.

CONCLUDING REMARKS

In terms of the proposed model, the range $\chi < \chi_*(\varphi_0)$ is distinguished on the plane (χ, φ_0) of thermodynamic states of the curing blend. In this region, any concentration heterogeneity in the completely cured blend relaxes via diffusion. A cure process starting from any thermodynamic state within this region must be realized without any structural instability. This region of stable homogeneous states of the cured blend seems to be unknown and not studied experimentally.

In the unstable region of spatially homogeneous states of the cured blend, the model predicts the existence of two subregions with qualitatively different character of relaxation of linear polymer concentration heterogeneity. These subregions are divided by a thermodynamic threshold $\chi_s(\varphi_0)$ that is the spinodal of the initial blend.

Moreover, the threshold of the reactive blend structural stability $k_*(\chi, \varphi_0)$ found in this work embraces a region of instability scales that is essentially wider than in the absence of chemical reaction. The peculiar structure of the plane of thermodynamic states for the cured blend (χ, φ_0) leads to an unusual temperature dependence of $k_*(\chi, \varphi_0)$.

In the region of unstable spatially homogeneous states of the initial blend for $\chi > \chi_s(\varphi_0)$, the most rapidly growing mode with the wave number $k_c/\sqrt{2}$ of concentration heterogeneities is distinguished at the linear stage of the process. This mode grows with time and approaches its limiting value of $k_*/\sqrt{2}$. The existence of time dependence for the fastest growth mode is one of the peculiar features of the proposed mechanism for microphase formation, in contrast to the classical SD. The predicted decrease in heterogeneities dimensions is a consequence of the model proposed; in experiments, the increase in heterogeneities size during the process was also observed.

Below the spinodal of the initial blend, for $\chi < \chi_s(\varphi_0)$, where spatially homogeneous states are stable, the most rapidly growing mode of concentration heterogeneities with a small wave number is separated after the characteristic delay time (before this moment, the blend remains spatially homogeneous). This mode grows with time and approaches a finite limiting value. Both the delay time and the limiting wave number depend on the distance between the initial thermodynamic state and the boundaries of either initial or cured blend thermodynamic stability (there is a simple relation between these two thermodynamic quantities).

One of the most important conclusions is the existence of characteristic exposure time when the so-called 'starting' state of the blend is formed. This state is characterized by a certain size distribution of concentration heterogeneities, conversion, and chain length of the reaction product. The blend deviation from its starting state is characterized by the spectrum of relaxation times for concentration heterogeneities, conversion and chain lengths depending on the wave number of heterogeneities as well as on the blend thermodynamic state. To estimate the evolution of purely concentration-type heterogeneities, the notion of heterogeneity development time is introduced in this paper. This time is defined as a sum of characteristic time of formation (exposure time) and growth of heterogeneity. An unexpected result of this model is the difference between the most rapidly growing mode and the mode with minimal development time.

Acknowledgments. This work was supported by the International Science and Technology Center (grant no. 358-96).

REFERENCES

1. Williams R.J., Rozenberg B.A., and Pascault J.-P., *Adv. Polym. Sci.*, **128**, 95 (1997).
2. Rozenberg B.A. and Irzhak V.I., *Macromol. Symp.*, **93**, 227 (1995).
3. Korolev G.V., Mogilevich M.M., and Golikov I.V., *Setchatye polyakrilaty: Mikrogeterogennye struktury, fizicheskie setki, deformatsionno-prochnostnye svoistva* (Network Polyacrylates: Microheterogeneous Structures, Physical Networks, Deformation and Strength), Khimiya, Moscow, 1995.
4. Yamanaka K. and Inoue T., *Polymer*, **30**, 662 (1989).
5. Yamanaka K., Takagi Y., and Inoue T., *Polymer*, **30**, 1839 (1989).
6. Lee H.-S. and Kyu T., *Macromolecules*, 23, 459 (1990).
7. Tanaka H., Suzuki T., Hayashi T., and Nishi T., *Macromolecules*, **25**, 4453 (1992).
8. Ohnaga T., Chen W., and Inoue T., *Polymer*, **35**, 3774 (1994).
9. Ginzburg V.V. and Clark N.A., *Phys. Rev. E*, in press.
10. Kyu T. and Lee J.-H., *Phys. Rev. Lett.*, **76**, 3746 (1996).
11. Shaginyan Sh.A. and Manevich L.I., *Polymer Science, Ser. A*, **39**, 908 (1997).
12. Irzhak V.I., Rozenberg B.A., and Enikolopyan N.S., *Setchatye polimery: Sintez, struktura, svoistva* (Network Polymers: Synthesis, Structure, Properties), Nauka, Moscow, 1979. Ch. 2.

13. Ryan M.E. and Dutta A., *Polymer*, **20**, 203 (1979).
14. de Gennes P.G., *Scaling Concepts in Polymer Physics*, Cornell University Press, Ithaca NY, 1979.
15. Binder K., *J. Chem. Phys.*, **79**, 6387 (1983).
16. Mitlin V.S., Manevich L.I., and Erukhimovich I.Ya., *Zh. Eksp. Teor. Fiz.*, **88**, 495 (1985).
17. Takenaka M. and Hashimoto T., *Macromolecules*, **27**, 6123 (1994).
18. Prigogine I., *The Molecular Theory of Solutions*, North-Holland, Amsterdam, 1957.
19. Glansdorff P. and Prigogine I., Thermodynamic Theory of Structure, Stability, and Fluctuations, Wiley, London, 1970.
20. Sigalov G.M. and Rozenberg B.A., *Polymer Science, Ser. A*, **37**, 1049 (1995).

Chapter 7

Two-Dimensional Model of Phase Separation during Polymerization of a Binary Polymer Blend

Avigeya N. IVANOVA[1]* and Leonid I. MANEVICH[2]

[1]*Institute of Problems of Chemical Physics, Russian Academy of Sciences, Chernogolovka, Moscow Region, 142432 Russia*
[2]*Semenov Institute of Chemical Physics, Russian Academy of Sciences, 4 Kosygin Street, Moscow, 117977 Russia*

*e-mail: ivanova@icp.ac.ru

ABSTRACT

For a large two-dimensional model of a system exhibiting cure reaction-induced microphase separation (CRIMPS), the evolution retardation scenario is realized independently of the initial conditions (either determinant or random). Long-lived structures play a decisive role in the process. Similarly to the one-dimensional case, the spinodal region is divided into sub-regions corresponding to soft and hard birth of spatial structures. Spinodal region upon rigid birth and metastable region may be considered within the same unifying description. Non-linear Cahn–Hilliard–de Gennes equation, which was earlier applied only to polymer blends in spinodal region, is shown to be suitable for the description of heterogeneous structures in the metastable region. The CRIMPS process proceeds according to a retardation scenario up to the gel point. The presence of the gel-point predetermines 'fixation' of an intermediate heterogeneous structure which seems to be thermodynamically stable.

INTRODUCTION

The dynamics of thermally and chemically induced phase separation of a binary polymer blend has been the subject of numerous experimental and theoretical studies during past years. The processes in metastable and spinodal regions are usually described by different initial equations. From the numerical study of the one-dimensional (1D) model in the spinodal region, the linear stage of the process characterized by exponential growth of long-wave fluctuations of the concentration (the Cahn modes) is followed by consecutive formation of long-lived unstable structures. Respectively, the fluctuation concentration changes fast and abruptly upon transition from one structure to another [1, 2].

From numerical analysis of two- and three-dimensional (2D, 3D) models [3, 5], dynamic behavior of the polymer blend was shown to depend essentially on the initial distribution of concentration fluctuations. For their random distribution, a fluent evolution of the system, instead of abrupt transitions, is observed. It should be noted that rather small systems were considered, with few or no unstable structures. For these systems, as distinguished from the 1D case, the dynamic process could not be always followed up to thermodynamic equilibrium, because of time limitations. In this work we have first analyzed the dynamics for 2D model in larger time ranges.

As far as heterogeneous structures (both thermodynamically stable and unstable) are important for the phase separation, conditions of their formation and lifetime should be clarified. For the 1D case, this was [6, 7] based on bifurcation analysis of a non-linear diffusion equation and detection of the leading eigenvalue for the operator linearized near stationary spatial structures. In a spinodal region, sub-regions were found that correspond to both a soft and rigid birth of spatial structures.

The non-linear Cahn–Hilliard–de Gennes equation, which was earlier used only for polymeric blends in a spinodal region, was shown to be suitable for description of heterogeneous structures in a metastable region. In this work, all results of papers [6, 7] are summarized for application to the 2D model.

Formulation of the Problem

The dynamics of separation of polymers A and B binary blend upon one of the components (A) polymerization is described by the Cahn–Hilliard–de Gennes equation:

$$\frac{\partial \varphi}{\partial t} = \operatorname{div}\left(\Lambda(\varphi) \nabla \frac{\delta F}{\delta \varphi} \right), \tag{1}$$

where

$$\Lambda(\varphi) = \frac{N_e \varphi(1-\varphi)}{\tau_A N_A (1-\varphi) + \tau_B N_B \varphi}. \tag{2}$$

is the Onzager diffusion coefficient, N_e is the average number of polymer chain segments between entanglements, N_A and N_B are numbers of segments in the molecules of A and B; $\delta F/\delta\varphi$ is the variation derivative of the specific free energy F for the Flory–Huggins–de Gennes model, with its local density determined by:

$$\frac{F}{k_B T} = \widetilde{F} = \widetilde{F}_0 + \widetilde{F}_1, \quad \widetilde{F}_0 = \frac{\varphi \ln \varphi}{N_A} + \frac{(1-\varphi)\ln(1-\varphi)}{N_B} + \chi\varphi(1-\varphi), \tag{3}$$

$$\widetilde{F}_1 = K(\varphi)(\nabla\varphi)^2, \qquad K(\varphi) = [36\varphi(1-\varphi)]^{-1}. \tag{4}$$

Equation (1) may be rewritten as follows:

$$\frac{\partial \varphi}{\partial t} = \nabla(\Lambda \nabla \mu_0) - \nabla\left[\Lambda \nabla\left(2K\nabla^2\varphi + K_\varphi (\nabla\varphi)^2\right)\right],$$

where

$$\mu_0 = \frac{1}{N_A}\ln\varphi - \frac{1}{N_B}\ln(1-\varphi) + \frac{1}{N_A} - \frac{1}{N_B} + \chi(1-2\varphi).$$

is the system chemical potential.

Suppose that boundary conditions for equation (1) are periodical within a rectangular cell with sides (L_1, L_2). The chain length for the polymerizing component in equation (1) is a function of the polymerization conversion which grows with time. In this case equation (1) describes the process until gel formation, and no stationary solutions and thermodynamic equilibrium may be considered. However, the influence of the chemical reaction on the blend separation can be understood if only the thermally initiated diffusion process (without reaction) is analyzed, i.e., in the case of N_A, N_B = const. It follows from 1D consideration, that the properties of stationary solutions as functions of parameters L_1, L_2, χ, φ_0, N_A, N_B also provide some information about the dynamics. Therefore, stationary solutions of equation (1) for two-dimensional case and the way of the solutions birth upon change of the above parameters will be consider below first of all.

RESULTS AND DISCUSSION

Stationary Solutions

According to equation (1) and periodical boundary conditions, the integral over the cell $\iint \varphi(x,y)dxdy = \varphi_0$ remains constant during the process. Therefore, we consider stationary solutions for a set of functions determined by this condition. Generally, many stationary solutions for equation (1) are available for this set. But if dimensions are small enough, there are no stationary solutions except the uniform one $\varphi(x,y) \equiv \varphi_0$. However, if parameters L_1, L_2, χ, φ_0, N_A, N_B belong to the spinodal region, there are heterogeneous solutions for the cell dimensions L_1 and L_2 close to some critical values L_1^*, L_2^*. These critical values correspond to the vanishing eigenvalue for the operator linearized on the homogeneous stationary solution. This is the necessary condition of branching. One or a few pairs of heterogeneous stationary solutions, depending on multiplicity of the zero eigenvalue, branch from the homogeneous solution. Branching can be supercritical or subcritical, similar to the homogeneous case [7]. The branching type may be determined by non-linear bifurcation analysis.

Bifurcation Analysis

By substituting variables, we introduce the cell parameters L_1, L_2 into the equation:

$$x' = x/L_1, \quad y = y/L_2; \quad \rho = 1/L_1^2, \quad \alpha = L_1^2/L_2^2.$$

Then the linearized operator over solution $\varphi = \varphi_0$ for problem (1) is

$$L(z)=\Lambda\left(\mu_0'\rho\left(\frac{\partial^2 z}{\partial x^2}+\alpha\frac{\partial^2 z}{\partial y^2}\right)-2K\rho^2\left(\frac{\partial^4 z}{\partial x^4}+2\alpha\frac{\partial^4 z}{\partial x^2\partial y^2}+\alpha^2\frac{\partial^4 z}{\partial y^4}\right)\right). \quad (5)$$

Eigenfunctions of operator (6) for periodical boundary conditions for the cell (0, L_1), (0, L_2) and corresponding eigenvalues are as follows ($m \neq n$):

$$z = A^+\cos(2\pi nx+2\pi my)+A^-\cos(2\pi nx-2\pi ym)+A_1^+\sin(2\pi nx+2\pi my)+$$

$$+A_1^-\sin(2\pi nx-2\pi my)+B^+\cos(2\pi mx+2\pi ny)+B^-\cos(2\pi mx-2\pi ny)+$$

$$+B_1^+\sin(2\pi mx+2\pi ny)+B_1^-\sin(2\pi mx-2\pi ny)\,, \quad (6)$$

$$\lambda_{n,m}=-4\pi^2L_1^{-2}\Lambda(\varphi_0)(n^2+\alpha m^2)\left(\mu_0'(\varphi_0)+8\pi^2L_1^{-2}K(\varphi_0)(n^2+\alpha m^2)\right). \quad (7)$$

It is obvious that if $\mu_0'(\varphi_0)>0$ then all eigenvalues are negative and the homogeneous solution is stable. Critical values of parameters are determined by the following relationship:

$$\mu_0'+8K\pi^2\rho(n^2+\alpha\, m^2)=0\,. \quad (8)$$

If $\rho=1/L_1^2$ is a bifurcation parameter, its critical values are as follows:

$$\rho_{n,m}^*=-\mu_0'(\varphi_0)\left[8\pi^2K(\varphi_0)\left(n^2+\alpha^2m^2\right)\right]^{-1}. \quad (9)$$

Expansion of the stationary solution $\varphi\,(x,y)$ over powers of the parameter characterizing deviation of ρ from its critical value yields

$$\varphi(x,y,\rho)=\varphi_0+\varepsilon\varphi_1(x,y,\rho^*)+\varepsilon^2\varphi_2(x,y,\rho^*)+\varepsilon^3\varphi_3(x,y,\rho^*)+o(\varepsilon^*)\,,$$

$$\rho=\rho^*+\varepsilon\rho_1+\varepsilon^2\rho_2+\varepsilon^3\rho_3+o(\varepsilon^3)\,. \quad (10)$$

Let us substitute these expressions into the stationary equation for φ, expand non-linear functions of φ in the vicinity of φ_0, and assume the terms with equal powers of ε to be zero. This yields recurrent relationships for φ_i, ρ_i. The first order terms have φ_1 as eigenfunction of the linearized operator corresponding to zero eigenvalue.

$$L(\varphi_0,\rho^*)\,\varphi_1(x,y,\rho^*)=0\,. \quad (11)$$

This equation does not define coefficients of the linear combination in (7). These coefficients are arbitrary at this stage. The second order terms determine the appearance of function $\varphi_2(x,y,\rho^*)$ and ρ_1:

$$L(\varphi_0,\rho^*)\,\varphi_2(x,y,\rho^*)=\Phi(\varphi_1,\rho_1)\,. \tag{12}$$

After lengthy transformations, omitted here, one can find that the right-hand side (r.h.s.) of equation (12) may be presented in the form

$$f_1(\varphi_0,\rho^*)\;\varphi_1^2+\rho_1\,f_2(\varphi_0,\rho^*)\,\varphi_1\,.$$

Hence, the necessary condition of existence of a solution to equation (12) is orthogonality of the r.h.s. with respect to φ_1; therefore, $\rho_1 = 0$. Then the function φ_2 from (13) is determined as a linear combination of cosines and sines of the arguments that are present in the expression for φ_1^2 after replacing cosine and sine products by functions of the arguments sum and difference. In particular, if $n = 1$, $m = 1$, then:

$$\begin{aligned}\varphi_2(x,y,\rho^*)&=c^+\cos[4\pi(x+y)]+c^-\cos[4\pi(x-y)]+s^+\sin[4\pi(x+y)]+\\&+s^-\sin[4\pi(x-y)]+c_x\cos 4\pi x+c_y\cos 4\pi y+s_x\sin 4\pi x+s_y\sin 4\pi y\,.\end{aligned} \tag{13}$$

The coefficients are given by substitution of (13) into (12) and setting the coefficients of linearly independent functions to zero.

The third order terms yield the following equation:

$$L(\varphi_0,\rho^*)\,\varphi_3(x,y,\rho^*)=\Phi_2(\varphi_2,\rho_2)\,. \tag{14}$$

Suppose that the r.h.s. of this equation is orthogonal to the linearized operator eigenfunctions corresponding to zero eigenvalue. Then we obtain equations for linear combination coefficients in (7) and for the ρ_2 value (for $n = 1$, $m = 1$, there are four such functions which are additives in (7)). In general case, this system of equation is non-linear. For $n = m = 1$ it is as follows:

$$\begin{aligned}&\rho_2\left(8\pi^2\Lambda(\varphi_0)\left(32\rho^*K(\varphi_0)+\mu_0'\right)\right)A^+=A^+\left(p\tilde{A}^++q\tilde{A}^-\right)\\&\rho_2\left(8\pi^2\Lambda(\varphi_0)\left(32\rho^*K(\varphi_0)+\mu_0'\right)\right)A^-=A^-\left(q\tilde{A}^++p\tilde{A}^-\right)\\&\tilde{A}^+=(A^+)^2+(A_1^+)^2,\;\tilde{A}^-=(A^-)^2+(A_1^-)^2\,.\end{aligned} \tag{15}$$

If A^+ is replaced by A_1^+, and A^- by A_1^-, two more equations are obtained. Here p and q are functions of coefficients from (1) and of their derivatives, including the third order ones. Two types of solutions are possible: either (i) one of the coefficients is not zero, and others are zero, or (ii) the coefficients satisfy the equation $\tilde{A}^+ = \tilde{A}^-$. These amplitudes can always be considered to be unity, because their variation corresponds to renormalization of ε and does not change the solution. In the first case, the structures that appear are virtually one-dimensional because they are invariant along lines $x + y = \text{const}$ or $x - y = \text{const}$. In this case the equation for ρ_2 is the same as in the 1D case at $N_A = N_B$:

$$\rho_2 = \frac{9(\Psi - \Psi_1)(\Psi - \Psi_2)}{4\pi^2 N^2 \mu_0'(\varphi_0)}, \quad \Psi_{1,2} = z^{-1}\left[1 - z \pm \sqrt{\frac{1}{3} + z\left(z - \frac{4}{3}\right)}\right]. \tag{16}$$

In the case (ii), expressions for ρ_2 are the same for all solutions, with accuracy to renormalizing with respect to the amplitude. Assuming $N_A =$ $= N_B$,

$$\rho_2 = \frac{9(40z - 7)(\Psi - \Psi_1)(\Psi - \Psi_2)}{4zN^2\pi^2\mu_0'(\varphi_0)}. \tag{17}$$

Here

$$z = \varphi_0(1 - \varphi_0), \; \Psi = N_A\chi, \; \Psi_{1,2} = \frac{13z - 1 \pm \frac{1}{6}\sqrt{321z^2 - 264z + 51}}{2z(7 - 40z)}$$

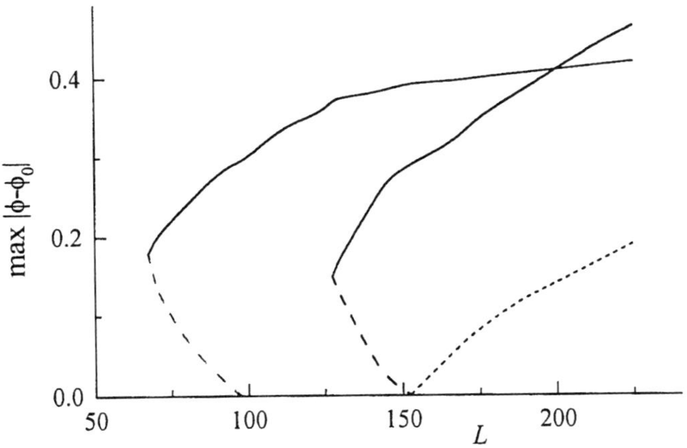

Figure 1. Heterogeneity amplitude max $|\varphi - \varphi_0|$ *vs.* system size L.

Expanding (11), we get $\varepsilon = \sqrt{(\rho - \rho^*)/\rho_2} + o(\varepsilon)$. Then, if $\rho_2 > 0$, there exists a solution for $\rho > \rho^*$ and hence for $L < L^*$. In this case the bifurcation over L is subcritical. Rigid birth of a pair of heterogeneous 2D structures occurs at a certain $L = L^{**}$ as well as in the 1D case. However, there is another possibility: a 2D structure can appear from a 1D structure via secondary bifurcation because of loss of stability relative to 2D perturbations. Fig. 1 shows a feasible bifurcation diagram. It corresponds to the points on the phase diagram shown in Fig. 2 in the region between the spinodal and the curve defined by $\Psi = \Psi_1$.

As the system size increases, the first bifurcation to occur corresponds to rigid birth of a pair of 1D structures in subcritical region at $L^{**} < L^* = (\rho^*_{1,0})^{-1/2}$. As L increases further and reaches $L^*_{1,1} = (\rho^*_{1,1})^{-1/2}$, an unstable 2D structure with two positive eigenvalues appears subcritically via rigid birth. This new structure is born as a pair with an unstable structure that has one positive eigenvalue. The latter structure exchanges its stability with a stable 1D structure that has appeared earlier. Thus, the 2D structure becomes stable. A numerical calculation of stability for 1D structures with respect to 2D perturbations [6] implies a possibility of stability loss by the 1D structure.

Figure 2 shows the phase diagram where curve $\Psi = \Psi_S(\varphi)$ is a spinodal, and $\Psi = \Psi_1(\varphi)$ corresponds to the second radical in (17). In the area between these curves a rigid birth of 1D structures occurs in a subcritical region over L. Curve $\Psi = \Psi_2(\varphi)$, where Ψ_2 is the second radical in (18), separates the region of a subcritical birth of 2D structures from the region of

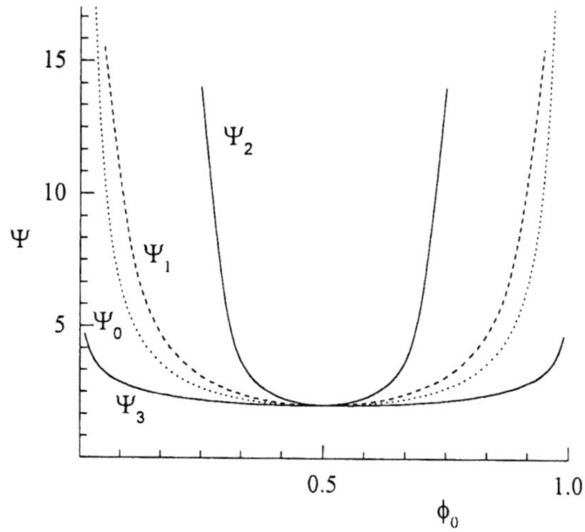

Figure 2. Phase diagram (Ψ, φ_0).

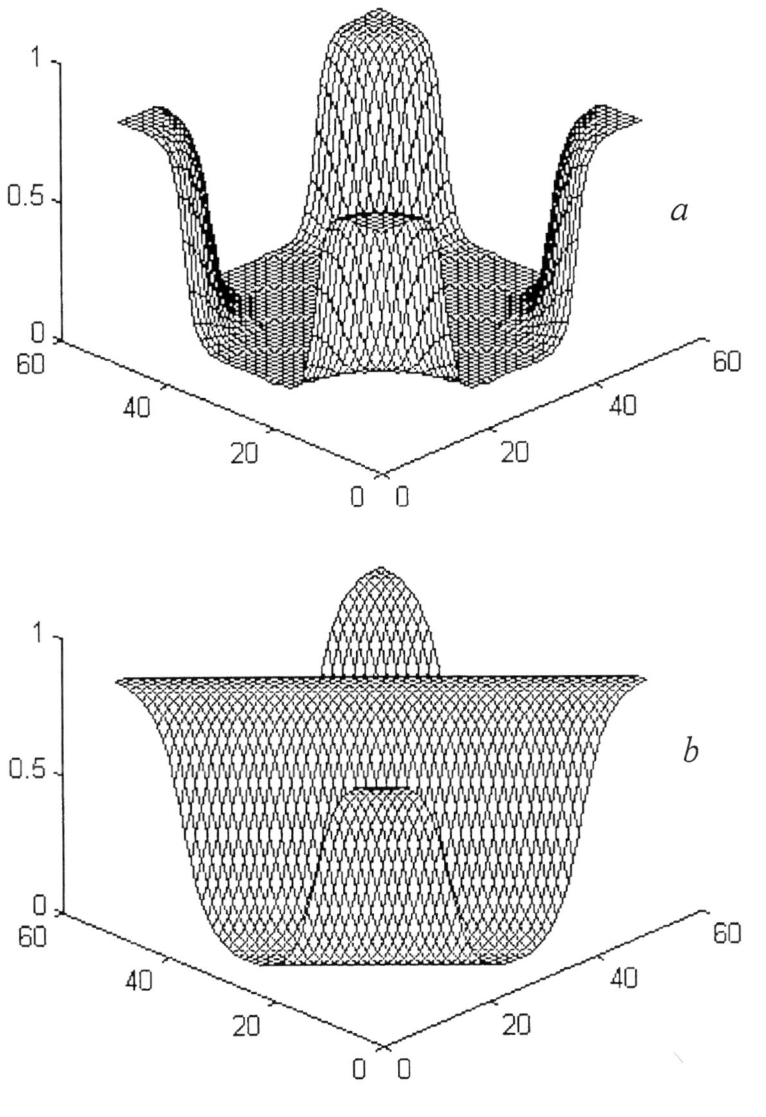

Figure 3. Spatial distribution of volume fraction for component *A*: (*a*) $\varphi_0 = 0.4$, $L_1 = L_2 = 200$; and (*b*) $\varphi_0 = 0.5$, $L_1 = L_2 = 200$.

the supercritical birth. In this region, the birth of 1D structures, their profile being constant along the lines $x+y$=const, $x-y$=const, occurs supercritically, and they are unstable. For $m=n\neq 1$, the regions of subcritical birth are the same. For $m\neq n$, analytical expressions are complicated and will not be presented here.

Numerical Modeling of 2D Structures

Numerical modeling of the 2D problem was performed with the help of an implicit difference scheme. The Newton method was used at each time step. Each linear problem was solved by the method of alternative directions, and the linear boundary value problem for each direction was solved by the cyclic sweep method, similar to 1D case [6]. An alternative time step was chosen automatically from the given criteria of approximation accuracy and iteration convergence. Based on the phase diagram obtained, the parameters values favorable for 2D or 1D structures were predicted. The values of $\chi = 0.003$, $N_A = N_B = 1000$, $\varphi_0 = 0.3$ or $\varphi_0 = 0.4$ correspond to subcritical bifurcation of 2D structures. Meanwhile, for $\varphi_0 = 0.5$ all 2D structures appear supercritically and must be unstable, and 1D structures are realized. These considerations are confirmed by the numerical results presented in Fig. 3.

In all cases the initial values were given by random perturbations relative to φ_0 level. The perturbations were simulated by 100 terms of a 2D Fourier series with random amplitudes, all below 0.01.

Numerical modeling for 2D case, as well as for 1D case, confirms the existence in the binodal region of structures revealed in the framework of the Cahn–Hilliard–de Gennes model. Stepwise behavior of the free energy during relaxation to equilibrium, typical of 1D case [1–3] and associated with slowing down of the system evolution in the vicinity of unstable stationary states, is also observed for 2D case. The number of the ‘steps’ is determined by the number of unstable stationary states lying on the system pathway to thermodynamic equilibrium. The stairs widths correlate with the inverse of the leading positive eigenvalue of the operator linearized on this structure. In 1D case, this correlation may be illustrated by lifetime comparison given in Table 1.

Such correlation exists also in 2D case. Numerical calculations for $\chi = 0.0025$, $\varphi_0 = 0.3$, $L_1 = L_2 = 300$ yield $\lambda = 5.97 \cdot 10^{-4}$, $\tau \approx 1/\lambda \approx 1600$, and the step width is ca. 1400 s. The larger is the system, the larger is the number of steps.

Modeling of CRIMPS

To perform modeling of phase separation during polycondensation, we took into account the change of molecular weight during the reaction. For a stoichiometric reaction of diepoxide and diamine, conversion degree $\alpha = 4A_0kt/(1+4A_0kt)$, and dependence of the average chain length on the conversion degree may be expressed as follows:

$$N(\alpha) = 1 + 4\left(\frac{\alpha}{1-3\alpha^2}\right)\left(\frac{M_{da}(1+\alpha) + M_{de}(1+3\alpha)}{M_{da} + 2M_{de}}\right),$$

Table 1. Comparison of unstable structure lifetimes estimated with the help of non-stationary calculation ($1/\lambda$) and from the linear approximation (τ)

N	L	λ	$1/\lambda$	τ
2	300	$6.2 \cdot 10^{-3}$	$1.6 \cdot 10^{2}$	$2.0 \cdot 10^{2}$
2	400	$4.5 \cdot 10^{-4}$	$2.2 \cdot 10^{3}$	$2.2 \cdot 10^{3}$
2	500	$4.1 \cdot 10^{-5}$	$2.5 \cdot 10^{4}$	$3.0 \cdot 10^{4}$
2	600	$4.1 \cdot 10^{-6}$	$2.5 \cdot 10^{5}$	$3.1 \cdot 10^{5}$
3	600	$2.0 \cdot 10^{-3}$	$5.0 \cdot 10^{2}$	$6.2 \cdot 10^{2}$

where A_0 is the initial concentration of component A, M_{da} and M_{de} are molecular weights of diamine and diepoxide, and k is the bimolecular reaction rate constant [8, 9].

Numerical calculations performed for the following parameter values: $N_e = 4$, $N_A(0) = 4$, $N_B = 12$, $4A_0k = 2 \cdot 10^{-5}$ s^{-1}, did not show much difference from the 1D case. The process peculiarities depend considerably on the ratio of characteristic times of reaction and diffusion, and on the position of the initial state on the phase diagram. For $\chi = 0.35$, $L_1 = L_2 = 100$, $\varphi_0 = 0.5$, $\tau_A = \tau_B = 0.05$, the initial point lies in the spinodal region, the reaction time is one order of magnitude larger than the diffusion time, and the phase separation is finished before the reaction is completed.

For $\tau_A = \tau_B = 0.5$, $L_1 = L_2 = 300$, the diffusion time is one order of magnitude larger than the reaction time, and the gelation point is reached before the phase separation is over.

CONCLUSIONS

For a large 2D system, the retardation scenario is realized, with a decisive role of long-lived structures. The process does not depend on whether or not the initial conditions are random.

The lifetime for unstable structures is determined with a good accuracy by increments of solutions for diffusion equations linearized in the vicinity of these structures.

Similarly to 1D case, the spinodal region is divided into sub-regions corresponding to a soft and rigid birth of spatial structures. The same description is possible for spinodal (upon rigid birth) and metastable regions.

In the case of CRIMPS, the process proceeds according to retardation scenario up to the gel point. The presence of the gel-point predetermines "fixation" of an intermediate heterogeneous structure, which seems to be thermodynamically stable.

Acknowledgments. This work was supported by the International Science and Technology Center (grant no. 358-96).

REFERENCES

1. Mitlin V.S., Manevich L.I., and Erukhimovich I.Ya., *Zh. Eksp. Teor. Fiz.*, **88**, 495 (1985).
2. Mitlin V.S. and Manevich L.I., *J. Polym. Sci., Part B: Polym. Phys.*, **28**, 1 (1990).
3. Prostomolotova E.V., Erukhimovich I.Ya., and Manevich L.I.,. *Polym. Sci.*, **39**, 682 (1997).
4. Kotnis M.A. and Muthukumar M., *Macromolecules*, **25**, 1716 (1992).
5. Brown G. and Chakrabarti A., *J. Chem. Phys.*, **98**, 2451 (1993).
6. Ivanova A.N., Manevich L.I., and Tarnopol'skii B.L., *Comp. Maths. Math. Phys.*, **39**, 284 (1999).
7. Ivanova A.N. and Tarnopol'skii B.L., *Comp. Maths. Math. Phys.*, **39**, 624 (1999).
8. Irzhak V.I., Rozenberg B.A., and Enikolopyan N.S., *Setchatye polimery: Sintez, struktura, svoistva* (Network Polymers: Synthesis, Structure, and Properties), Nauka, Moscow, 1979. Ch. 2.
9. Manevich L.I., Rozenberg B.A., and Shaginyan Sh.A., *this book*, p. 65.

Chapter 8

Theory and Simulation for Dynamics of Polymerization-Induced Phase Separation in Reactive Polymer Blends

Thein KYU*, Hao-Wen CHIU, and Jae-Hyung LEE

Institute of Polymer Engineering, The University of Akron, Akron OH, 44325-0301 USA

ABSTRACT

Mechanisms and dynamics of phase decomposition following polymerization-induced phase separation (PIPS) of thermoset/thermoplastic blends have been investigated. The phenomenon of PIPS is a non-linear dynamic process that involves competition between reaction kinetics and phase separation dynamics. The mechanism of PIPS has been thought to be a nucleation and growth (NG) originally, however, newer results indicate spinodal decomposition (SD). In PIPS, the coexistence curve generally passes through the reaction temperature at off-critical

*e-mail: tkyu@uakron.edu

points, thus phase separation must be initiated first in the metastable region where nucleation occurs. When the system further drifts from the metastable to the unstable region, the NG structure transforms to the SD bicontinuous morphology. The crossover behavior of PIPS may be called nucleation initiated spinodal decomposition (NISD) so that it can be distinguished from the conventional SD. The formation of newer domains between the existing ones is responsible for the early stage of PIPS. Since PIPS is a non-equilibrium kinetic process, it would not be surprising to discern either NG or SD textures.

INTRODUCTION

In recent years, the field of polymerization induced phase separation (PIPS) in reactive prepolymer/polymer blends has gained considerable interest because of development of unusual equilibrium and/or non-equilibrium patterns [1, 2] and also for practical purposes [3]. In general, liquid–liquid phase separation occurs in polymer blends either by thermal quenching into an unstable region from an initially homogeneous state or through polymerization. While thermally induced phase separation (TIPS) has been extensively investigated for quenched binary systems, there are only limited studies on phase separation driven by polymerization [3–12], although this process may be at least equally important. When a polymer blend is brought from an initially homogeneous state into an unstable spinodal region, various modes of concentration fluctuations develop and are amplified simultaneously by virtue of thermal fluctuations, resulting in an irregular two-phase structure [1, 13]. However, if thermal fluctuations were suppressed fully, a single selective mode grows predominantly creating a more regular structure. In the case of reaction induced phase separation, the instability in the system is driven by a continuous increase in molecular weight of one or both components [6–9]. Once this kind of chemical reaction has been initiated, there will be a competition [6–12] between phase separation dynamics and reaction kinetics that determines the final non-equilibrium structure. Understanding the governing mechanism(s) of polymerization-induced phase separation is therefore of paramount importance in order to gain insight into development of the final blend morphology.

The mechanism of nucleation and growth (NG) has been perceived to be prevalent in the polymerization-induced phase separation because of frequent observation of a globular structure (i.e., spherical domains that are often interconnected) in microscopic investigations of the post-cured blends [5]. Time-resolved light scattering studies [6] on PIPS have shown that phase separation occurs through spinodal decomposition (SD) that casts some doubt on the assignment of the NG mechanism to PIPS. Recently we found that the

PIPS mechanism is more complex than hitherto reported by others [5–9], i.e., phase separation occurs in the metastable region via a nucleation and growth process due to the asymmetric movement of the upper critical solution temperature during polymerization. The system then enters into the unstable regions with progressive polymerization, resulting in a crossover in behavior from the NG to the SD. In this article, we introduce recent theoretical advances and two-dimensional numerical simulations on PIPS with emphasis on structure development and coarsening dynamics of the PIPS.

THEORETICAL MODELING

The dynamics of phase separation driven by polymerization may be treated as a reaction-diffusion process [1, 10, 11]. The system under consideration is a binary blend such as polymer dispersed liquid crystal (PDLC) prepared via polymerization induced phase separation of low molar mass liquid crystals (in an isotropic state)/epoxy mixtures or of liquid rubber/epoxy blends. However, only one component (i.e., epoxy) undergoes polymerization and/or crosslinking reactions, which may be represented by

$$\mathrm{M}+\mathrm{C}\xrightarrow{\mathrm{S}}\mathrm{P}\,, \tag{1}$$

where S stands for solvent (e.g., non-reacting component such as isotropic liquid crystals or prepolymers such as liquid rubber), M is the reacting monomer, C is the crosslinking agent, and P is the resulting polymer. The diffusion process for this system is expressed according to the time-dependent Ginzburg–Landau equation [1, 10]

$$\frac{\partial \phi_1(r,t)}{\partial t} = -\nabla J_1 + \eta(r,t)\,, \tag{2}$$

where J_1 is the flux and $\eta(r, t)$ is thermal fluctuation that satisfies the fluctuation-dissipation theorem [1] and $\phi_1(r,t)$ is the volume fraction of the non-reacting component at position r and reaction time t. When the polymerization rate is slow compared to the kinetics of phase separation, a sizable amount of monomers would remain unreacted at a given time. In principle, the emerging polymer can segregate from the residual monomer as well as from the non-reacting component (i.e., polymer solvent). Hence, such a reacting blend should be treated as a three-phase system because it contains the residual monomer, the emerging polymer, and the polymer solvent. The change of monomer concentration (volume fraction) for such a three-phase system may be described through the reaction-diffusion equation [10, 11], viz.,

$$\frac{\partial\phi_m(r,t)}{\partial t} = -\nabla J_m - \alpha(T,t)\phi_m(r,t) + \eta_m(r,t), \tag{3}$$

$$\frac{\partial\phi_p(r,t)}{\partial t} = -\nabla J_p + \alpha(t)\phi_m(r,t) + \eta_p(r,t), \tag{4}$$

where η_m and η_p are the thermal fluctuations produced by the reacting monomer and the resulting polymer. The monomer concentration, ϕ_m, can be related to the volume fraction of the emerging polymer (ϕ_p) in terms of the incompressibility condition $\phi_1 + \phi_2 = 1$ with $\varphi_2 = \varphi_m + \varphi_p$. Further, the rate of polymerization reaction at a given reaction temperature, $\alpha(T, t) = dp(t)/dt$ is given as [15,16]

$$\alpha(T,t) = k(T)p(t)^m[1-p(t)]^n, \tag{5}$$

where $k(T)$ is the reaction rate constant with m and n being the reaction exponents to characterize the consumption of monomer and the emergence of polymer, respectively. Further the degree of conversion $p(t)$ can be related [15] to the increasing degree of polymerization $N(t)$ according to $1 - f_{av}p(t)/2 = 1/N(t)$, where f_{av} is the average functionality.

When the polymerization rate is slow as compared to the kinetics of phase separation, a significant amount of monomer would remain unreacted at a given time. In principle, the emerging polymer could segregate from the residual monomer as well as from the non-reacting LC component. Hence, such a reacting blend should be treated as a three-phase system because it contains the residual monomer, the emerging polymer, and the liquid rubbers. The pattern forming aspects for such a three-phase system may be modeled by numerically solving equations (2) and (3) simultaneously. On the other hand, if the polymerization rate is faster than the kinetics of phase separation, most of the monomers will be consumed during polymerization. It can be anticipated that the emerging polymers may result in a wide distribution of molecular weights. As is well known, the molecular weight distribution exerts profound effect on the establishment of thermodynamic phase diagrams [13]. However, the polydispersity plays an insignificant role in the phase separation dynamics of the thermal-quenched case [17]. Hence, it is reasonable to assume that the influence of molecular weight distribution on the dynamics of PIPS may be inconsequential.

Assuming that the residual oligomers and the emerging polymers are completely miscible, the polymerizing component may be treated as a single component (hereafter designated as component 2 with ($\phi_2 = \phi_m + \phi_p$), which further simplifies the treatment of the polymerizing system as a pseudo two-phase blend. From the incompressibility condition $\phi_1 + \phi_2 = 1$, equation (3) leads to

$$\frac{\partial\phi_2(r,t)}{\partial t} = -\nabla J_2 + \eta_2(r,t)\,. \tag{6}$$

It is evident that equation (6) is complementary to (2). Hence, it should be sufficient to solve equations (2) and (5) simultaneously in describing the dynamics of phase separation in a PDLC in which only one component is reactive [11].

The thermal fluctuation force, $\eta(r,\ t)$, is customarily expressed according to the fluctuation-dissipation theorem [1] as

$$\langle \eta(r,t)\eta(r',t')\rangle = -2k_B T\Lambda\nabla^2\delta(r-r')\delta(t-t')\,, \tag{7}$$

where k_B is the Boltzmann constant and T temperature. Λ is defined as the mutual diffusion coefficient having the property of the Onsager reciprocity [17, 18]. For a two-phase system, Λ depends on changing blend composition and increasing degree of polymerization as follows:

$$\frac{1}{\Lambda} = \frac{1}{D_1 N_1 \phi_1} + \frac{1}{D_2 N_2(t)\phi_2}\,, \tag{8}$$

where D_j are the self-diffusion coefficients of the components j. N_1 represents the degree of polymerization of the non-polymerizing component 1 and $N_2(t)$ is that of the polymerizing component 2. D_j are further related to N_j, viz., $D_j = D_{j0}N_j^{-2}$ for a reptation model [18] or $D_j = D_{j0}N_j^{-1}$ for a Rouse model [17]. Here, we adopt the former model. The diffusion flux, J_1, is given by

$$J_1 = -\frac{\Lambda}{k_B T}\nabla\left[\frac{\delta G(\phi_1)}{\delta\phi_1}\right], \tag{9}$$

where $G(\phi)$ is the total free energy of the mixture. Further, $G(\phi)$ can be expressed in the form of the Cahn–Hilliard–de Gennes expression [19–21]:

$$\frac{G}{k_B T} = \int_V (g_\ell + g_{\text{grad}})\,dV \tag{10}$$

in which $g = g_\ell + g_{\text{grad}}$ denotes the local and non-local free energy densities. It is customary to describe the local free energy density of a binary blend in terms of the Flory–Huggins lattice model [22]:

$$g_\ell = \frac{\phi_1}{N_1} ln\,\phi_1 + \frac{\phi_2}{N_2(t)} ln\,\phi_2 + \chi\phi_1\phi_2\,, \tag{11}$$

where χ is known as the Flory–Huggins interaction parameter. In general, χ is assumed to be a function of reciprocal absolute temperature, i.e., $\chi = A + B/T$, where A and B are constants. Note that equation (11) needs to be modified for a three-phase system. The second term in equation (10), g_{grad}, represents the free energy density arising from the concentration gradient defined [20] as

$$g_{\text{grad}} = \kappa\left|\nabla\phi_i\right|^2, \tag{12}$$

where $i = 1$ or 2 and κ is a coefficient relating to the segmental correlation length and the local concentration. For an asymmetric polymer–polymer mixture, κ is given [21] by

$$\kappa = \frac{1}{36}\left[\frac{a_1^2}{\phi_1} + \frac{a_2^2}{\phi_2}\right], \tag{13}$$

where a_1 and a_2 are the correlation lengths of polymer segments of the component 1 and 2, respectively. The equation of motion has been customarily expressed by combining equations (2), (9), and (10) as follows [11]:

$$\frac{\partial\phi_1(r,t)}{\partial t} = \nabla\left[\Lambda\nabla\left\{\frac{\partial g}{\partial\phi_1} - \nabla\left(\frac{\partial g}{\partial(\nabla\phi_1)}\right)\right\}\right] + \eta(r,t), \tag{14}$$

where

$$\frac{\partial g}{\partial\phi_1} = \frac{\ln\phi_1 + 1}{N_1} - \frac{\ln\phi_2 + 1}{N_2(t)} + \chi(1 - 2\phi_1)\,, \tag{15}$$

$$\nabla\left(\frac{\partial g}{\partial\nabla\phi_1}\right) = \frac{1}{18}\left[\frac{a_1^2}{\phi_1} + \frac{a_2^2}{\phi_2}\right](\nabla^2\phi_1) - \frac{1}{36}\left[\frac{a_1^2}{\phi_1{}^2} - \frac{a_2^2}{\phi_2{}^2}\right](\nabla\phi_1)^2\,. \tag{16}$$

From equations (2), (14)–(16), the pattern forming aspects of phase separation during polymerization may be investigated. It should be pointed out that the molecular diffusion is simply coupled with the polymerization reaction through the time dependence of the molecular weight of the polymerizing component, $N_2(t)$. In the event of a three-phase system consisting of the residual monomer (oligomer), the emerging polymer, and the non-reacting solvent (e.g., liquid crystals or liquid rubber), the reaction and diffusion processes are coupled through both $N_2(t)$ and $\alpha(T, t)$ of equation (5). Hence the temporal change of concentration fluctuations will be dominated by both the change in the local free energy density (or chemical potential) associated with the progressive polymerization as well as by the coupling term involving the reaction rate, $\alpha(t)$, and the monomer concentration.

Next, equation (14) may be rewritten in Fourier space to determine the temporal evolution of structure factor, $s(q, t)$, i.e.,

$$s(q,t) = F[\phi_1(r_1,t)\phi_1(r_2,t)], \tag{17}$$

where F represents the Fourier transform and q is the scattering wavenumber defined as $q = (4\pi/\lambda)\sin(\theta/2)$ where λ and θ are wavelength of incident light and scattering angle, respectively. Comparing the temporal change of the calculated structure factor with the experimental results of the time-resolved scattering studies, the validity of equation (17) may be tested. Numerical calculation was performed on a two-dimensional square lattice (128×128) using a finite difference scheme for spatial steps and an explicit method for temporal steps with a periodic boundary condition.

RESULTS AND DISCUSSION

In PIPS, establishment of a temperature–composition phase diagram of the starting mixture is indispensable to guide polymerization reaction for controlling morphology development and PIPS dynamics. Figure 1 shows the cloud point phase diagram of the starting blends of diglycidyl ether bisphenol-A epoxy (BADGE) and carboxyl terminated butadiene acrylonitrile (CTBN), showing a UCST-type coexistence curve with a convex maximum at 60°C and about 12.5 wt % CTBN. The addition of methylene dianiline (MDA) curing agent in the equivalent amount to the epoxy tends to suppress the UCST curve. Polymerization was initiated in a single-phase temperature denoted by X in the figure. Upon polymerization, the molecular weight of the reacting epoxy increases which makes the system unstable. This instability drives the coexistence curve to move up to a higher temperature and asymmetrically to a higher CTBN side.

Eventually the UCST curve surpasses the reaction temperature at off-critical points (Figure 1). In view of the asymmetric shift of the UCST, phase separation is believed to occur in the metastable region, and then the system enters into the unstable region with progressive polymerization. Since phase separation is initiated in the metastable region then enters into the unstable region, there is a crossover in behavior from nucleation and growth to spinodal decomposition. Another interesting observation was that the length scale or the average size of the domains decreases due to increasing supercooling, ΔT. Similar behavior was also observed independently by Inoue *et al.* [6–8] and later by Chan and Ray [12]. It should be pointed out that the decrease in the length scale is observable only in the early stage of PIPS where the reaction kinetics predominates over the structural growth associated with thermal relaxation. This mechanism, termed nucleation initiated spinodal decomposition (NISD), is completely different from the linear growth of fluctuations observed in some thermally quenched systems near the critical point [16–18] where the early stage of SD is characterized by a linear growth. To account for the NISD phenomenon, we analytically and numerically demonstrated in the linear limit that the length scale is reduced [11] due to increasing supercooling and/or the development of the newer domains between those existing.

Now, we shall extend our study to a two-dimensional simulation in order to elucidate the PIPS dynamics without linearization. The calculation was performed by assuming $A = -1$ that in turn gives $B = 550.72$ according to a criticality condition, viz., $\chi = A + (\chi_c - A) \cdot T_c / T$. Further, the initial conditions of the polymerization were set as $k = 0.001$, $m = 0.5$ and $n = 1.5$

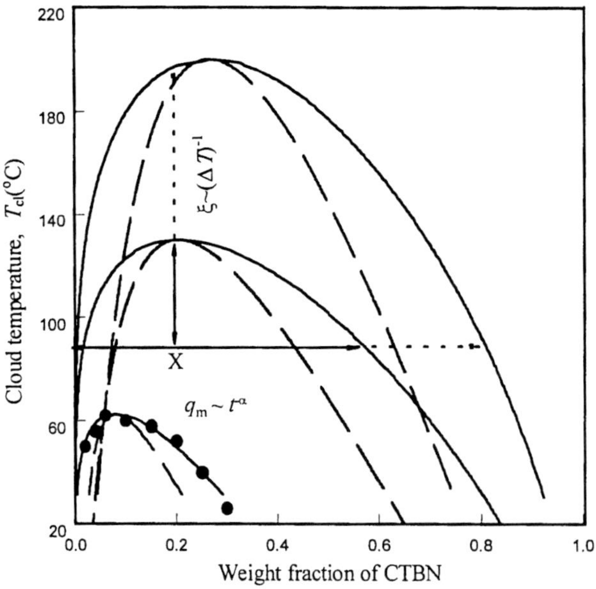

Figure 1. Temperature–composition phase diagram for the starting mixture of BADGE/CTBN and the snapshots of the coexistence curve with the progression of polymerization. The reaction temperature is indicated by X.

with $a_1 = 1.5$, $a_2 = 1.5$, $D_{10} = 2$, and $D_{20} = 98$ in dimensionless units. The reaction was initiated at a single-phase temperature of 90°C. As the polymerization advances, the UCST curve moves progressively to a higher temperature but noticeably to a lower composition of the polymerizing component at later times (Fig. 1). When the UCST surpasses the reaction temperature, phase separation begins in the blends.

Figure 2 shows the temporal evolution of the phase separated domain structures during the progressive polymerization. The smaller thermal fluctuations diminish much faster than the larger ones during the so-called induction period. When the UCST curve catches up with the reaction temperature, phase separation starts in the metastable region and enters rapidly to the unstable region. In liquid-liquid phase separation, it is well known that spinodal decomposition is an unstable process. Hence, even small concentration fluctuations can grow spontaneously. In the metastable region, all small modes of concentration fluctuations tend to diminish during the induction period. The nucleation process is a natural occurrence, and thus thought to be the preferred mechanism for the polymerization-induced phase separation. However, as the system drifts from the metastable to the unstable region these concentration fluctuations grow in magnitude while newer fluctuations develop in between those already present and eventually transform into a so-called bicontinuous structure reminiscent of a spinodal texture.

To appreciate the formation of the newer fluctuations more clearly, the two-dimensional matrix (128 × 128) was reduced to (64 × 64) space steps and subsequently sliced into one-dimension. Note that the width of the slice was the average of 3 tracks. The resulting temporal change of the concentration fluctuation profiles is depicted in Fig. 3. The small thermal fluctuations decay rapidly during the induction period (1000 time steps) leaving behind predominantly the larger ones. These larger fluctuations could decay further if the gap between the critical and the reaction temperatures were large. When the system reaches the metastable region, the amplitudes of these large fluctuations increase (t = 2000). Subsequently, newer fluctuation peaks (indicated by arrows) develop between the existing ones leading to the reduction of the inter-domain distances (i.e., peak-to-peak distance of the fluctuations). The domain size (half-width) decreases while the amplitude (contrast of electron density or concentration) continues to increase (t = 2500–3500), resulting in the sharp interface. The sharpening of the interface domain boundary during polymerization is consistent with that reported by Glotzer and co-workers [10] for their simulation of reaction-induced phase separation.

With continued polymerization, the system is thrust deeply into the unstable region. The magnitude of fluctuations increases (note the change of ordinate scale at t = 2500, 3000, and 3500), while newer fluctuations develop (shown by arrows). As the system has entered from the metastable to the unstable spinodal region, the fluctuating domain structure gets sharper and becomes more regular. This crossover in behavior from nucleation to spinodal decomposition driven by polymerization [11] is strikingly similar to that in a slowly cooled system [6].

Figure 4 shows the temporal evolution of the corresponding scattering patterns obtained by Fourier transforming the domain structures (patterns) of Fig. 2. The structure factor initially shows a diffused scattering pattern without a clear maximum, suggestive of a heterogeneous nucleation process (e.g. see t = 100). Later, it transforms into a scattering ring, while the diameter increases with progressive polymerization (t = 1500). The increase in diameter of the scattering ring at t = 2000 may be attributed to the formation of newer fluctuations as opposed to the Ostwald ripening observed in some thermal quenched systems. Another possibility is that the difference between the coexistence point and the reaction temperature (i.e., supercooling) becomes larger due to the progressive shift of the UCST to a higher temperature (or the LCST to a lower temperature) by virtue of increasing molecular weight. The PIPS tends to afford smaller domain sizes because the larger the supercooling the smaller the domain size, i.e., $\xi \propto 1/\Delta T$ (Figure 1). The increase of the intensity (structure factor) may be caused by the increasing number of fluctuations (scattering centers) as well as by the increase in the magnitude of the fluctuations (scattering contrast).

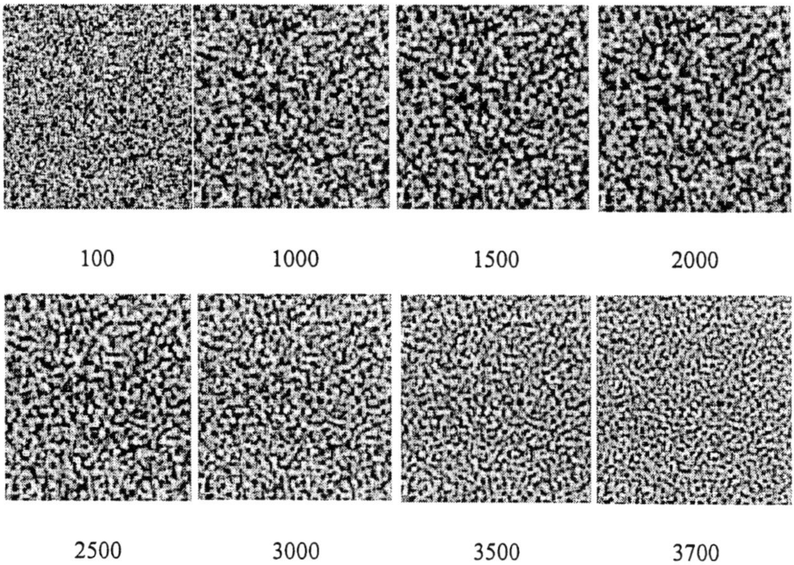

Figure 2. Temporal evolution of phase separated domains with the progression of polymerization.

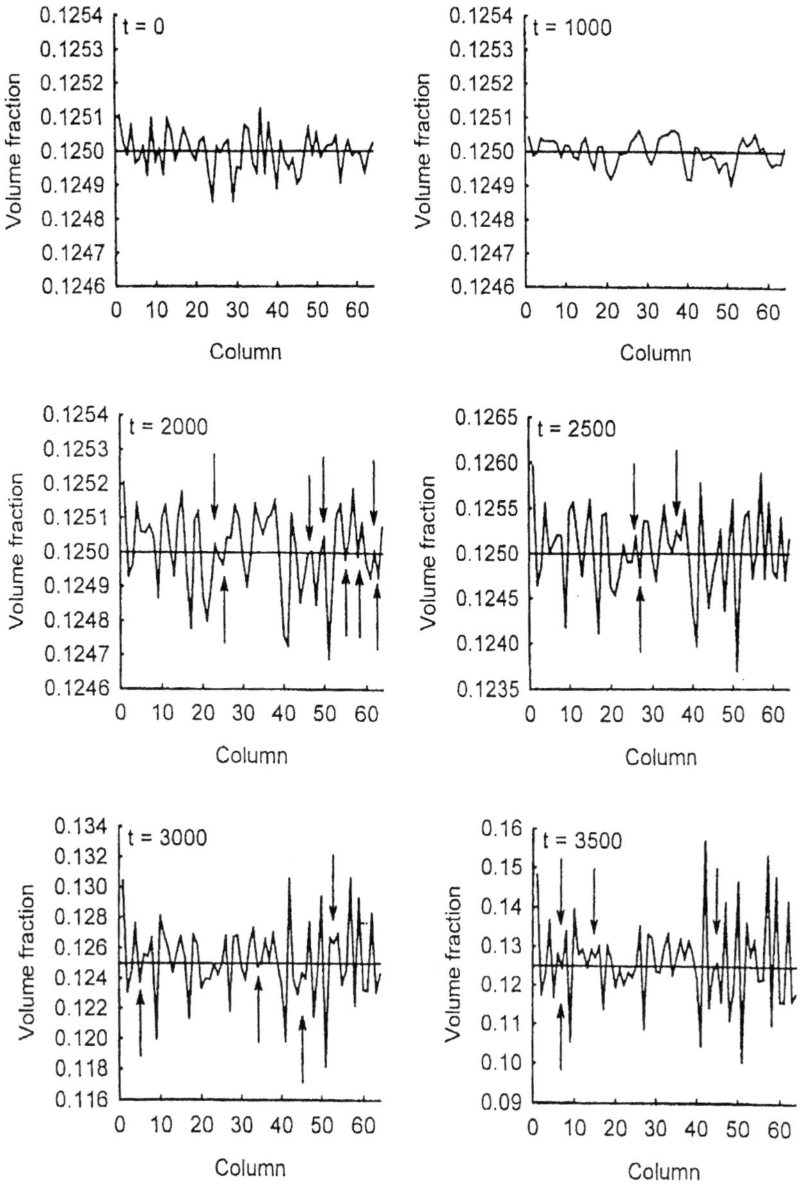

Figure 3. Temporal change of concentration fluctuation profiles during phase separation driven by progressive polymerization, displaying initial decay of small fluctuations and subsequent formation of newer fluctuations with elapsed time. These calculated results were sliced in one dimension from the 128 × 128 matrix and reduced to the 64 space steps for clarity.

As the polymerization continues, the peak of the structure factor gets sharper while moving to a wider angle.

Figure 5a shows the log q_m versus log$(t - t_i)$ plot in which t_i represents the induction time. It is striking to observe a discrete variation of the wavenumber maximum with time at a relatively fast reaction rate (e.g., k = 0.005). At slower reaction rates, this behavior is more gradual. It is tempting to speculate that when newer fluctuations are formed between those already existing, the inter-domain distances may be shorten which is exactly what was seen in the simulation (Figure 3). Later, it follows a power law behavior with an exponent that depends on the choice of m values. At a given set of constants, m and n, the slope seemingly remains unchanged with increasing reaction rate (k). Hence, it is reasonable to conclude that the onset of the temporal change of the wavenumber maximum, q_m, increases with increasing k. Another interesting feature is that the final length scale is reduced with increasing k, i.e., the faster the reaction rate, the smaller the domain size. This behavior is reminiscent of the domain structures developed in the slowly cooled (or shallow quench) system to be larger than that in the rapidly quenched (or deep quench) blends.

Figure 5b shows the influence of the n values on the time dependent behavior of PIPS. For a given k and m values, the onset of the reaction time as well as the slope of log q_m versus log$(t - t_i)$ plot appear nearly the same regardless of the n values. Ignoring the order of reaction, the m value is varied simply from 0 to 1. As shown in Figure 5c, the m value exerts significant effects on both the slopes as well as the onset of phase separation

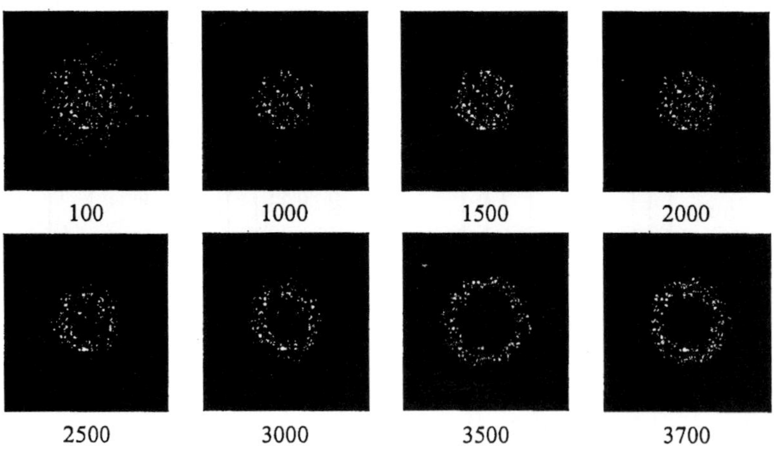

Figure 4. Temporal evolution of Fourier-transformed scattering patterns during phase separation driven by progressive polymerization, showing a change from a diffused scattering pattern without a maximum (nucleation) to a clear scattering ring (spinodal).

time. The sigmoidal variation of wavenumber maximum becomes steeper with increasing m and eventually levels off due to crosslinking.

When the reactivity of the curing agent is low, the reaction rate will be slow relative to the dynamics of phase separation. For the case of a slow polymerization reaction, it can be anticipated that the domains would grow as opposed to the early stage of PIPS where the length scales get smaller with elapsed time. This process would be reminiscent of the late stages of SD of the conventional TIPS, which may be scaled according to the power law[1], i.e., $q_m(t) = 1/\xi(t) \propto t^{-\alpha}$, where $\xi(t)$ is the length scale. The classical TIPS predicts the growth exponent of −1/3 for the intermediate stage crossing over to the late stages of SD with the value of −1 where hydrodynamics dominates. However, the wavelength selection rule predicts a smaller value of −1/4 for the PIPS process [10].

As demonstrated above, the growth exponents determined experimentally could vary from 1/2 to −1 depending on the reactivity of the curing agent, its amount and curing temperature, and blend composition. As shown in Figure 1, the progressive shift of the UCST to a higher temperature (or the LCST to a lower temperature) will drive the PIPS to afford smaller

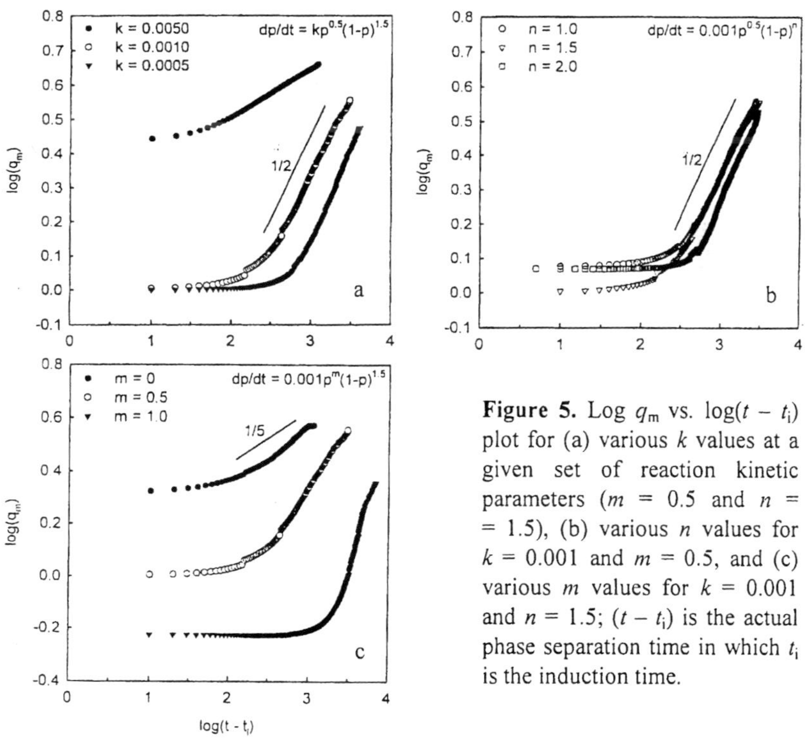

Figure 5. Log q_m vs. $\log(t - t_i)$ plot for (a) various k values at a given set of reaction kinetic parameters (m = 0.5 and n = = 1.5), (b) various n values for k = 0.001 and m = 0.5, and (c) various m values for k = 0.001 and n = 1.5; $(t - t_i)$ is the actual phase separation time in which t_i is the induction time.

domain sizes because the larger the supercooling (ΔT), the smaller the domain size, whereas the structural growth due to the coalescence driven by thermal relaxation will drive the average size to increase in time. When supercooling is dominant, q_m increases with time and then levels off (Figure 6*a*). In the event that the coarsening process prevails (Figure 6*c*), the growth dynamics would resemble that of the thermal quench case. If the two competing processes were comparable the q_m in the initial period would appear invariant like a linear regime (Figure 6*b*). Hence, these two opposing mechanisms would naturally give a growth exponent between the limiting scaling exponents of 1/2 for the length scale reduction due to the supercooling effect to −1 for the coarsening in the hydrodynamic regime due to thermal relaxation. Moreover, the increase in molecular weight will increase viscosity and hence slow diffusion; therefore the domain growth must slow down. This prediction is exactly what one observed experimentally for the polymerization induced phase separation of the BADGE/CTBN mixtures. It should be pointed out that the NISD structures strongly depend on the magnitude of thermal noise introduced initially to the system as well as on the temperature gap. The most crucial findings in the polymerization induced phase separation are the finer average domain size, the reduced inter-domain distances, and the uniform dispersion of these domains, which are undoubtedly important for the improvement of the materials properties.

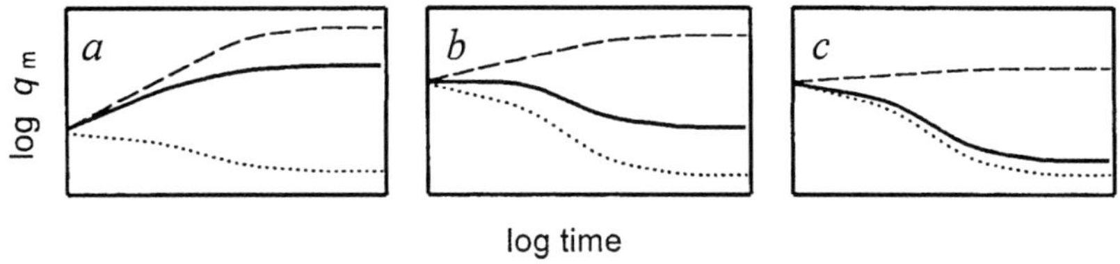

Figure 6. Predicted scaling laws for the growth dynamics resulting from the competition between the reduction of length scale due to increasing ΔT (i.e., supercooling) driven by progressive polymerization and domain coarsening due to thermal relaxation: (*a*) the supercooling is dominant, (*b*) the supercooling and coarsening are comparable, and (*c*) the coarsening is dominant.

CONCLUDING REMARKS

The initial reduction in the scale caused by the increase of the degree of conversion is unique to the early stage of phase separation driven by polymerization, which may be attributed to the formation of newer fluctuations as well as the reduction in size of fluctuations due to increasing supercooling. Another important point is that phase separation was initiated in the metastable region before drifting to the spinodal unstable region with progressive polymerization. As a consequence, there is a change in texture from the sea-and-island type (NG) to the bicontinuous structure (SD), which is referred to as nucleation initiated spinodal decomposition (NISD) in order to differentiate it from the conventional NG or SD of the thermal quenched system. This mechanism is definitely different from the early stage of thermal quench-induced spinodal decomposition, where the gradient of fluctuations grows without involving the movement of the scattering peak, and also from the Ostwald ripening mechanism. The coupling of the nucleation and spinodal decomposition is the dominant mechanism as the system drifts from the metastable to unstable regime during the course of polymerization. It is striking to observe that the formation of newer fluctuations between those existing resulted in a decrease of the inter-domain distances. Furthermore, the progressive shift of the UCST to a higher temperature (or the LCST to a lower temperature) will drive the PIPS to afford smaller domain sizes because the larger the supercooling the smaller the domain size, whereas the structural growth due to the domain coalescence driven by thermal relaxation will drive the average size to increase in time. The onset of phase separation time is greatly influenced by both the kinetic rate constant (k) and the kinetic exponent m, but it is less sensitive to n. The most important characteristics of PIPS are the reduced fluctuation size (domain size), the shorter inter-domain distances, and the finer distribution of the domains, which should have significant influence on mechanical and physical properties of reactive blends. Such fine domain structures are achievable if the domain coarsening driven thermal relaxation can be fully suppressed.

Acknowledgments. The research described in this paper was made possible by the support of National Science Foundation, DMR 95-29296 and the NSF-ALCOM through Grant No. DMR 89-20147. We thank Nwabunma Domasius and Andy Guenthner for their helpful comments and suggestions.

REFERENCES

1. Gunton J.D., San Miguel M., and Sahni P.S., in *Phase Transitions and Critical Phenomena*, Domb C. and Lebowitz J.L., Eds., Academic Press, New York, Ch. 3, 1983.
2. *Dynamics of Ordering Processes in Condensed Matters*, Komura S. and Furukawa H., Eds., Plenum Press, New York, 1988.
3. Doane J.W., in *Liquid Crystals: Applications and Usages*, Vol. 1, Bahadur B., Ed., World Scientific, Singapore, 1990.

4. Smith G.W., *Int. J. Mod. Phys.*, **B 7**, 4187 (1991).
5. *Rubber-Toughened Plastics*, Riew C.K., Ed., Adv. Chem. Series, **222**, 1989.
6. Inoue T., *Prog. Polym. Sci.*, **20**, 119 (1995).
7. Yamanaka K., Takagi Y., and Inoue T., *Polymer*, **30**, 1839 (1989).
8. Ohnaga T., Chen W., and Inoue T., *Polymer*, **35**, 3774 (1994).
9. Kim J.Y., Cho C.H., Palffy-Muhoray P., Mustafa M., and Kyu T., *Phys. Rev. Lett.*, **71**, 2232 (1993).
10. Glotzer S.C. and Coniglio A., *Phys. Rev. E*, **50**, 4241 (1994); *Phys. Rev. Lett.*, **74**, 2034 (1995).
11. Kyu T. and Lee J.H., *Phys. Rev. Lett.*, **76**, 3746, (1996).
12. Chan P.K. and Ray A.D., *Macromolecules*, **29**, 8934 (1996).
13. Olabisi O., Robeson L.M., and Shaw M.T., *Polymer–Polymer Miscibility*, Academic Press, New York, 1979.
14. Lee H.S. and Kyu T., *Macromolecules*, **23**, 459 (1990).
15. Odian G. *Principles of Polymerization*, Wiley, New York, 1981.
16. Ryan M.E. and Dutta A., *Polymer*, **20**, 203 (1979).
17. Takenaka M. and Hashimoto T., *Phys. Rev. E*, **48**, 47 (1993).
18. Doi M. and Edwards S.F., *Theory of Polymer Dynamics*, Academic Press, New York, 1986.
19. Cahn J.W. and Hilliard J.E., *J. Chem. Phys.*, **28**, 258 (1958).
20. de Gennes P.-G., *J. Chem. Phys.*, **72**, 4756 (1980).
21. Binder K., *J. Chem. Phys.*, **79**, 6387 (1983).
22. Flory P.J., *Principles of Polymer Chemistry*, Cornell University Press, Ithaca NY, 1953.

Chapter 9

Phase Diagrams of Multicomponent Reacting Polymer Systems Undergoing Phase Decomposition

Elina L. MANEVITCH, Grigori M. SIGALOV*, and Boris A. ROZENBERG

Institute of Problems of Chemical Physics, Russian Academy of Sciences, Chernogolovka, Moscow Region, 142432 Russia

*e-mail: sigalov@icp.ac.ru

ABSTRACT

The problem of calculating the phase diagrams for multicomponent reacting and non-reacting polymer systems has been formulated. To solve this problem, the concept of the free energy of phase separation (FEPS) was introduced. Calculation of phase diagrams implies global minimization of FEPS. A numerical algorithm for FEPS minimization and appropriate computer program were developed. This procedure was used to calculate phase diagrams for the ternary systems styrene/PS/PMMA, chloroform/PS/polybutadiene, and styrene/PS/polydimethyl-siloxane. The calculated data were found to agree with experiment. We also calculated a phase diagram for a reacting multicomponent system (diglycidyl ether of bisphenol A cured by 2,6-diaminopyridine in the presence of polyurethane). Up to 1000 intermediate reaction products that appear during cure were taken into account as individual components. The van Krevelen theory was applied to calculating the solubility parameter and molar volume for the system components. Reasonable qualitative agreement between experimental data on the cloud point and numerical results was reached.

INTRODUCTION

In order to predict the final structure of materials formed upon cure-reaction-induced microphase separation (CRIMPS) in multicomponent thermoset–polymer blends, one has to know the phase diagrams of these systems. For binary polymer solutions, the Flory–Huggins (FH) equation [1] is known to apply:

$$\Delta g_m = RT\left(\frac{\rho_A}{M_A}(1-\phi)\ln(1-\phi)+\frac{\rho_B}{M_B}\phi\ln\phi+\chi\phi(1-\phi)\right), \qquad (1)$$

where Δg_m is specific free energy of mixing, ρ_A and ρ_B are the specific weights of the components, M_A and M_B are their molecular weights, ϕ is the volume fraction of component B, and χ is the binary interaction parameter.

Quasi-Binary Approach

The two-component equation (1) is also applicable to the description of multicomponent thermoset–additive blends. This approximation is referred to as a quasi-binary approach and involves the following assumptions:

- a complex blend of monomers (e.g., epoxy and curing agent) and intermediate products of their reactions are considered as one component, A;
- a modifying agent (e.g., liquid rubber) is also considered as a single component, B, even if its MWD is not too narrow;

- in the FH equation, weight averaged MW for A and B are used.

Equation (1) has only one parameter, χ, that has to be assessed from experimental data, for example, from the cloud point measurements. For every particular χ and reaction conversion, the binodal and spinodal points may be found from the free energy plot with the help of simple geometric considerations (see Fig. 1).

The quasi-binary approach is widely used because of the following advantages:

- easy experimental measurement of $\overline{M}_w$ for the components;
- simplicity of equation leads to easy calculation of the phase diagram.

However, the quasi-binary approach has also its drawbacks:

- polydispersity of the components is neglected;
- quasi-binary approach is not applicable in case of chain polymerization;
- differences in pair interaction between individual components are not considered;
- differences in the specific volume of individual components are neglected; therefore, thermoset shrinkage at cure may not be allowed for;
- some experimental data do not fit the two-component FH equation at any choice of parameters.

For better description of phase separation in multicomponent blends, one should use the multicomponent FH equation.

Multicomponent Flory–Huggins Equation

A natural extension of the binary FH equation to multicomponent solutions is the multicomponent FH equation:

$$\frac{\Delta g_m}{RT} = \sum_i \frac{1}{V_i}\phi_i \ln \phi_i + \frac{1}{V_r}\sum_{i \neq j} \chi_{ij}\phi_i\phi_j \ . \tag{2}$$

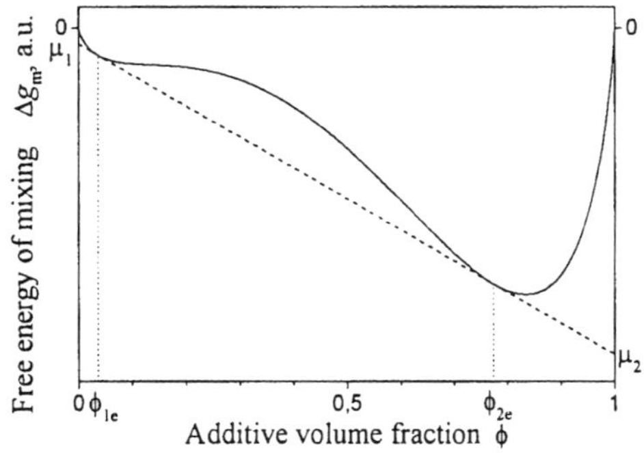

Figure 1. Common tangent to the conditional minima of the free energy plot yielding the binodal points [1].

Equation (2) involves a set of pair interaction parameters χ_{ij} that may be calculated from solubility parameters of the components:

$$\chi_{ij} = \chi_s + \frac{V_r}{RT}(\delta_i - \delta_j)^2. \tag{3}$$

In turn, the solubility parameters δ_i may be found (by using the additive group contribution approach [2]) from the published data. The molar volume of components, V_i, may be measured whenever possible or calculated using the theory of additive group contribution. In particular, the molar volume of intermediate products of polycondensation reaction may be estimated by using the molar volume of monomers and data on the shrinkage–conversion relation. An entropy-related contribution to the pair interaction parameters, χ_S, is usually assumed to be zero. However, calculation of phase diagrams with the help of the multicomponent FH equation is a non-trivial problem. To simplify the solution, the notion of free energy of phase separation is introduced.

Free Energy of Phase Separation

Note that the free energy of mixing given by the FH equation is the change of the system free energy upon formation of a multicomponent homophase solution from a number of individual components. In contrast, *the free energy of phase separation* is the change in the system free energy upon formation of a stable two-phase system from a homophase solution.

Let n components to form a homophase solution of total volume V. In this case, the free energy of mixing is given by:

$$\Delta G_m^1 = V\,\Delta g_m(\phi_1,\ldots,\phi_n), \tag{4}$$

where the specific free energy $\Delta g_m(\phi_1,\ldots,\phi_n)$, or, in shorter notation, $\Delta g_m(\phi)$, is given by equation (2). If this system separates into two phases, α and β, then the free energy of mixing of the heterophase system can be written as

$$\Delta G_m^2 = \Delta G_m^\alpha + \Delta G_m^\beta = V^\alpha\,\Delta g_m(\phi_1^\alpha,\ldots,\phi_n^\alpha) + V^\beta\,\Delta g_m(\phi_1^\beta,\ldots,\phi_n^\beta). \tag{5}$$

The specific free energy of phase separation (FEPS) may be formally written as the normalized difference between (5) and (4):

$$\Delta\tilde{g}_{ps} = (\Delta G_m^2 - \Delta G_m^1)/V = \Phi\Delta g_m(\phi^\alpha) + (1-\Phi)\Delta g_m(\phi^\beta) - \Delta g_m(\phi), \tag{6}$$

where $\Phi = V^\alpha / V$ is the volume fraction of α-phase. If the phase separation is allowed by thermodynamics, then FEPS must be negative and have a

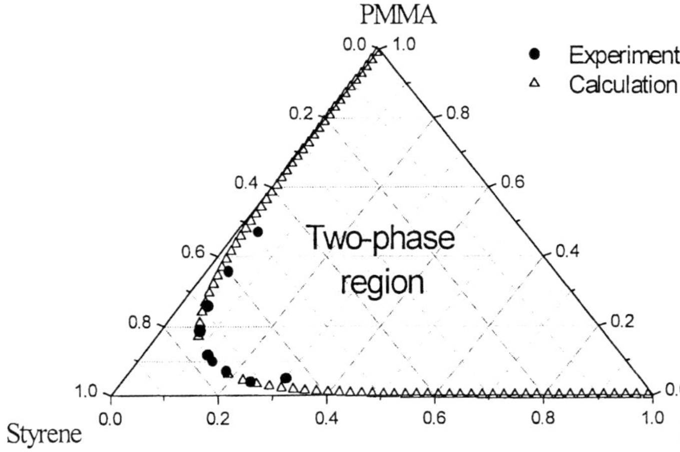

Figure 2. Ternary phase diagram for styrene/PS/PMMA system. Styrene: $\delta = 8.96$; PS: $\overline{M}_w = 2350$, $\delta = 8.83$; PMMA: $\overline{M}_w = 33 \cdot 10^3$, $\delta = 9.3$; $T = 20°C$, $\chi_s = 0$.

global minimum. Therefore, the unknown independent variables† $\Phi, \phi_1^\beta, \phi_2^\beta, \ldots, \phi_{n-1}^\beta$ from equation (6) can be found by minimization of $\Delta\tilde{g}_{ps}$:

$$\Delta g_{ps} = \min_{\Phi,\phi_i^\beta}(\Delta\tilde{g}_{ps}). \tag{7}$$

Equation (7) gives the final expression for the specific free energy of phase separation. At the same time, minimization yields the volume fraction of the two phases and their detailed composition. In general case, minimization of (7) cannot be performed analytically. However, numerically it is quite straightforward.

Algorithm for Numerical Calculation of Phase Diagrams

Preliminary testing of different standard minimization procedures has shown that the only method that turned out applicable to solution of equation (7) is the Powell technique [3]. Nevertheless, even this method, when applied to the problem under consideration, is too sensitive to the initial conditions for the numerical minimization procedure. To overcome this obstacle, the Monte-Carlo technique was applied to choose the initial conditions, followed by the Powell minimization procedure. The use of

†If the variables $\Phi, \phi_1^\beta, \phi_2^\beta, \ldots, \phi_{n-1}^\beta$, are assumed independent, then the volume fraction of the α-phase components ϕ_i^α may be easily found from the condition of material balance $\phi_i = \Phi\phi_i^\alpha + (1-\Phi)\phi_i^\beta$ and normalization condition $\sum\phi_i = 1$ for all of the phases.

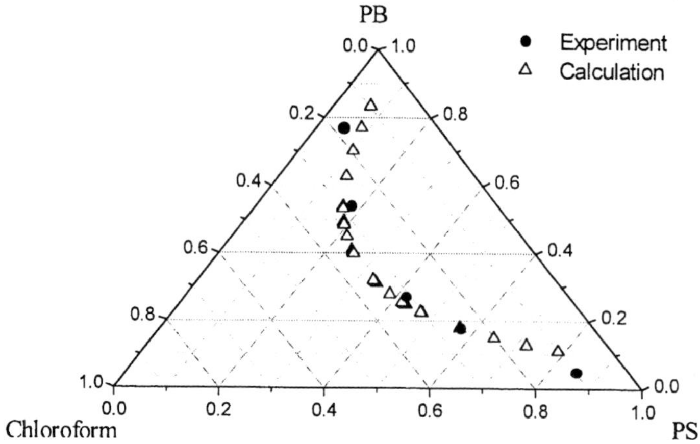

Figure 3. Ternary phase diagram for chloroform/PS/PB system. Chloroform: $\delta = 9.3$; PS: $\overline{M}_w = 3570$, $\delta = 8.83$; PB: $\overline{M}_w = 2250$, $\delta = 8.44$; $T = 20°C$, $\chi_s = 0.06$.

random initial conditions was justified because only the global minimum was sought. On the basis of the proposed algorithm, a computer program has been developed in the Turbo C language. The program allowed calculating the phase diagrams for systems with an arbitrary number of components.

RESULTS AND DISCUSSION

Phase Diagrams of Ternary Systems

To check the proposed algorithm, the phase diagrams calculated for a number of the ternary systems 'solvent/polymer A/polymer B' have been compared to he available experimental data. When the solvent is the A-type monomer, the phase diagrams may describe a polymerizing system that contains polymer B and undergoes phase separation in the course of polymerization. In calculations, χ_s was used as a fitting parameter.

The solubility parameter δ for all of the components was calculated using the additive group contribution approach [2]. The data obtained are given in the figure captions along with other characteristics and best-fit values of parameter χ_s.

Phase diagrams have been calculated for the 'styrene/polystyrene (PS)/polymethylmethacrylate (PMMA)', 'chloroform/PS/polybutadiene (PB)', and 'styrene/PS/polydimethylsiloxane (PDMS)' ternary systems. As follows from Figs. 2–4, agreement between the calculated and experimental data is satisfactory. As expected, the best agreement is observed in the

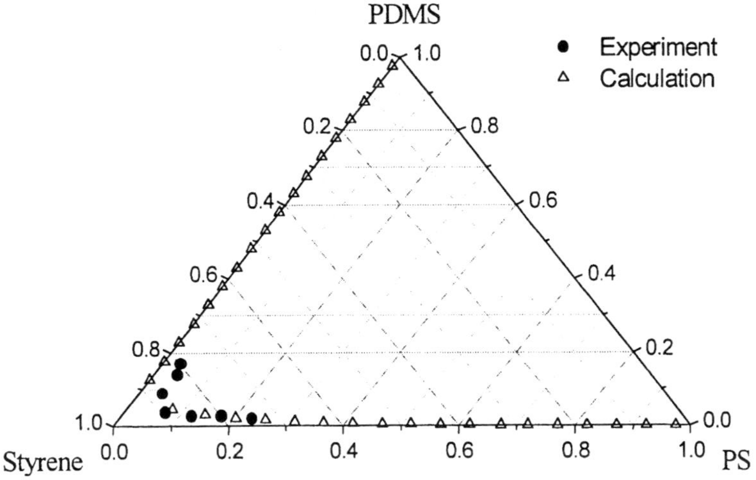

Figure 4. Phase diagram for the styrene/PS/PDMS system. Styrene: $\delta = 8.96$; PS: $\overline{M}_w = 3.1 \cdot 10^5$, $\delta = 8.83$; PDMS: $\overline{M}_w = 1.3 \cdot 10^4$, $\delta = 9.53$; $T = 20°C$, $\chi_s = 0$.

region of relatively low polymer content. Upon variation in χ_s, one can fit the calculated data to the measured critical point. Meanwhile, the branches of the theoretical binodal curve may markedly deviate from the measured coexistance boundary. This behavior may be attributed to the fact that the FH theory is less accurate at low concentration of blend components.

The best-fit value of parameter χ_s was found to be zero for the styrene/PS/PMMA and styrene/PS/PDMS systems (calculation error for χ_s was about 0.01). This testifies the random character of these blends. As expected, the chloroform/PS/PB system exhibits a higher χ_s ($\chi_s = 0.06$) due to stronger specific interactions leading to cluster formation and, therefore, a lower enthropy-related contribution to the free energy of mixing.

Phase Diagram of Curing Multicomponent System

A practical aim of our work was to model CRIMPS in curing epoxy systems modified by a non-reactive rubber additive. Numerical modeling was performed for a system containing diglycidyl ether of bisphenol A (**E**), 2,6-diaminopyridine (**A**), intermediate products of their reaction, and an additive (S13) of the following structure:

$$X–O–Y–O–[–Z_n–Y–O–]_m–X,$$

where n = 13, m is determined by the molecular weight of S13, and structural units are as follows:

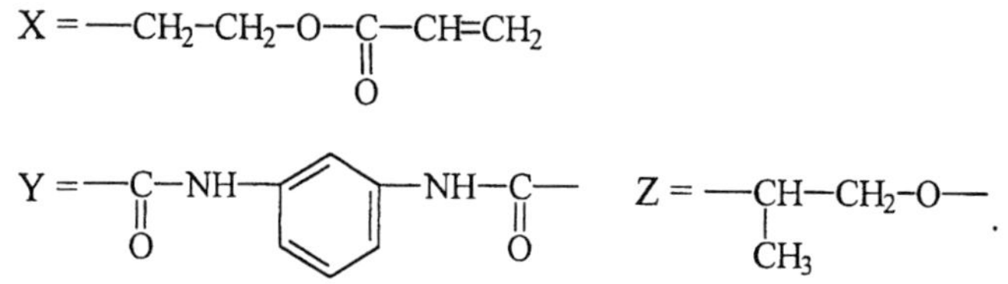

Intermediate products of the polycondensation reaction, **EA**, $\mathbf{E_2A}$, $\mathbf{E_2A_2}$, $\mathbf{E_2A_2}$, etc., were also taken into account as individual components of a multicomponent blend.

The monomers, principal products, and the additive used in our experiments are characterized in Table 1. The van Krevelen method was used to find the solubility parameter δ for all of the components, as well as the specific weight ρ for all components, except for monomers and additive.

Our procedure is applicable systems with an arbitrary number of components. Its applicability is only restricted by computer potentialities. We used a Pentium II computer. The ultimate number of individual components taken into account was ca. 1000, so that calculations took several hours.

The calculation procedure was as follows. A system of conventional kinetic equations for the polycondensation reaction of **E** and **A** was integrated by using the fourth order Runge–Kutta method to yield a tabulated dependence of the component concentrations on reaction conversion, α, and additive concentration, ϕ. Then the FEPS was calculated for the multicomponent system corresponding to given ϕ and α. The FEPS was calculated from (7) by using a combination of the Monte-Carlo and Powell minimization methods. For every (ϕ,α), we found whether the equilibrium state of the system was homophase or not (the global minimum of FEPS is zero for a homophase system and negative for a heterophase one). In this way, the boundary value of α corresponding to the onset of phase separation for a given ϕ was found. A detailed mathematical description of the procedure will be given elsewhere.

Table 1. Characterization of blend components

Compound	$\overline{M}_w$	ρ	δ
E	340	1.15	8.52
A	210	0.97	9.10
EA	550	1.27	10.59
$\mathbf{E_2A}$	890	1.23	9.98
$\mathbf{EA_2}$	760	1.32	11.59
$\mathbf{E_2A_2}$	1100	1.27	10.75
S13	5520	1.05	8.43

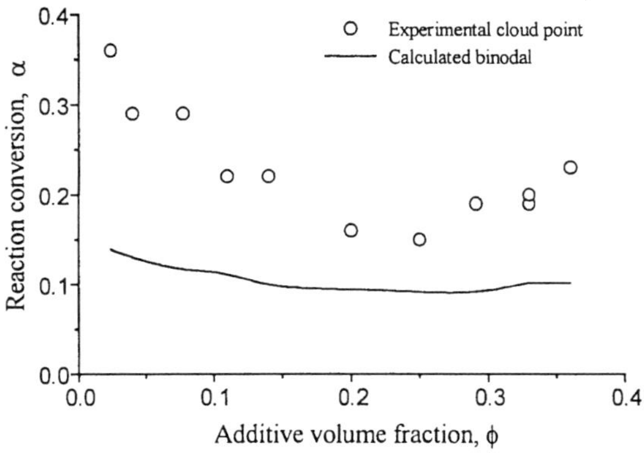

Figure 5. Phase diagram for the multicomponent curing polymer system (see text for details).

The phase diagram of the multicomponent reacting system was plotted as the dependence of α on ϕ [4–6].

The results of numerical modeling are compared in Fig. 5 to the experimental data obtained by cloud point measurements. Although the theoretical and measured binodal curves do not coincide, the agreement between seems satisfactory, especially if we take into account the following considerations.

First, the cloud point is always observed at somewhat higher conversions than those corresponding to actual onset of phase separation. The theoretical binodal is calculated on the basis of quasi-stationary thermodynamic consideration. The ideal binodal may be experimentally measured only for reversible thermally induced phase separation, since the temperature in this case may be varied as slowly as needed. In the case of reaction-induced phase separation, the reaction rate is often higher than the rate of phase separation [7–9]. Therefore, the cloud points can be expected to occur above the theoretical binodal, due to the non-equilibrium conditions of CRIMPS.

Second, it is generally recognized that the FH theory is more accurate in the vicinity of the critical point, that is, at low component concentrations. This may account for the discrepancy between the data points and theoretical curves at lower ϕ for both the ternary (Figs. 2–4) and multi-component (Fig. 5) systems.

CONCLUSIONS

The Flory–Huggins (FH) approach is applicable to numerical modeling of phase diagrams for both non-reacting and reacting (in particular, curing) multicomponent polymer systems. Although a system that exhibits CRIMPS never reaches exact thermodynamic equilibrium, the multicomponent FH equation may be used for analysis of a curing system at any reaction conversion. Calculation of the equilibrium phase state of a multicomponent system involves global minimization of the free energy of phase separation (FEPS). Proper combination of (i) the cure kinetic equations, (ii) the van Krevelen approach to retrieving lacking parameters forthe FH equation, (iii) FEPS as a goal function for global minimization, and (iv) the Monte-Carlo/Powell optimization algorithm yields reasonable prediction of the phase structure for curing multicomponent systems.

REFERENCES

1. Hiemenz P.C., *Polymer Chemistry: The Basic Concepts*, Marcel Dekker, New York, 1984.
2. van Krevelen, D.W., *Properties of Polymers*, Elsevier, Amsterdam, 1990.
3. Press W.H., Teukolsky S.A., Vetterling W.T., and Flannery B.P., *Numerical Recipes in C: The Art of Scientific Computing*, Cambridge University Press, Cambridge, 1992.
4. Rozenberg B.A. and Sigalov G.M., *Polym. Adv. Technol.*, **7**, 356 (1996).
5. Rozenberg B.A. and Sigalov G.M., *Macromol. Symp.*, **102**, 329 (1996).
6. Rozenberg B.A. and Sigalov G.M., in *Chemical and Physical Networks: Formation and Control of Properties*, Mijs W.J. and te Nijnhuis K., Eds., Wiley, New York, 1998. Vol. 1, p. 209.
7. Rozenberg B.A., *Makromol. Chem., Macromol. Symp.*, **41**, 1648 (1991).
8. Rozenberg B.A. and Sigalov G.M., *Polym. Sci., Ser. A*, **37**, 1049 (1995).
9. Bogdanova L.M., Dzhavadyan E.A., Sigalov G.M., and Rozenberg B.A., this book, p. 121.

Section III

KINETICS AND MECHANISM OF CURE REACTIONS AND REACTION-INDUCED MICROPHASE SEPARATION

Chapter 10

Cure Rate and CRIMPS Mechanism

Lyudmila M. BOGDANOVA, Emma A. DZHAVADYAN, Grigori M. SIGALOV, and Boris A. ROZENBERG*

Institute of Problems of Chemical Physics, Russian Academy of Sciences, Chernogolovka, Moscow Region, 142432 Russia

ABSTRACT

The dependence of cloud point at cure reaction-induced microphase separation (CRIMPS) on the reaction rate for multicomponent curing system (diepoxide + tertiary amine + additive) was studied. Three distinct regions associated with the change in the phase separation mechanism were detected. For additive concentrations far from the critical point in the phase diagram, the nucleation-and-growth mechanism with a low cure rate is found responsible for the phase separation. The mechanism of spinodal decomposition is typical of rapid cure.

*e-mail: rozen@icp.ac.ru

A mixed mechanism may be expected in case of intermediate cure rates. Near the critical additive concentration, the spinodal decomposition is realized at any cure rate. These conclusions are confirmed by the kinetic data obtained.

INTRODUCTION

Numerous studies have dealt with elucidation of the microphase separation mechanism in multicomponent curing systems [1–4]. However, this important problem is still a subject of controversy. Much experimental data suggest that, for a modifier concentration much lower than the critical value (that is commonly met in practice), the following nucleation-and-growth (NG) mechanism of microphase separation is taking place [1, 5–16]:

- Particles of the new phase are perfectly spherical. But at late stages of spinodal decomposition (SD), the particles are often nearly spherical, (but by no means ideal) because of high system viscosity. The latter factor inhibits complete decomposition of the co-continuous phase morphology initially formed during spinodal decomposition.
- Using small-angle X-ray scattering [5], dispersed spherical particles 30–500 Å in size were shown to appear in the curing system well before the cloud point. Before the cloud point the particles grow gradually, while after the cloud point their dimension jumps up to a certain value of a few micrometers. This occurs well before the gel point.
- The interdiffusion coefficient measured during phase separation is positive [6, 7].
- For epoxy–rubber blends, interphase surface tension in the metastable region is rather low (0.1–0.3 mN/m at 80–140°C) [8]. As shown in [9], this is a reason for NG mechanism of phase separation.
- A drastic decrease in the volume fraction and dimension of the dispersed particles with the increase of cure rate occurs. Moreover, the particle size distribution (PSD) may become polymodal. Polymodal PSD can be considered as a specific feature of NG mechanism in curing systems [6, 10–15].
- There is a good qualitative consistency between regularities of morphology formation for heterophase polymers (those formed upon the phase separation of the curing system) predicted by NG mechanism-based theory [9, 16], and extensive experimental data for the phase separation in epoxy-rubber formulations.

However, it was claimed [2, 3] that the NG mechanism is fundamentally impossible for microphase separation in the polymer–oligomer systems under consideration because of a very low rate of homogeneous nucleation.

The rate of homogeneous nucleation is known [17] to be strongly determined—as $\exp(-a\sigma^3)$—by the interphase surface tension, σ. Since $d\sigma/dT < 0$, σ may be infinitely small and the nucleation rate may be then infinitely high (it may be limited due to other factors) [9]. Binder suggested another argument against homogeneous nucleation [18]. He assumes that nucleation in polymers was always nonhomogeneous because, due to existence of some molecular weight distribution, the largest molecules act as nuclei. Recently, the idea to use a soluble high-molecular fraction as nuclei was tested by using the epoxy–amine system with castor oil and a small amount of polyester synthesized from castor oil and oxalic acid as an additive [19]. This small change results in drastic changes in the morphology of resultant heterophase polymer: particles of the dispersed phase are much bigger, and the PSD profile changes from unimodal to bimodal.

It is supposed that for the case of high-molecular polymer instead of low-molecular rubber as a modifying additive and its higher concentration, the SD mechanism is the more probable [1].

Experimental evidence of spinodal decomposition in curing systems is usually based on the morphological structure of synthesized heterophase polymers, i.e., the presence of two continuous phases at early stages of spinodal decomposition, irregular shape of the dispersed phase particles at later stages of SD (nearly spherical for low viscosity medium), as well as on light scattering evolution during phase transformation (maximum of the scattering intensity *vs.* scattering angle curve is shifted toward smaller angles for higher cure reaction conversions) [2–4]. Recently, it was shown that, for high enough cure rates, the scattering intensity maximum might drift toward larger angles with increasing conversion [20]. A mixed mechanism of the phase separation (nucleation-initiated spinodal decomposition) was assumed to be responsible for this phenomenon. Using the Monte-Carlo computer simulation [21], similar behavior was also found for the systems characterized by the NG mechanism of phase separation (at early stages). Only for later stages, a monotonic decrease in scattering intensity *vs.* scattering angle is observed, such a behavior being typical of NG. On the other hand, completely cured polymer was used for morphology studies. The results of such research are not sufficient for understanding the mechanism of phase separation because, at later stages of SD (if it occurs in low-viscous media), the morphological structure is usually a dispersion similar to that obtained via the NG mechanism. In this context, only simultaneous real-time monitoring of phase separation by various methods—such as the kinetics of phase morphology and angular dependence of scattering intensity change during CRIMPS—can provide a reliable background for understanding the mechanism of microphase separation. A few studies of this type have been performed so far [4, 5].

Therefore, development of new methods for elucidation of microphase separation mechanism in curing polymer systems is of a great importance.

In this work, the mechanism of phase separation was studied using the following system: the diglycidyl ether of diphenylpropane cured under the action of a tertiary amine (TA, dimethylbenzylamine) in the presence of polypropyleneglycol-*bis*-(toluylenediurethanethyleneacrylate) (PBUA). The cure rate was varied by changing the TA concentration. A particular emphasis was put on examining the effect of the cure rate on the conversion degree at the onset of phase separation. The conversion degree can be easily determined from the cloud point of the curing system by the light scattering technique. Variation in the reaction rate within a wide range changes the mechanism of phase separation, while insignificant changes in the cure kinetics virtually does not affect the thermodynamics of the process. Note that such a separation of kinetics and thermodynamics is impossible for polycondensation where the reaction rate may only be varied at the expense of temperature change.

Using the above approach, we managed to shed light on the effect of major process parameters, additive concentration and temperature on the mechanism of phase separation.

EXPERIMENTAL

Benzyldimethylamine was distilled at 100°C/65 mm Hg. Other chemicals were used without further purification.

The physicochemical characteristics of starting materials are presented in Table 1.

Table 1. Physicochemical characteristics of the compounds under study

Compound	$\overline{M}_n$	$\overline{M}_w$	d^{20}, g/cm^3	T_b, °C
Diglycidyl ether of bisphenol A (DGEBA), Dow Chemicals Co.	380	—	1.16	—
Polypropyleneglycol-bis(toluylene-diurethanethyleneacrylate) (PBUA-1), Air Products & Chemicals	2540	3410	1.10	—
Polypropyleneglycol-bis(toluylene-diurethanethyleneacrylate) (PBUA-2), Air Products & Chemicals	2040	5520	1.05	—
Benzyldimethylamine (TA)	121	—	0.896	185

The molecular weight distribution was determined by SEC using a liquid microcolumn in a Milikhrom-1 chromatograph.

The cure kinetics was studied by isothermal calorimetry (DAK-1-1 calorimeter).

The phase separation kinetics was monitored by light scattering technique (TOP-1 apparatus, λ = 546 nm, fixed scattering angle 90°). The reaction blend was pre-degassed in vacuum. Polymer morphology was studied with a Philips scanning electron microscope SEM-599 (100 kV).

RESULTS AND DISCUSSION

Kinetics of Light Scattering

Figure 1 shows the kinetic curves of change in relative intensity of light scattering during phase separation. For low additive volume fractions ($\phi_2 < 0.01$), the kinetic curves exhibit a broad maximum with subsequent decay to a plateau. For larger additive concentrations, a sharp maximum is observed, its amplitude increasing with increasing additive concentration. The presence of the maximum indicates rearrangement in the initial structure of the formed dispersed phase via different mechanisms (Lifshitz–Slyozov mechanism and direct coalescence). Their role rises with an increase in the additive volume fraction.

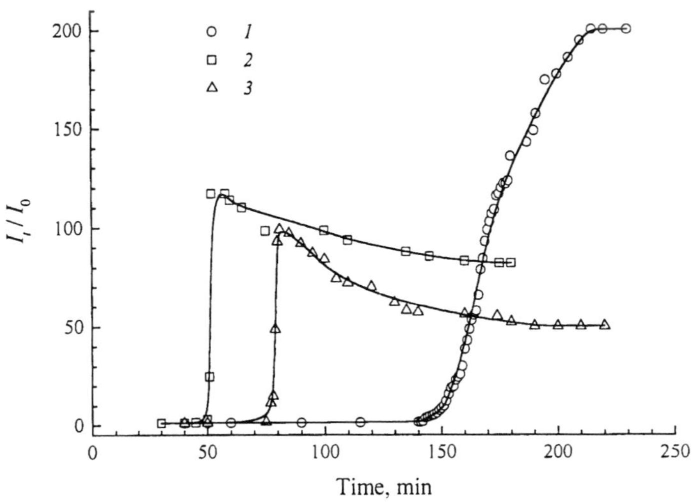

Figure 1. Change of the relative light scattering intensity in the course of phase separation at different additive concentrations for DGEBA/TA/PBUA-2 system. Isothermal cure at 50°C, [TA] = 0.22 m/l, ϕ_2 = (*1*) 0.005; (*2*) 0.05, (*3*) 0.15.

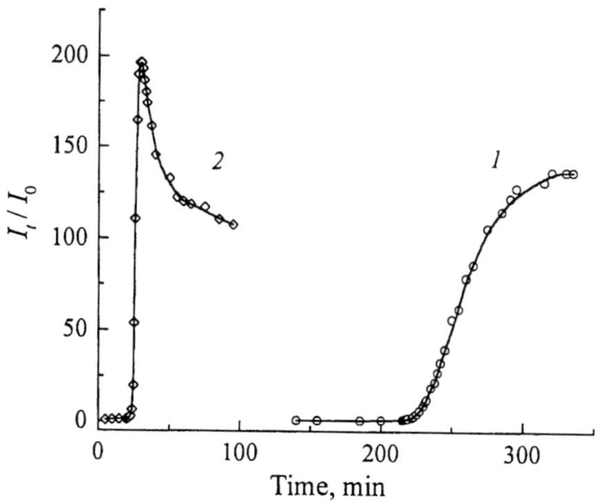

Figure 2. Change of the relative intensity of light scattering during phase separation at different amine concentrations. DGEBA/TA/PBUA-2 system, isothermal cure at 70°C, $\phi_2 = 0.01$. [TA] = (*1*) 0.04 and (*2*) 0.47 mol/l.

The reaction rate determined by the catalyst concentration is seen to markedly affect the kinetics of light scattering (Fig. 2).

Influence of the Reaction Rate

The kinetics and mechanism of the cure reaction have been studied by the authors previously [22]. It should be noted that, upon introduction of additive into the system, the cure rate slightly decreases because of the effect of dilution but the reaction mechanism does not change.

Figure 3 shows the dependence of the phase separation starting moment upon the value of the reaction rate. The phase separation starting moment is determined as the onset of drastic growth in the intensity of light scattering. Figure 3 suggests that the cloud point is an exponential function of the catalyst concentration. For low reaction rates, the exponent equals –1, while for larger reaction rates it is about –0.4. The reaction is of the second order, hence, for low reaction rates, the phase separation starts at a virtually constant conversion, independent on the reaction rate. This means that the phase separation is a quasi-stationary process, and the reaction rate is much less than the phase separation rate. The volume fraction of the additive is only 1 %, which is well below the critical concentration. Therefore, the SD mechanism for quasi-stationary phase separation is not feasible. Hence, the left part of the curve in Fig. 3 corresponds to NG mechanism of the phase separation.

For larger reaction rates, a conversion corresponding to the cloud point increases with increasing reaction rate. Therefore, the characteristic times of reaction and phase separation are close in this range of the process parameters. The higher is the reaction rate, the larger is the delay of the phase separation onset.

As follows from these data, the cloud point conversion can be plotted against the cure rate for different systems (Fig. 4a). It can be seen that for the additive concentration differing considerably from the critical value, the curves have three distinct areas. For low (area I) and high (area III) cure rates, conversion value corresponding to the phase separation onset does not depend on the reaction rate (though the values are different for these areas). For the third area (area II), the cloud point monotonically grows with the cure rate. This means that in this area the phase separation delays while the reaction rate grows because of fast drop in the interdiffusion coefficient for the additive molecules in curing system. While the additive volume fraction approaches the critical one, the delay decreases because of reduction of the conversion interval between spinodal and binodal. For the additive volume fraction being close to the critical one, area II degenerates completely. As one would expect, the cloud point becomes independent on the reaction rate. Similar regularities are observed upon changing the cure temperature (Fig. 4b) and the additive molecular weight characteristics. As expected, the increase in the additive polydispersity leads to earlier onset of phase separation (Figs. 4b, 4c). It should be taken into account that the number-average molecular weights of PBUA-1 and PBUA-2 are close. In the curve showing the cloud point–reaction rate relationship, area II shifts toward lower reaction rates at cure temperature growth (Figs. 4a, 4b).

The above regularities are easily tractable. In area I (low cure rates) quasi-equilibrium phase separation occurs via the NG mechanism. In area II (moderate reaction rates) bordering area I with weak metastability, NG phase separation also takes place. Meanwhile, area II that borders area III is

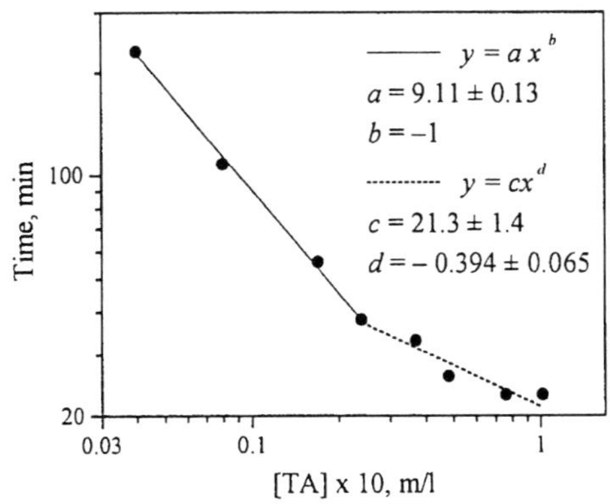

Figure 3. Dependence of the cloud point on the catalyst concentration. DGEBA/TA/PBUA-2 system; isothermal cure at 70°C; $\phi_2 = 0.01$.

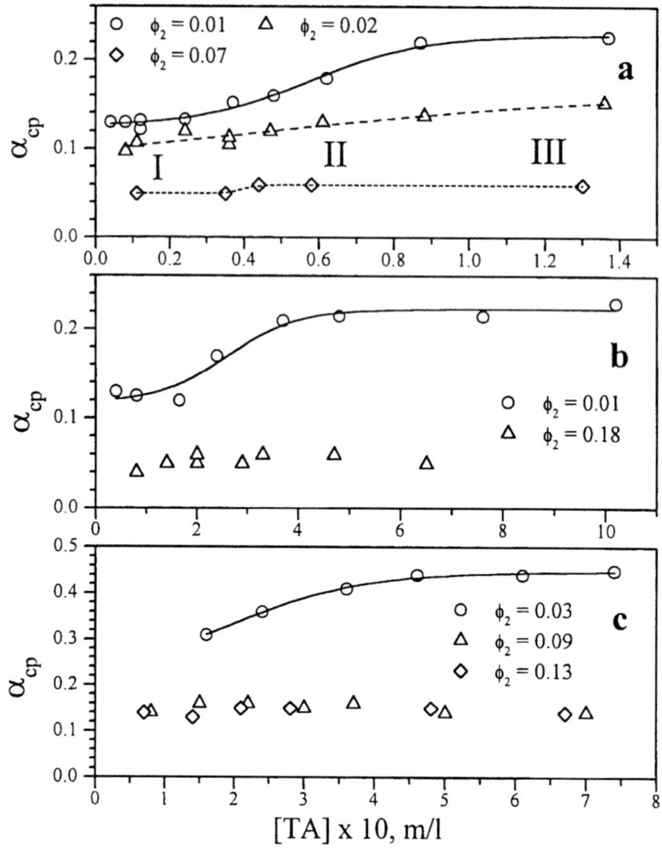

Figure 4. Dependence of the cloud point on the catalyst concentration for isothermal cure: (a) DGEBA/TA/PBUA-2, 50°C; $\phi_{2cr} = 0.08$; (b) DGEBA/TA/PBUA-2, 70°C; $\phi_{2cr} = 0.23$; (c) DGEBA/TA/PBUA-1, 70°C; $\phi_{2cr} = 0.21$.

characterized by strong metastability, and here both NG and mixed mechanism (NG and SD) can be realized. Area III (high reaction rates) is characterized by SD mechanism for any additive volume fractions.

Polymer Morphology

Figure 5 shows typical morphological structures of heterophase polymers for low, moderate, and high cure rates. It can be seen that these structures are very similar, and, as one would expect, conclusions on mechanism of the phase separation cannot be based on these pictures alone. For rather high cure rates and the additive concentration being nearly critical, when

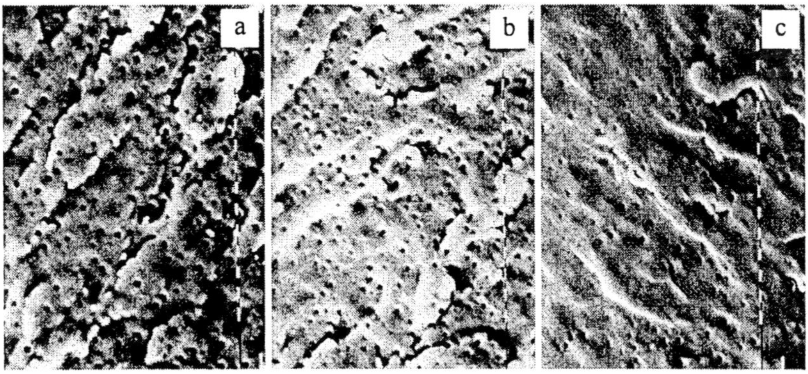

Figure 5. Morphology of samples of cured DGEBA/TA/PBUA-2 systems. Cure schedule: 10 h at 70°C and 5 h at 150°C. $\phi_2 = 0.02$, [TA] = (a) 0.11 mol/l; (b) 0.62 mol/l; (c) 0.87 mol/l. White segment of the scale bar corresponds to 1 μm.

SD mechanism is realized, there are no co-continuous phase structures typical of the early stages of spinodal decomposition, which would be an unambiguous evidence for the SD mechanism. This suggests that the reaction rate is still not high enough to hinder further evolution of the heterophase polymer structure.

CONCLUSIONS

The results as presented in Fig. 4 provide easy discrimination between the phase separation mechanisms. Such presentation is a unique kind of phase diagram, and in its explicit form involves a kinetic factor (kinetics of change of the CRIMPS driving factor). It provides a direct experimental estimation of the conditions responsible for quasi-equilibrium phase separation in curing systems. Generally, this is assumed without proof when plotting the phase diagram and estimating thermodynamic parameters of phase separation [1].

In areas II and III, the phase separation is non-equilibrium and is described by a phase trajectory declining from binodal [21].

Acknowledgements. This work was supported by the International Science and Technology Center (grant no. 358-96) and the Russian Foundation for Basic Research (project no. 96-03-32027). Authors are grateful to Prof. A.E. Chalykh and Dr. V.K. Gerasimov for the SEM measurements.

REFERENCES

1. Williams R.J.J., Rozenberg B.A., and Pascault J.-P., *Adv. Polym. Sci.*, **128**, 95 (1997).
2. Inoue T., *Progr. Polym. Sci.*, **20**, 119 (1995).
3. Yamanaka K., Takagi V., and Inoue T., *Polymer*, **30**, 1839 (1989).
4. Okada M., Fujimoto K., and Nose T., *Macromolecules*, **28**, 1795 (1995).
5. Chen D., Pascault J.-P., Sautereau H., and Vigier G., *Polym. Int.*, **32**, 369 (1993).
6. Volkov V.P., Roginskaya G.F., Chalykh A.E., and Rozenberg B.A., *Usp. Khim.*, **51**, 1733 (1982).
7. Rozenberg B.A., Makromol. Chem., Macromol. Symp., **41**, 165 (1991).
8. Vazques A., Rojas A.J., Adabbo H.E., Borrajo J., and Williams R.J.J., *Polymer*, **28**, 1156 (1987).
9. Moshiar S.M., Riccardi C.C., Williams R.J.J., Verchere D., Sautereau H., and Pascault J.-P., *J. Appl. Polym. Sci.*, **42**, 717 (1991).
10. Roginskaya G.F., Volkov V.P., Chalykh A.E., Matveev V.V., Rozenberg B.A., and Enikolopyan N.S., *Dokl. Akad. Nauk SSSR*, **252**, 402 (1980).
11. Roginskaya G.F., Volkov V.P., Dzhavadyan E.A., Zaspinok G.S., Rozenberg B.A., and Enikolopyan N.S., *Dokl. Akad. Nauk SSSR*, **290**, 630 (1986).
12. Nikitin O.V. and Rozenberg B.A., *Polym. Sci., Ser. A*, **34**, 365 (1992).
13. Nikitin O.V. and Rozenberg B.A., *Polym. Sci., Ser. A*, **38**, 883 (1996).
14. Rozenberg B.A. and Sigalov G.M., *Polym. Adv. Technol.*, **7**, 356 (1996).
15. Rozenberg B.A. and Sigalov G.M., *Macromol. Symp.*, **102**, 329 (1996).
16. Williams R.J.J., Borrajo J., Rojas A.J., and Adabbo H.E., in *Rubber Modified Thermoset Resins*, Riew C.K. and Gillham J.K., Eds., *Adv. Chem. Ser.*, **208**, 405, Amer. Chem. Soc., Washington DC, 1984.
17. Lyubov V.A., *Kineticheskaya teoriya fazovykh prevrashchenii* (Kinetic Theory of Phase Transformations), Metallurgiya, Moscow, 1969.
18. Binder K., *J. Chem. Phys.*, **79**, 6387 (1983).
19. Ruzeckaite R., Fasce D.P., and Williams R.J.J., *Polym. Int.*, **30**, 297 (1993).
20. Kyu T., Chiu H.W., and Lee J.-H., *this book*, p. 93.
21. Jo W.H. and Ko M.B., *Macromolecules*, **27**, 7815 (1994).
22. Dzhavadyan E.A., Bogdanova L.M., Irzhak V.I., and Rozenberg B.A., *Polym. Sci., Ser. A*, **39**, 383 (1997).

Chapter 11

The Role of Intermolecular Interactions in Polyurethane Formation

Elena V. STOVBUN, Vera P. LODYGINA,
Elmira R. BADAMSHINA*, Valentina A. GRIGOR'EVA,
Irina V. DORONINA, and Sergei M. BATURIN

Institute of Problems of Chemical Physics, Russian Academy of Sciences, Chernogolovka, Moscow Region, 142432 Russia

ABSTRACT

The effect of molecular organization on the kinetics of urethane formation was investigated for the reaction of 2,4-toluelenediisocyanate with different oligo-butadienediols in the presence/absence of a catalyst (tin dibutyldilaurate).

---

*e-mail: badamsh@icp.ac.ru

The kinetic parameters for this reaction were found to depend on relative amount and type of associated hydroxyl groups in starting oligobutadienediols (autoassociated and π-conjugated) and reaction products (autoassociated, π-conjugated, and associated with urethane groups). Mathematical modelling of the reaction was performed. The concentrations of all hydroxyl associates were calculated numerically. For the oligomeric systems under study, we investigated the effect of associates on the polyurethane (PU) formation in the presence/absence of catalysts.

INTRODUCTION

The importance of organized structures, homo- and heteroassociates, in the kinetics and mechanism of liquid-phase reactions is postulated by various authors [1–5]. The type of associates formed upon hydrogen bonding, donor–acceptor interaction, and van der Waals interaction is defined by the chemical nature of starting reactants and reaction products [6–9].

The reaction mechanism suggested for formation of low-molecular urethanes can be readily applied (under some assumptions) to the kinetics of polyurethane formation.

This work aims at investigating the effect of different associates on the kinetics and mechanism of polyurethane formation in the presence/absence of a catalyst.

EXPERIMENTAL

In our experiments, we used oligobutadienediols (OBDs) of different molecular weight (MW), asymmetric 2,4-toluylenediisocyanate (TDI), and tin dibutyldilaurate (TBL) (solution in heptane) as a catalyst.

OBDs were purified by precipitation with ethanol from dilute benzene solutions and then dried at 333 K/13.3 Pa. Lithium content was below 10^{-4} g-eqv/l. The OBDs are characterized in Table 1.

TDI was distilled at 323–325 K/12 Pa ($d_4^{20} = 1.2178$, $n_D^{20} = 1.5669$) and stored in sealed ampoules. The reaction kinetics in the range 313–349 K was monitored by microcalorimetry and IR spectroscopy [10]. Reaction rate constants were determined within 10 % accuracy. IR spectra were recorded with a Specord M82 spectrophotometer.

The molecular parameters of OBDs were determined with a Waters GPC instrument [10].

Table 1. Characterization of OBDs

OBD	$\overline{M}_e^* \times 10^{-3}$	$\overline{M}_n \times 10^{-3}$	$\overline{M}_w / \overline{M}_n$	[OH]**, %
I	1.3	2.2	1.12	1.31
II	1.7	3.2	1.09	0.98
III	2.2	4.1	1.08	0.79
IV	3.2	6.3	1.19	0.51
V	3.5	6.8	1.09	0.49
VI	4.1	8.3	1.06	0.41
VII	5.5	11.4	1.12	0.31

* Equivalent molecular weight M_e = 1700/[OH].

** Hydroxyl groups were determined by chemical analysis and IR spectra [10].

RESULTS AND DISCUSSION

The kinetics of catalyzed and uncatalyzed reactions of OBD with TDI can be rationalized in terms of the classical reaction scheme [11] taking into account interaction between free and urethane-substituted TDI with the hydroxyl groups of OBD.

This reaction can also be described by using a simplified scheme involving two consecutive reactions with rate constants k_{1ef} and k_{2ef}, where k_{1ef} is the effective rate constant for the reaction of free isocyanate and k_{2ef} is that for the reaction of urethane-substituted isocyanate with hydroxyl groups of OBD.

The presence in TDI of two isocyanate groups with different reactivity manifests itself as a breakpoint in the anamorphose of the kinetic curve plotted in the second-order reaction coordinates. The slopes of linear segments of this curve yield k_{1ef} and k_{2ef}.

Uncatalyzed Reaction

For uncatalyzed urethane formation, k_{1ef} was found to depend on the molecular weight of OBD, k_{2ef} be virtually constant with the given range of OBD molecular weights (Table 2).

The end OH-groups of OBD were found [12] to form two types of associates: autoassociates, [OH$\cdots$OH], and associates with π-electrons of the double bond in the butadiene unit, [OH$\cdots\pi$]. The relative amount of these associates depends on the total amount of OH groups and related equivalent molecular weight, M_e of OBD. The amount of more reactive [OH$\cdots$OH] associates gradually decreases with increasing M_e, which explains the observed dependence of k_{1ef} on M_e.

The constancy of k_{2ef} (Table 2) might be attributed to the domination of relatively less reactive [OH$\cdots\pi$] associates at this stage. But we have to

Table 2. The values of k_{1ef} and k_{2ef} for different OBDs

OBD	T, K	Effective rate constant, l/(g-eqv·s)		β*
		$k_{1ef}\times10^3$	$k_{2ef}\times10^4$	
II	313	8.5	1.5	56.6
	337	12.5	2.8	44.6
III	317	7.7	1.4	55.0
VI	313	2.3	1.1	20.9
	335	4.9	2.9	16.9
VII	317	—	1.6	—
	337	4.5	2.8	16.1

*$\beta = k_{1ef}/k_{2ef}$.

keep in mind that the reaction yields highly reactive urethane groups that may form associates with the OH groups of OBD [13]. In this case, the constancy of k_{2ef} may take place only if the strength of these associates is close to that of the [OH ··· π] associates.

The above data can be adequately rationalized only upon account of all associates formed in the system and all possible equilibria between them. However, the required relevant information is difficult to obtain even for model compounds. Therefore, we have performed mathematical modelling of urethane formation by solving the direct kinetic problem. The adopted reaction scheme involved 20 equations for reaction of the OH groups in OBD with the NCO groups of TDI and the set of equilibria established in the system. We also made the following simplifying assumptions:

- the reactivity of the OH group does not depend on the length of a given OH-carrying oligomer chain;
- [OH ··· OH] associates are linear dimers;
- carbonyl of the urethane group reacts only with the OH group of OBD;
- solvation by surrounding molecules with π-donating bonds is neglected.

A reaction scheme was regarded as adequate when the predicted behavior became coincident with that observed.

The adopted reaction scheme is shown in Fig. 1. Here S is the unassociated OH group, d stands for [OH ··· OH], P stands for the [OH ··· π] associates involving double bonds of OBD, US stands for the OH groups associated with the carbonyl of the urethane group, [OH ··· CO]; U is the unassociated urethane group, X_1 is the free diisocyanate, X_3 is the p-urethane-substituted diisocyanate, X_4 is the o-urethane-substituted diisocyanate, π denotes the double bonds in OBD, k_1, k_{-1}, k_2, k_{-2}, k_3, k_{-3} are

$$(1)\ S + S \underset{k_{-1}}{\overset{k_1}{\rightleftharpoons}} d \qquad (2)\ S + P \underset{k_{-2}}{\overset{k_2}{\rightleftharpoons}} P \qquad (3)\ S + U \underset{k_{-3}}{\overset{k_3}{\rightleftharpoons}} US$$

$$(4)\quad S + X_1 \rightarrow X_4 + U \quad (k_4) \qquad (5)\quad S + X_1 \rightarrow X_3 + U \quad (k_5)$$

$$(6)\quad d + X_1 \rightarrow X_4 + U + S \quad (k_6) \qquad (7)\quad d + X_1 \rightarrow X_3 + U + S \quad (k_7)$$

$$(8)\quad d + X_3 \rightarrow U + S \quad (k_8) \qquad (9)\quad d + X_4 \rightarrow U + S \quad (k_9)$$

$$(10)\ P + X_1 \rightarrow X_4 + U + \pi \quad (k_{10}) \qquad (11)\ P + X_1 \rightarrow X_3 + U + \pi \quad (k_{11})$$

$$(12)\ P + X_3 \rightarrow U + \pi \quad (k_{12}) \qquad (13)\ P + X_4 \rightarrow U + \pi \quad (k_{13})$$

$$(14)\ US + X_1 \rightarrow X_4 + U + U \quad (k_{14}) \qquad (15)\ US + X_1 \rightarrow X_3 + U + U \quad (k_{15})$$

$$(16)\ US + X_3 \rightarrow U + U \quad (k_{16}) \qquad (17)\ US + X_4 \rightarrow U + U \quad (k_{17})$$

Figure 1. Adopted reaction scheme (see text for details).

the rate constants for formation and decay of appropriate associates, and k_4–k_{17} are the rate constants for appropriate reactions.

For oligomeric systems, the direct experimental determination of equilibrium constants, rate constants for direct and reverse reactions, and amount of associates is difficult. For this reason, the data needed for calculations were partially taken from the published data [14] and partially estimated from the heat of mixing measured for some model compounds.

We investigated the relation between the total concentration of hydroxyl groups in the system, $[OH]_0$, and reaction kinetics of urethane formation.

Calculations were carried out for the following three cases:

(1) the rate constants for reaction of isocyanate with OH are identical for all three types of associates: $[OH \cdots OH]$, $[OH \cdots \pi]$, and $[OH \cdots CO]$;
(2) the rate constants for reactions of $[OH \cdots OH]$ and $[OH \cdots \pi]$ are different while those for $[OH \cdots \pi]$ and $[OH \cdots CO]$ are identical;
(3) the rate constant for $[OH \cdots CO]$ is lower than that for $[OH \cdots \pi]$.

The reactivity ratio of free to urethane-substituted isocyanate was always assumed to be the same. In all three cases we calculated the concentration profiles, plotted the anamorphoses of the kinetic curves in the second-order reaction coordinates, and analyzed variation in the rate constants (and their ratios) with $[OH]_0$.

In case (1), the values k_{2ef} and $\beta = k_{1ef}/k_{2ef}$ were found to be independent of $[OH]_0$ (Fig. 2, curve *1*).

In case (2), variation in $[OH]_0$ had no influence on k_{2ef} but affected β, the effect becoming more pronounced with increasing difference between the reactivity of the $[OH \cdots OH]$ and $[OH \cdots \pi]$ associates (Fig. 2, curves *2*, *3*).

In case (3), variation in $[OH]_0$ affected both k_{2ef} and β.

Comparing the calculated data presented in Fig. 2 with the experimental data given in Table 2, we can conclude that the case (1) is purely hypothetical and unrealistic for the system under study.

As follows from Table 2, the values of k_{2ef} remain virtually unchanged over the entire range of $[OH]_0$ while β diminishes with decreasing $[OH]_0$. This corresponds to case (2) when the behavior of β as a function of $[OH]_0$ is governed by the difference in reactivity of $[OH \cdots OH]$ and $[OH \cdots \pi]$ associates (the reactivity of $[OH \cdots OH]$ is one order of magnitude higher than that of $[OH \cdots \pi]$). The concentration of $[OH \cdots OH]$ associates was found to be significant for degrees of conversion, α, below 50%. For higher α, the OH groups are present in the form of the $[OH \cdots \pi]$ and $[OH \cdots CO]$ associates that agrees with IR spectra [13].

For the system under investigation, case (3) is also unrealistic.

Our reaction scheme was tested on real oligomer systems with different MW. Figure 3 shows the calculated concentration of different associates and free OH group as a function of α for a low-molecular OBD. In this case, autoassociates, d, are present in significant amounts.

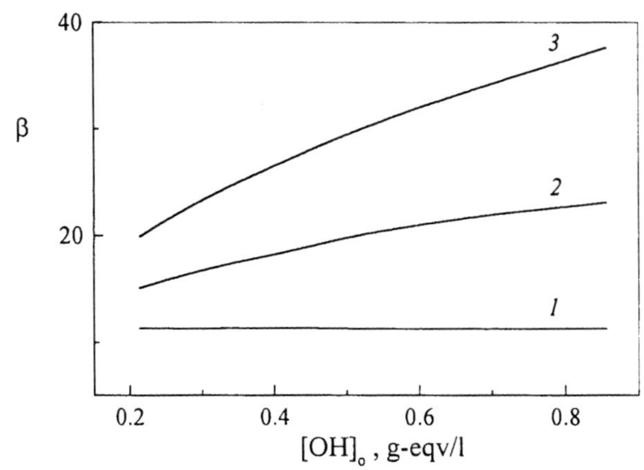

Figure 2. Calculated dependence of $\beta = k_{1ef}/k_{2ef}$ on $[OH]_0$: (*1*) case (1) and (*2, 3*) case (2) (see text for details).

Table 3. Calculated rate constants for reaction of different OH associates with NCO groups in free and urethane-substituted diisocyanates

T, K	OBD	OH group types	Rate constant $k\times10^3$, l/(g-eqv·s) position of NCO in TDI				k_d/k_p
			1*	2*	3*	4*	
313	**VI**	[OH ··· OH]	20.52	3.42	5.40	0.90	10
313	**II**	[OH ··· OH]	30.78	5.13	8.10	1.35	15
313	**II**	[OH ··· π]	2.05	0.34	0.54	0.09	15
313	**II**	[OH ··· CO]	2.05	0.34	0.54	0.09	15
313	**II**	free [OH]	0.20	0.03	—	—	—
333	**VI**	[OH ··· OH]	46.8	7.80	14.40	2.40	10
333	**VI**	[OH ··· π]	4.68	0.78	1.44	0.24	10
333	**II**	[OH ··· CO]	4.68	0.78	1.44	0.24	10
333	**II**	free [OH]	0.468	0.078	—	—	10

*1 and 3 indicate *p*-position while 2 and 4 stand for *o*-position of NCO in the free and urethane-substituted TDI.

For [OH] corresponding to low-molecular OBD, autoassociates may be present in the form of di-, tri-, tetra-, or *n*-mers. The calculated rate constants for reaction of OH with the NCO groups in TDI are collected in Table 3. It follows from these data that the [OH ··· OH] associates are 10-fold more reactive than the [OH ··· π] associates (except for low-molecular OBD where the difference is 15-fold). Conversely, the autoassociates are present largely in the form of dimers in case of high-

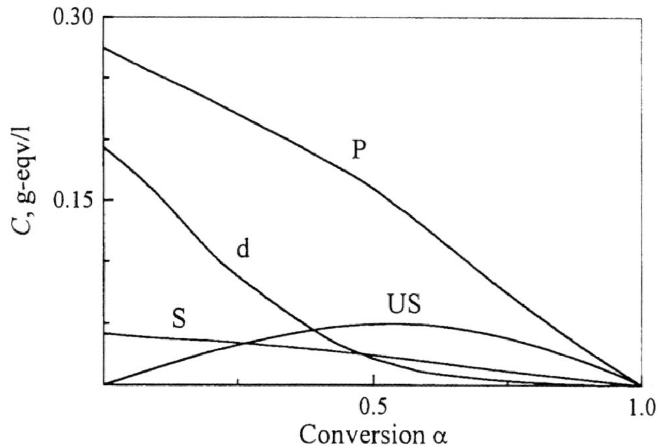

Figure 3. Calculated concentration profiles for all types of OH groups in the **II**–TDI system (313 K). For details, see the reaction scheme (Fig. 1).

molecular OBD [13]. This affects the reactivity of OH groups and, accordingly, changes the reactivity ratio of the auto- to [OH ··· π] associates.

Catalyzed Reaction

Since in the presence of catalyst (TBL) the urethane formation proceeds much faster, we could measure only the value of $k_{2ef\ cat}$. This parameter as a function of catalyst concentration, [TBL], is given in Fig. 4. In contrast to uncatalyzed reaction where k_{2ef} = const, $k_{2ef\ cat}$ is seen to increase with the MW of oligomer.

For the oligomeric systems under study, we detected the effect of saturation (with respect to catalyst) for [TBL] = (6–9)·10^{-5} M (Fig. 4). Since OH groups are activated upon electron–donor interaction with catalyst molecules, the stage of intermediate complex formation is strongly accelerated [15]. Consequently, transformation of this complex to urethane becomes the rate-controlling stage. This explains the above saturation.

The saturation threshold was found to shift towards lower [TBL] with increasing MW of oligomer (Fig. 4) and temperature. Our calculations show that $\beta_{cat} = k_{1ef\ cat}/k_{2ef\ cat}$ is constant and equal to 21. This allowed us to calculate the $k_{1ef\ cat}$ values that cannot be measured in experiments. In contrast to k_{1ef} for uncatalyzed reaction, $k_{1ef\ cat}$ was found to increase with increasing MW of oligomer. This can be associated with a catalyst-induced change in the reactivity of all OH associates. The reactivity of unassociated OH groups is the largest, and the reactivity of associates is decreasing with increasing strength of associates: [OH ··· π] > [OH ··· CO] > [OH ··· OH]. An increase in the amount of active associates with increasing oligomer MW

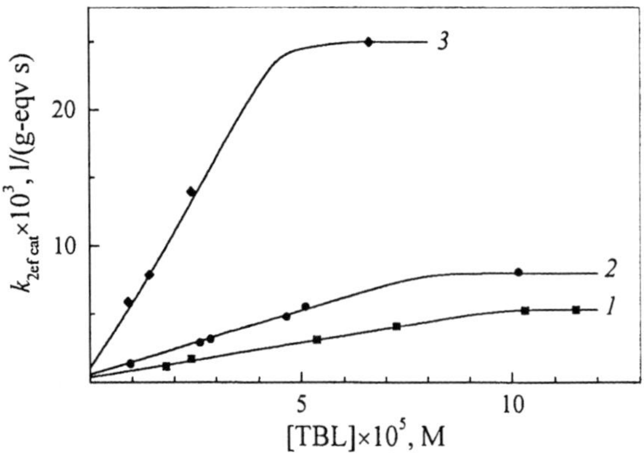

Figure 4. The value of $k_{2ef\ cat}$ *vs.* [TBL] for OBDs with different MW: (*1*) **I**, (2) **II**, and (*3*) **VI**. T = 313 K.

defines the behavior of rate constants for both the stages and explains the shift of the saturation threshold. It cannot also be excluded that formation of a donor–acceptor complex between catalyst and urethane groups results in partial passivation of the catalyst.

CONCLUSIONS

The kinetics and mechanism of uncatalyzed and catalyzed urethane formation were quantitatively related to the reactivity of all kinds of OH associates involved in the reaction.

Kinetic investigations of the reaction were found to be useful for elucidating the subtle details of intermolecular interactions taking place during urethane formation.

Acknowledgments. This work was supported by the International Science and Technology Center (grant no. 358-96).

REFERENCES

1. Entelis S.G. and Tiger R.P., *Kinetika reaktsii v zhidkoi faze* (Kinetics of Liquid-Phase Reactions: Quantitative Account of Environment Effects), Khimiya, Moscow, 1973.
2. Pal'm V.A., *Osnovy kolichestvennoi teorii organicheskikh reaktsii* (Fundamentals of the Quantitative Theory of Organic Reactions), Khimiya, Leningrad, 1967.
3. Oleinik N.M., Litvinenko L.M., Sadovskii Yu.S., Piskunova Zh.P., and Popov A.F., *Zh. Org. Khim.*, **16**, 1469 (1980).
4. Rozenberg B.A., *Adv. Polym. Sci.*, **75**, 113 (1986).
5. Rozenberg B.A., *Usp. Khim.* **60**, 1473 (1991).
6. Egorochkin A.N. and Skobeleva S.E., *Usp. Khim.*, **48**, 2216 (1979).
7. Thiele L., *Monatsh. Chem.* **123**, 536 (1992).
8. Gur'yanova E.N., Goldstein I.P., and Romm I.P., *Donorno–aktseptornya svyaz'* (Donor–Acceptor Coupling), Khimiya, Moscow, 1973.
9. Lodygina V.P., Stovbun E.V., and Baturin S.M., *Vysokomol. Soedin., Ser. A*, **27**, 921 (1985).
10. Stovbun E.V., Lodygina V.P., Kuzaev A.I., Romanov A.I., and Baturin S.M., *Vysokomol. Soedin., Ser. A*, **26**, 1449 (1984).
11. Saunders J.H and Frish K.C., *Polyurethanes*, Wiley, New York, 1965, Ch. 3.
12. Gafurova M.P., Lodygina V.P., Grigor'eva V.A., Chernyi G.I., Komratova V.V., and Baturin S.M., *Vysokomol. Soedin., Ser. A*, **24**, 858 (1972).
13. Stovbun E.V., Badamshina E.R., Grigor'eva V.A., Lodygina V.P., and Baturin S.M., *Vysokomol. Soedin., Ser. A*, **40**, 1286 (1998).

14. Tiger R.P., *Doctoral (Chem.) Dissertation*, Inst. Chem. Phys., Moscow, 1979.
15. Chirkov Yu.N., Tiger R.P., Entelis S.G., and Tondeur J.J., *Kinet. Katal.*, **36**, 612 (1995).

Chapter 12

Formation of Spatial Dissipative Structures during Synthesis of Polyurethanes

Lev P. SMIRNOV* and Evgenii V. DEYUN

Institute of Problems of Chemical Physics, Russian Academy of Sciences, Chernogolovka, Moscow Region, 142432 Russia

ABSTRACT

The formation of spatial dissipative structures in reactive mixtures containing compounds with hydroxyl and isocyanate functional groups is theoretically analyzed

*e-mail: lsmirn@icp.ac.ru

for different kinetics and mechanism of urethane formation. Nonuniform distribution of alcohol associates was found to take place when (*i*) the reaction involves heteroassociates of reactants or (*ii*) alcohol tetramers are formed in catalytic reaction. Owing to nonuniform distribution of associates, the local rate for urethane formation differs from the mean value. This explains observed microinhomogeneity in the structure and properties of polyurethanes.

INTRODUCTION

The supramolecular structure of network polymers is known to exhibit microinhomogeneities normally related to poor solubility of polymer in reactive mixture [1–2]. The microinhomogeneity of structure may be caused not only by thermodynamic but also other factors. Fluctuations in a multicomponent single-phase medium are assumed to give rise to nonuniform distribution of intermediate products, thus forming the so-called spatial dissipative structure. Consequently, the local polymerization rate and extent of conversion in some small volumes begin to differ from the mean values. Because of impurities, reactive shrinkage, and other factors, microinhomogeneity does not vanish even upon completion of polymerization, so that resultant polymer exhibits micro-inhomogeneity in supramolecular structure and related properties [2, 3].

Thermodynamic aspects of structural microinhomogeneity in network polymers have been investigated in numerous works. Meanwhile, the conditions for formation of spatial dissipative structures during polycondensation remain (to our knowledge) unclear.

In this work, we investigated the conditions for the buildup of nonuniform distribution of reactants during synthesis of polyurethanes in different kinetic schemes of reactant associations and urethane formation. Supramolecular structures (associates of different origin and composition) play important part in the kinetics and mechanism of polymerization and structure formation in resultant polymers [4, 5].

Kinetic Schemes for Association of Reagents during Urethane Formation

In polycondensation, both the starting and reactive mixtures exhibit well-pronounced supramolecular structure [4, 5]. Owing to donor–acceptor interactions, reagents form the auto- and heteroassociates. Moreover, reactive mixture may also contain *n*-meric forms with different degree of association. In many aspects, association actually reminds equilibrium polycondensation, so that the processes of association should be analyzed with regard to dynamic equilibrium between formation and decomposition of associates.

Formation of the urethane group in reaction of isocyanates with alcohols, RN=C=O + HOR' → RNHCOOR', is a multistage process whose details have not been specified so far. The measured reaction rate, k_{meas}, is known to depend on alcohol (A) concentration. Its nonmonotonic dependence on [A] can be rationalized in terms of a reaction scheme that involves alcohol associates A_n [6] or inhomogeneities caused by formation of associate clusters in concentrated solutions [7].

The available experimental data and quantum-chemical calculations show that monomeric alcohol A is inactive in noncatalytic reaction of urethane formation [8, 9]. For simplicity, we will take into account only the dimers A_2 and tetramers A_4 that are formed in the following reactions:

$$2A \xleftrightarrow{K_1} A_2, \tag{1}$$

$$2A_2 \xleftrightarrow{K_2} A_4, \tag{2}$$

where $K_i = k_i/k_{-i}$ is the equilibrium constant for the i-th reaction (i = 1, 2, 3, ...) and k_i, k_{-i} are the rate constants for the direct and reverse reaction, respectively.

Formation of n-mers by scheme (1) and (2) is not the only possibility. In models suggested [10] for prebiological systems, the concept of the matrix catalysis of n-merization is being widely used. The mechanism of matrix catalysis is defined by the structural organization of reacting system. Without going into details, let us consider the following two routes for association on a matrix:

(a) formation of dimer on a dimer matrix:

$$A + A_2 \xleftrightarrow{K_3} Y, \tag{3}$$

$$A + Y \xrightarrow{k_4} 2A_2, \tag{4}$$

where Y is the short-lived transition complex;

(b) formation of tetramer on a tetramer matrix

$$A_2 + A_4 \xleftrightarrow{K_3} Y, \tag{3'}$$

$$A_2 + Y \xrightarrow{k_4} 2A_4. \tag{4'}$$

Since dimers A_2 are reactive only at elevated temperatures [7], let us investigate the spatial stability of reactive system under the assumption that the urethane group U is formed in reaction between isocyanate B and tetramer A_4:

$$A_4 + B \xrightarrow{k_5} U + A_2 + A. \tag{5}$$

Again, this is not the only reaction route. For compounds capable of forming ordered structures through hydrogen bonding, nucleophilic attachment and substitution are assumed to proceed via cyclic transition states by synchronous (concert-like) rearrangement of heteroassociates that are present in the so-called optimized microreactors [11]. Taking account of different auto- and heteroassociates, the reaction scheme can be written in the following general form [11]:

$$A_i + B_j \xleftrightarrow{K_{ij}} [A_iB_j] \xrightarrow{k_{ij}} \text{product}.$$

Accordingly, formation of urethanes (at low concentration of B) may be described by the following scheme:

$$A_4 + B \xleftrightarrow{K_6} [A_4B], \tag{6}$$

$$[A_4B] \xrightarrow{k_7} U + 3A \text{ (or} \rightarrow U + A_2 + A). \tag{7}$$

At high concentrations of B, formation of B_2 can also be expected:

$$2B \xleftrightarrow{K_8} B_2, \tag{8}$$

$$A_4 + B_2 \xleftrightarrow{K_6} [A_4B_2], \tag{6'}$$

$$[A_4B_2] \xrightarrow{k_7} U + B + 3A \text{ (or} \rightarrow U + B + A + A_2). \tag{7'}$$

RESULTS

Stability of Reactive Mixture

Some details of the reaction mechanism being unknown, further analysis was performed by several models. A distinctive feature of these models is the assumption of only two intermediate products, A_2 and A_4.

The equations describing evolution of open chemical systems are known to adopt not only homogeneous but also inhomogeneous stationary solutions that correspond to spatial dissipative structures [10, 12]. Mathematical tools for analyzing the stability of chemical systems (applicable also to polycondensation) are well developed. Finding the conditions for buildup of dissipative structures is based on the analysis of the characteristic equation for linearized operator of a kinetic model with regard to mathematical conditions for stability [12, 13]. Stability of stationary solutions to the system of differential equations is analyzed by using the technique of perturbed stationary solutions. Suppose that local concentrations $C_i(r, t)$ satisfy the field equations

$$\partial C_i(r, t)/\partial t = F_i(C_i) + D_i\nabla^2 C_i(r, t)$$

that contain the flux terms $D_i\nabla^2 C_i(r, t)$, then for deviations $c_j(r, t)$ from the steady state we obtain:

$$\partial c_i(r, t)/\partial t = \Sigma a_{ij}c_j(r, t) + D_i\nabla^2 c_i(r, t), \qquad (9)$$

where $a_{ij} = [\partial F_i/\partial C_j]$ and $C_k = C_{k,s}$.

In order to determine the stability of the solution to periodic (spatial) perturbation (normal mode) with wavelength λ, let us substitute the expression $c_i(r, t) = A_\lambda \exp[i2\pi r/\lambda + \omega t]$ in (9). In this case, we obtain:

$$\sum_{i=1}^{f} [a_{ij} - D_i z^2 \delta_{ij} - \omega\delta_{ij}]c_j = 0, \quad |z| = 2\pi/\lambda. \qquad (10)$$

Parameter ω can be found from the characteristic equation:

$$\mathrm{Det}(a_{ij} - D_i z^2 \delta_{ij} - \omega\delta_{ij}) = 0. \qquad (11)$$

For $z \neq 0$, there exist some additional instabilities termed diffusion instabilities [12]. For a system with two degrees of freedom ($f = 2$), the Hurwitz criterion gives the following inequalities [12]:

$$a_{11} + a_{22} < 0, \qquad (12)$$

$$a_{11}a_{22} > a_{12}a_{21}, \qquad (13)$$

$$a_{11}a_{22} + [D_1D_2z^4 - (D_1a_{22} + D_2a_{11})z^2] < a_{12}a_{21}, \qquad (14)$$

$$(D_1a_{22} + D_2a_{11}) > D_1D_2z^2 > 0, \qquad (15)$$

$$\mathrm{sgn}\, a_{11} \neq \mathrm{sgn}\, a_{22}, D_1 \neq D_2. \qquad (16)$$

Therefore, the evolution of a microheterogeneous reactive system can be described only with due regard for diffusion of reactants and system size. For the sake of certainty, let the system have the linear size h. Stability of the system will be analyzed by assuming 1D diffusion of reactants. The effect of conversion degree will be neglected.

Autoassociates[1]

The models given by equations (1)–(5), (1)–(2), (3')–(5) take into account only species A_2 and A_4 that is consistent with reported data [14, 15] on the association of alcohols in inert media. But these models are applicable only to relatively low concentrations of A. For the autoassociates model (1–5), the steady-state treatment with respect to complex Y leads to the following system of kinetic equations:

$$\partial A_2/\partial t = k_1 A^2 - k_{-1}A_2 - 2k_2A_2^2 + k_3k_4A^2A_2/(k_{-3} + k_4A) + \\ + 2k_{-2}A_4 + k_5A_4B + D_1\nabla^2A_2,$$

$$\partial A_4/\partial t = k_2A_2^2 - k_{-2}A_4 - k_5A_4B + D_2\nabla^2A_4, \qquad (17)$$

where t is time and A, A_n are the concentrations of monomeric and n-meric alcohol, respectively.

Introducing new variables $x = 2k_2A_2/k_{-2}$, $y = 4k_2A_4/k_{-2}$, $\tau = k_{-2}t$ and parameters $\alpha = 2k_1k_2A^2/k_{-2}^2$, $\beta = k_{-1}/k_{-2}$, $\gamma = k_3k_4A^2/[k_{-2}(k_{-3} + k_4A)]$, $c = k_5B/2k_{-2}$, $D_1^* = D_1/k_{-2}$, $D_2^* = D_2/k_{-2}$, we can simplify (17):

$$\partial x/\partial\tau = \alpha - \beta x - x^2 + \gamma x + y + cy + D_1^*\nabla^2x,$$

$$\partial y/\partial\tau = x^2 - y - 2cy + D_2^*\nabla^2y. \qquad (18)$$

A unique homogeneous stationary solution to (18) has the form:

$$x_s = (1 + 2c)\{\gamma - \beta + [(\gamma - \beta)^2 + 4\alpha c/(1 + 2c)]^{1/2}\}/2c,$$

$$y_s = x_s^2/(1 + 2c). \qquad (19)$$

Let us consider the possibility of diffusion instability for system (18). In this case, the coefficients of characteristic equation (11) have the following values: $a_{11} = \gamma - \beta - 2x_s$, $a_{22} = -1 - 2c$, $a_{12} = 1 + c$, and $a_{21} = 2x_s$. As follows from condition (13), $(\beta + 2x_s - \gamma)(1 + 2c) > 2x_s(1 + c)$. This inequality is in

---

[1]The figures indicate the reactions taken into account in terms of a given model.

contradiction with condition (16) which gives $\beta + 2x_s - \gamma < 0$. This means that all vibration modes eliminate perturbations arising in the system (1–5).

Similar result was also obtained for the model (1,2,3'–5) (formation of A_4 from A_2 on a tetramer matrix) that takes into account only the A_2 and A_4 autoassociates.

Heteroassociates

Now let us consider the models that take into account both auto- and heteroassociation of A and B. For the model (1,2,3',4',6,7) the steady-state treatment with respect to A_4B leads to the following kinetic equations:

$$\partial A_2/\partial t = k_1A^2 - k_{-1}A_2 - 2A_2^2[k_2 + k_3A_4/(k_{-3}/k_4 + A_2)] + \\ + 2k_{-2}A_4 + k_6A_4B/(k_{-6}/k_7 + 1) + D_1\nabla^2A_2,$$

$$\partial A_4/\partial t = k_2A_2^2 - k_{-2}A_4 + k_3A_2^2A_4/(k_{-3}/k_4 + A_2) - \\ - k_6A_4B/(k_{-6}/k_7 + 1) + D_2\nabla^2A_4.$$

After transformations, we have:

$$\partial x/\partial\tau = \alpha - \beta x - x^2 + y - fx^2y/(g + x) + cy + D_1^*\nabla^2x,$$

$$\partial y/\partial\tau = x^2 - y + fx^2y/(g + x) - 2cy + D_2^*\nabla^2y, \qquad (20)$$

where $x = 2k_2A_2/k_{-2}$, $y = 4k_2A_4/k_{-2}$, $\tau = k_{-2}t$, $\alpha = 2k_1k_2A^2/k_{-2}^2$, $\beta = k_{-1}/k_{-2}$, $c = k_6B/2k_{-2}(k_{-6}/k_7 + 1)$, $f = k_3/2k_2$, $g = 2k_2k_{-3}/k_4k_{-2}$, $D_1^* = D_1/k_{-2}$, and $D_2^* = D_2/k_{-2}$.

In equilibrium, we obtain from (20) for dimensionless concentrations (x_s, y_s) of A_2 and A_4:

$$y_s = (\alpha - \beta x_s)/c,$$

$$(c - \beta f)x_s^3 + [cg + \alpha f + \beta(1 + 2c)]x_s^2 + \\ + (1 + 2c)(\beta g - \alpha)x_s - \alpha g(1 + 2c) = 0. \qquad (21)$$

Now let us evaluate the possibility of diffusion instability for system (20). First, let us determine the number of stationary solutions. Given that $c \neq \beta f$, the second equation in (21) is cubic,

$$a_0x^3 + a_1x^2 + a_2x + a_3 = 0. \qquad (22)$$

The number of real (i.e., stationary) solutions to (20) is defined [13] by the sign of the sum

$$Q = p^3 + q^2, \tag{23}$$

where $p = a_2/3a_0 - (a_1/3a_0)^2$, $q = a_3/2a_0 + (a_1/3a_0)^3 - a_1a_2/6a_0^2$.

For $Q > 0$, equation (22) has a unique real root. For $Q < 0$, the number of real roots is three: the middle one corresponds to unsteady stationary state. At $Q = p = q = 0$ (bifurcation point), the real roots coincide [12, 13]. From (21) и (23), it follows that the bifurcation point is defined by the equation $f^2\alpha + (\beta + 2\beta c + 8\beta fg - 8cg)\alpha - \beta g(cg + \beta + 2c) = 0$. For realistic rate constants, estimates show that we can expect only one positive root.

For system (20), coefficients a_{ij} in (11) have the following values: $a_{11} = -\beta - 2x_s - f(2g + x_s)x_sy_s/(g + x_s)^2$, $a_{22} = -1 - 2c + fx_s^2/(g + x_s)$, $a_{12} = 1 + c - fx_s^2/(g + x_s)$, and $a_{21} = 2x_s + f(2g + x_s)x_sy_s/(g + x_s)^2$. According to criteria (15), (16), diffusion instability arises for $fx_s^2/(g + x_s) > 1 + 2c$, provided that

$$D_1^*[fx_s^2/(g + x_s) - 1 - 2c] > D_2^*[\beta + 2x_s + f(2g + x_s)x_sy_s/(g + x_s)^2].$$

For $f = 0$ or $g \to \infty$ (i.e., $k_3 = 0$ or $k_4 = 0$, that is, no matrix catalysis), the reacting system remains homogeneous.

For the model (1,2,3',4',6',7',8), the kinetic equations (steady-state treatment with respect to B_2 и A_4B_2) can be written in the form:

$$\partial x/\partial\tau = \alpha - \beta x - x^2 + y - fx^2y/(g + x) + cy/(k + y) + D_1^*\nabla^2 x,$$

$$\partial y/\partial\tau = x^2 - y + fx^2y/(g + x) - 2cy/(k + y) + D_2^*\nabla^2 y, \tag{24}$$

where $c = 2k_2k_8B_2/k_{-2}^2$ and $k = 4k_2k_{-8}(k_{-6}/k_7 + 1)/k_{-2}k_6$, other notations being the same.

The type of reaction (7') does not affect the stability of the system. It affects only the bifurcation parameters and type of the linearized operator. Steady-state treatment of (24) leads to the following expressions for x_s and y_s:

$$(\alpha - \beta x_s) = cy_s/(k + y_s),$$

$$x_s^2 + fx_s^2y_s/(g + x_s) = y_s + 2cy_s/(k + y_s). \tag{25}$$

For system (24), coefficients a_{ij} in (11) have the following values:

$$a_{11} = -\beta - 2x_s - f(2g + x_s)x_s y_s/(g + x_s)^2,$$
$$a_{22} = -1 - 2ck/(k + y_s)^2 + fx_s^2/(g + x_s),$$
$$a_{12} = 1 + ck/(k + y_s)^2 - fx_s^2/(g + x_s),$$
$$a_{21} = 2x_s + f(2g + x_s)x_s y_s/(g + x_s)^2.$$

According to (15), (16), the diffusion instability takes place when

$$x_s^2/(g + x_s) > 1 + 2ck/(k + y_s)^2 \tag{26}$$

for

$$D_1[fx_s^2/(g + x_s) - 1 - 2c] > D_2[\beta + 2x_s + f(2g + x_s)x_s y_s/(g + x_s)^2].$$

Again, the stage of matrix catalysis is a necessary condition. Expression (25) is a fourth-order algebraic equation with respect to x_s. Its analytical solution is too bulky. Analytical estimates can be simplified if we take into account inequality (26). In this case, we find that $x_s < \alpha/[(0.5c)^{0.5} + \beta]$ while $y_s > k/[\beta(2c)^{0.5} + c - \alpha]$.

Given that $y_s >> k$, we find from (25), (26) the lower boundary for microredistribution in the reacting system:

$$\alpha > c + \beta[1 + (1 + 4fg)^{0.5}]/2f.$$

The upper boundary is found from the condition $y_s > 0$. Therefore, we come to the following relationship between dimensionless system parameters:

$$c + \beta[1 + (1 + 4fg)^{0.5}]/2f < \alpha < \beta(2c)^{0.5} + c.$$

In the dimensional form, it can be written as

$$k_8 B_s^2/k_1 + 0.5K_1 k_{-2}[1 + (1 + 4k_3 k_{-3}/k_{-2}k_4)^{0.5}]k_3 < A_s^2 <$$
$$< k_8 B_s^2/k_1 + (k_8/k_2)^{0.5} B_s/K_1. \tag{27}$$

Apparently, these inequalities hold true when

$$B_s > B^* = 0.5K_1^2 k_2^{0.5} k_{-2}\,[1 + (1 + 4k_3 k_{-3}/k_{-2}k_4)^{0.5}]/k_3 k_8^{0.5}. \tag{28}$$

Using (27), (28), we can find the least value of starting concentrations A^* and B^* at which nonuniform distribution of intermediate products may arise. According to [7], the measured values of equilibrium constants, at 40–50°C, are $K_1 \approx 0.7\ \mathrm{M}^{-1}$ and $K_2 \approx 10^2\ \mathrm{M}^{-1}$. Since other constants are unknown, we tentatively assumed that $K_3 \approx 10^3\ \mathrm{M}^{-1}$, $k_8 \approx k_1$, $k_{-1}/k_{-2} \approx 1$, $k_3 \approx k_4$, and $k_{-3}/k_{-2} \approx 2$. In this case, $\mathrm{A}^* = 0.03$ M и $\mathrm{B}^* = 0.01$ M. Analysis shows that equilibrium concentrations of A_2 and A_4 are 2–3 orders of magnitude lower than those of A and B. Our estimates for A^* and B^* are difficult to compare with measured kinetic data. For this to be done, NMR or electron microscopic studies are needed on the effect of starting concentrations on the supramolecular structure of polymerizing system or resultant polymer. Note that aggregation of globules during synthesis of cross-linked polyurethanes starts at $\mathrm{A}_g = 0.5$ M [16, 17]. This is consistent with above estimates for A^* because A^* must clearly be below A_g.

Using (15), we may estimate the minimal size of microinhomogeneity λ_{min} by determining z from (15), since, according to (10), $\lambda = 2\pi/z$. Given that $D_1 >> D_2$, one obtains for $y_s >> k$ and $x_s << c$ that $\lambda_{min} > \pi(D_2/K_2k_8\mathrm{B}^2)^{1/2}$. At $D_2 = 10^{-6}\ \mathrm{cm}^2/\mathrm{s}$, $\lambda_{min} = 10^{-6}$ cm, which agrees with the reported [17] diameter of globular associates in cross-linked polyurethane: as measured by electron microscopy, it ranges between 150 and 500 Å.

DISCUSSION

Synthesis of cross-linked polyurethanes proceeds in a reacting system that contains not only hydroxyl (diols, triols) and isocyanate groups (diisocyanate) but also their associates. Even at moderate temperatures, the estimated concentration of associates is markedly lower than the concentration of starting reagents. In terms of widely adopted reaction schemes, the rate of urethane formation depends on the concentration of associates (in particular, hydroxyl-containing tetramer). When the characteristic time for physical relaxation of supramolecular system is longer than that for chemical reaction, uniform spatial distribution of reagents in the system is infringed, the effect being most pronounced for associates (because of their low amount).

Analysis shows that if reaction involved only autoassociates, fluctuation in local concentration would not have resulted in microphase separation of the system. A range of wave numbers whose amplitudes of spatial harmonics exponentially grow around steady-state concentration of associates does exist only when heteroassociates take part in reaction. This infringes uniform spatial distribution of intermediate products in reactive mixture. In some microvolumes ('grains'), the local reaction rate becomes higher than the average, which results in higher extent of conversion (compared to intergrain space). Within the range of diffusion kinetic instability, formation of urethanes becomes spatially inhomogeneous:

reaction proceeds as if in two different phases. A necessary condition for the onset of spatial inhomogeneity is the presence of both noncatalytic and catalytic routes for the formation of A_4.

A question arises whether the results obtained for open systems can be applied to closed systems most frequently used in experimental studies. In closed systems, steady-state treatment is applicable (for a limited period of time) only to intermediate products: the lower their concentration, the longer is the duration of the steady state. For the Belousov–Zhabotinsky reaction taken as an example, formation of dissipative structures in closed reactive systems involving low-molecular compounds was demonstrated experimentally [10, 12]. Spatial and transient dissipative structures were observed during synthesis of polymers [2, 3, 16–18]. Theoretical analysis of “rapid” variables at ‘frozen slow’ variable has a strict theoretical background. In this context, it can be inferred that the obtained data can be readily applied (at least, qualitatively) to closed systems (with no reagents motion through the system boundary).

Other explanations have also been suggested for formation of dissipative structures during synthesis of cross-linked polymers. In particular, distribution of structural elements in cross-linked polyimide (prepared by stepwise cyclization of linear polyamido acid) was explained [19] by the rupture of stressed chemical bonds and increasing molecular mobility of macromolecules with increasing temperature.

CONCLUSIONS

Theoretical analysis shows that synthesis of polyurethane is accompanied by the formation of structural microheterogeneities of a polymer formed. The homogeneity range of reactive system depends on the kinetic constants and concentration of starting reagents. Structural inhomogeneity during synthesis of polyurethane arises when (*i*) the reaction involves heteroassociates of reactants and (*ii*) alcohol tetramers are formed in catalytic reaction. The minimal size of microinhomogeneity was estimated.

Acknowledgments. This work was supported by the International Science and Technology Center (grant no. 358-96).

REFERENCES

1. Irzhak V.I., Rozenberg B.A., and Enikolopyan N.S., *Setchatye polimery: Sintez, struktura, svoistva* (Network Polymers: Synthesis, Structure, Properties), Nauka, Moscow, 1979.

2. Berlin A.A., Korolev G.V., Kefeli T.Ya., and Sivergin Yu.M., *Akrilovye oligomery i materialy na ikh osnove* (Acrylic Oligomers and Related Materials), Khimiya, Moscow, 1983.
3. Volkova N.N., Sosikov A.I., Berezin M.P., Korolev G.V., Erofeev L.N., and Smirnov L.P., *Vysokomol. Soedin., Ser. A*, **30**, 2133 (1988).
4. Rozenberg B.A., *Usp. Khim.*, **60**, 1473 (1991).
5. Rozenberg B.A., *Adv. Polym. Sci.*, **75**, 113 (1986).
6. Bondarenko S.P., Zaporozhskaya S.V., Tiger R.P., and Entelis S.G., *Khim. Fiz.*, **5**, 1264 (1986).
7. Tiger R.P., Tarasov D.N., and Entelis S.G., *Khim. Fiz.*, **15**, 11 (1996).
8. Bondarenko S.P., Chirkov Yu.N., Zaporozhskaya S.V., and Entelis S.G., *Kinet. Katal.*, **30**, 599 (1989).
9. Chernova E.A., Tiger R.P., and Tarakanov O.G., *Zh. Strukt. Khim.*, **27**, 19 (1986).
10. Nicolis G. and Prigogine I., *Self-Organization in Nonequilibrium Systems*, Wiley, New York, 1977.
11. Zelenyuk A.N., Berlin P.A., Tiger R.P., and Entelis S.G., *Kinet. Katal.*, **35**, 852 (1994).
12. Ebeling W. *Strukturbildung bei Irreversibilen Prozessen*, Teubner, Leipzig, 1976.
13. Mishina A.P. and Proskuryakov I.V., *Vysshaya algebra* (Higher Algebra), Nauka, Moscow, 1965.
14. Tucker E.E., Fornham S.B., and Cristian S.D., *J. Phys. Chem.*, **73**, 3820 (1969).
15. Duboc C., *Spectrochim. Acta A*, **30**, 431 (1974).
16. Lipatova T.E., Ivatchenko V.K., and Bezruk L.I., *Vysokomol. Soedin., Ser. A*, **13**, 1701 (1971).
17. Lipatova T.E., Babitch V.F., Sheinina L.S., Vengerovskii Sh.G., and Korzhuk N.I., *Vysokomol. Soedin., Ser. A*, **20**, 2051 (1978).
18. Ivanov P.V., Maslova V.I., Bondareva N.G., *et al.*, *Izv. Akad. Nauk, Ser. Khim.*, **46**, 2256 (1997)
19. Bovenko V.N. and Startsev V.M., *Vysokomol. Soedin, Ser. B*, **35**, 1004 (1994).

Section IV

STRUCTURE–PROPERTIES RELATIONSHIP

Chapter 13

Thermoset/Thermoplastic Blends with a Crosslinked Thermoplastic Network Matrix

Ying YANG, Tsuneo CHIBA, and Takashi INOUE*

Department of Organic and Polymeric Materials,
Tokyo Institute of Technology,
Ookayama, Meguro-ku, Tokyo, 152-8552 Japan

ABSTRACT

A thermoset/thermoplastic system, diallyl phthalate (DAP)/poly(2,6-dimethyl-1,4-phenylene ether) (PPE), was found to show upper critical solution temperature (UCST)-type phase behavior (UCST ≈ 145°C). The system was cured by organic peroxide just above the UCST. A light scattering peak and its growth at an early stage of curing suggested the reaction-induced spinodal decomposition caused by elevation of the UCST with the polymerization of DAP. Transmission electron

*e-mail: tinoue@o.cc.titech.ac.jp

microscopy showed that, even in the case of low PPE content (e.g., 20 wt%), PPE-rich phase could be the matrix of phase-decomposed material. Furthermore, it was shown that fine polyDAP domains of 10 nm diameter are dispersed in the matrix, and in the polyDAP-rich dispersed particles of μm diameter, fine PPE domains of 10 nm scale are occluded. The fine domains in both the dispersed particles and the matrix could be formed by successive spinodal decomposition under very deep quench after the micrometer-scale particle/matrix morphology had been arrested by partial cure. Tensile modulus, elongation at break, and fracture toughness (G_{IC}, K_{IC}) of the fully cured blends increased almost linearly with increasing PPE content. The toughening seems to be realized mostly by the unique morphology *via* the two-step spinodal decomposition to render the PPE-rich matrix even for the low PPE content system. Dynamic mechanical analysis, Fourier transform infrared spectroscopy, and solvent extraction experiment showed that polyDAP significantly grafted on PPE chains to yield a network matrix.

INTRODUCTION

Many combinations of dissimilar polymers are found to follow phase diagrams; i.e., the polymers are miscible within limited temperature and composition ranges but are immiscible outside of these ranges [1]. Spinodal decomposition is induced by a temperature jump from a miscible region to an immiscible region. The thermally-induced spinodal decomposition is an interesting subject in polymer physics in terms of the order-disorder transition [2]. Spinodal decomposition can be also driven by chemical reactions. The reaction-induced spinodal decomposition is interesting in an attempt to design new materials [3].

A typical example of the reaction-induced spinodal decomposition is seen in thermoset/thermoplastic system, e.g., epoxy/poly(ether sulfone) blend. The system is usually homogenous at the beginning of curing. With the proceeding of polymerization, i.e. with the increase of molecular weight, the system will be thrust into a two-phase region and the phase decomposition takes place either by the nucleation and growth or the spinodal decomposition mechanism. Then the morphology developed by the reaction-induced phase decomposition would be fixed at a certain stage of curing by gelation and/or vitrification. Even by the spinodal decomposition mechanism, the fixed morphology would be versatile, depending on the cure condition. A variety of morphologies have been found: dispersed domain structure with uniform domain size, interconnected globule structure and bimodal domain structure [3]. In this article, a new and sophisticated morphology will be demonstrated in a thermoset/thermoplastic blend, diallyl phthalate (DAP)/poly(2,6-dimethyl-1,4-phenylene ether) (PPE). Time-resolved light scattering and transmission electron microscopy were used to study the morphology development during curing. The phase-separated structure was further investigated by Fourier transform infrared spectroscopy and dynamic mechanical analysis. Tensile properties and fracture toughness of cured materials were measured and discussed in connection with the morphology.

EXPERIMENTAL

PPE was supplied by General Electric Co. ($M_w = 41.5 \times 10^3$). Diallyl phthalate (DAP) was supplied by Kanto Chemical Co. The cure agent for DAP was *a,a'-bis*(*t*-butylperoxy-*m*-isopropyl)benzene (PBP) from Japan Fat Co.

The phase diagram of the DAP/PPE system was measured by the cloud point method, i.e., a solution cast film of DAP and PPE on cover-glass was annealed at different temperatures for 12 h and then the structure in the film was observed under an optical microscope.

DAP/PPE blends loaded with PBP were prepared as follows. The PPE was first mixed mechanically with DAP monomer at room temperature. The mixture was heated with stirring at 170°C to dissolve PPE completely in DAP. Then the mixture was cooled down to 150°C and the PBP was added with rigorously stirring to ensure PBP dispersed uniformly in the mixture. Two curing procedures, one-step and two-step, were applied. In the one-step curing procedure, the mixture was injected into a mold made of two glass plates sealed by rubber gasket and cured at 150°C for 8 h. In the two-step curing procedure, the mixture was cured at 120°C for 12 h then 150°C for 5 h. Further, it was postcured at 180°C for 1 h and then at 250°C for 0.5 h under vacuum.

The light scattering profiles were obtained by a laser light scattering apparatus with a goniometer and a hot stage. The description of the apparatus in detail can be found elsewhere [4]. The radiation from a He-Ne gas laser of wavelength 632.8 nm was applied vertically to the film specimen. The goniometer traces of the scattering light were obtained under a V_V (parallel polarized) optical alignment. Thus during isothermal curing, the change in the light scattering intensity with time was recorded at appropriate intervals.

The morphology of the cured specimens was examined by a transmission electron microscope (JEM 100 CX, Jeol Co.) at 100 kV acceleration voltage. The ultrathin sections of thickness about 70 nm were obtained with an ultramicrotome (Ultracut N, Reichert-Nissei). The sections were stained by RuO_4 vapor for 10 min at room temperature.

The cure process was monitored with a JIR-600 Fourier transform infrared spectrometer (Jeol Ltd.) equipped with a hot stage.

Tensile stress-strain curves were measured by a tensile machine (Orientec RTC-1350A) at room temperature, using dumb-bell specimen (gauge length 20 mm; thickness 3 mm; width 3 mm) at a crosshead speed of 10 mm/min.

Fracture toughness was measured following ASTM D5045-93. The critical stress intensity factor K_{IC} and the critical strain energy release rate G_{IC} were obtained using a single-edge notched specimen (3 mm × 10 mm × × 50 mm) in a three point bending geometry. A fresh crack was created by a sharp razor blade.

Dynamic mechanical behavior was measured by a Toyoseiki Dynamic Mechanical Analyzer at 100 kHz and a heating rate of 2°C/ min. The dynamic loss and the storage modulus were obtained as a function of temperature.

Cured materials were immersed in trichloroethylene (good solvent for PPE) at the boiling point for 72 h. Weight loss with the extraction was measured.

RESULTS AND DISCUSSION

The phase diagram of DAP/PPE is shown in Fig. 1. The system exhibits upper critical solution temperature (UCST)-type phase behavior. The UCST locates at about 145°C at PPE content of 20 wt %. It is expected that the DAP/PPE system can be cured starting from a homogenous mixture, when a cure agent with high decomposition temperature such as PBP is used. It is

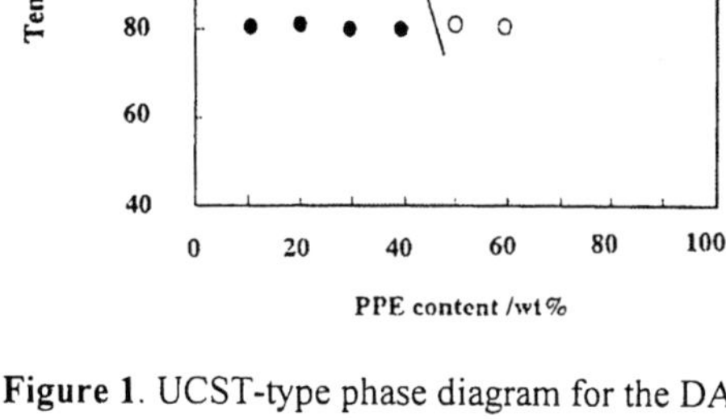

Figure 1. UCST-type phase diagram for the DAP/PPE system.

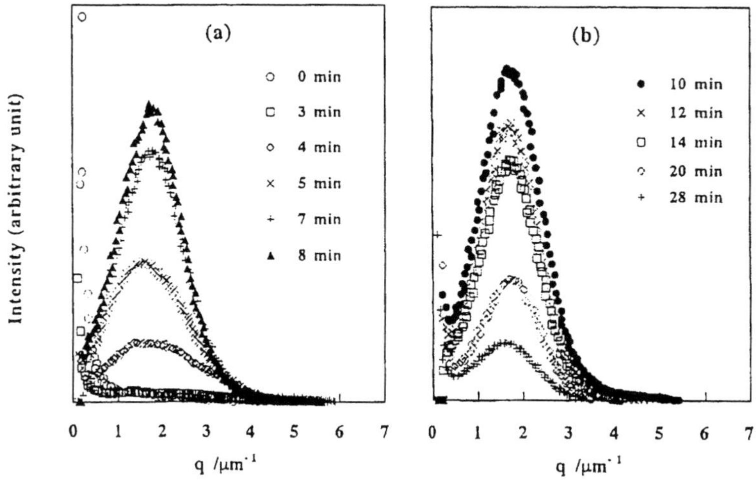

Figure 2. Time variation of light scattering profiles for a DAP/PPE/PBP 80/20/1.1 mixture during curing at 150°C: (a) early stages and (b) later stages.

also expected that, when curing proceeds, the phase boundary will shift upwards and the system will be thrust into a two-phase region.

Figure 2 shows the typical time-resolved light scattering profiles during curing at 150°C, where q is the scattering vector defined by q = $= (4\pi/\lambda)\sin(\theta/2)$; λ and θ being the wavelength of light in the specimen and the scattering angle, respectively. A DAP/PPE/PBP 80/20/1.1 mixture was a single-phase system at the curing temperature of 150°C. There was no appreciable light scattering detected from the mixture in the first few minutes. A peak in the light scattering profile appeared after about 3 min, then the scattering intensity increased with time, keeping the peak position almost constant (Figure 2a). About 10 min later, the peak intensity started to decrease (Figure 2b) and then the peak disappeared after 40 min.

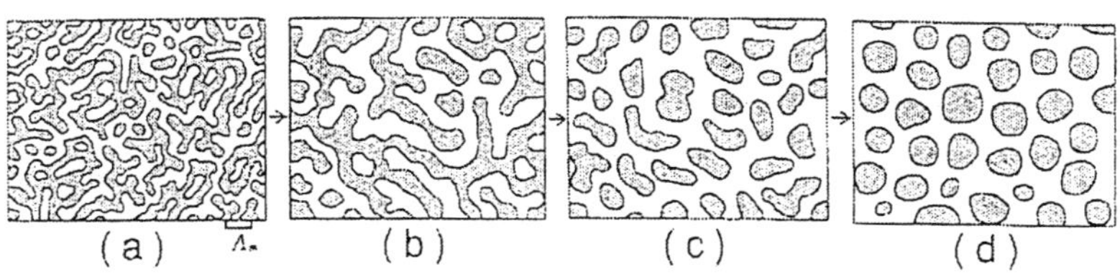

Figure 3. Schematic representation of the morphology development via spinodal decomposition: bicontinuous structure developed at early stage (a) grows up self-similarly to yield a similar structure with larger spacing (b) and the phase connectivity is interrupted to be converted to a fragmented structure (c) and then to a rather spherical domain structure (d).

The same tendency was observed at different levels of PBP loading (0.4–1.75 phr).

It is well known that the formation of phase-separated morphology *via* spinodal decomposition can be well characterized by light scattering profiles. The appearance of a scattering peak and the continuous increase of the scattering intensity are indications of the development of a regular phase-separated morphology as schematically shown in Fig. 3. After the intensity of the scattered light reaches the maximum value, the change of the scattering profile will be different depending on the further development of the morphology. When the phase-separated morphology is arrested by gelation or vitrification, the peak intensity will level off [5].

When the phase dissolution takes place [6] and the difference in the refractive index between the two phases becomes smaller with further reaction [7], the peak intensity would start to decrease. The decay of the scattering peak in the present system can be ascribed to a disordering in the mutual arrangement of dispersed particles; i.e. broader size distribution and more irregular arrangement of particles in Fig. 3d, as shown in TEM micrographs in Fig. 4.

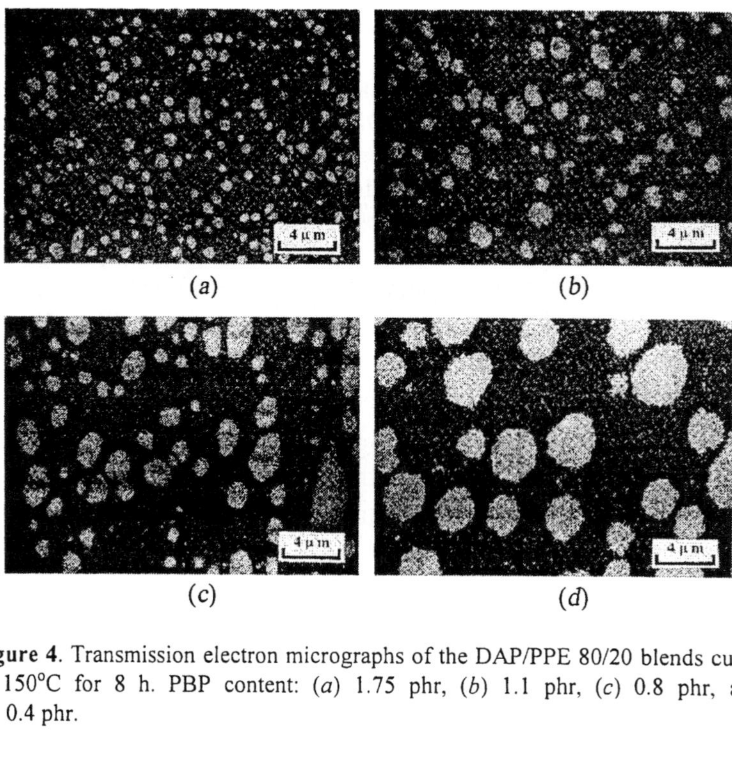

Figure 4. Transmission electron micrographs of the DAP/PPE 80/20 blends cured at 150°C for 8 h. PBP content: (*a*) 1.75 phr, (*b*) 1.1 phr, (*c*) 0.8 phr, and (*d*) 0.4 phr.

In Figure 4, the bright region shows polyDAP-rich particles, while the dark region shows PPE-rich matrix [8]. It can be seen that, in all the specimens, spherical polyDAP-rich particles are dispersed rather irregularly in the PPE-rich matrix. The lower PBP content system yields the larger polyDAP-rich particles. Regardless of higher or lower PBP content, the domains are not interconnected. This implies that the rate of phase separation was very fast, compared with the reaction rate. In other words, before the network formation of the polyDAP-rich phase, the phase connectivity had been broken and the polyDAP-rich phase had been converted to dispersed particles. Furthermore, the mutual arrangement of particles seems to become irregular by coarsening. The disordering may cause the decay of the scattering peak. Then the morphology at quite a late stage of spinodal decomposition seems to be fixed.

It should be noted that DAP has a unique gelation behavior. Usually free radical polymerization of divinyl compound yields a crosslinked network at only 1% conversion, while diallyl phthalate can be polymerized to 21% conversion without forming a network [9]. The delayed gelation of polyDAP may render the structure fixation at a very late stage of spinodal decomposition; without premature stopping at early stages in Fig. 3a–c. The lower PBP content will lead to the lower reaction rate; in turn, the slower increase of viscosity in the system, which could allow the dispersed particles to coarsen to the higher degree, so that the morphology at the later stage would be arrested.

A TEM micrograph at higher magnification is shown in Fig. 5. There are small bright domains in the matrix and small dark domains in the particles. The small domains in both the matrix and the dispersed particles may be formed by the second spinodal decomposition in individual regions after the micrometer scale structure was fixed. The situation is schematically

Figure 5. Transmission electron micrograph of the DAP/PPE/PBP 80/20/0.4 system (see Figure 4e) at high magnification.

described in Fig. 6. That is, long waves in the concentration profile shown by the broken line are created by the first spinodal decomposition to provide the dispersed particles of micrometer diameter and then shorter waves are formed by the second spinodal decomposition to yield fine domains of the order of 10 nm in both the matrix and the dispersed particles. The morphology shown in Fig. 4 and 5 is a new one in thermoset/thermoplastic blends.

Note that the term second spinodal decomposition in the above discussion is used for convenience to describe the phenomenon. It is just an apparent one, because the thermodynamic quench depth may increase continuously with time during curing and one can not expect any particular new instability to cause another decomposition. That is, the decomposition should not proceed step-wise but continuously. However, such apparent two-step spinodal decomposition is actually seen in the computer simulation of spinodal decomposition under non-isoquench depth conditions; i.e., when the quench depth increases with time [10] (as in the present case). In the simulation, as the quench depth increases with time, the long waves in the concentration profile develop to certain levels of amplitude and wavelength (as shown by the broken lines in Fig. 6) and then short waves suddenly start to appear at a certain stage. This apparently looks like the second-step decomposition. The apparent two-step spinodal decomposition could be expected to take place especially in the thermoset system, because the

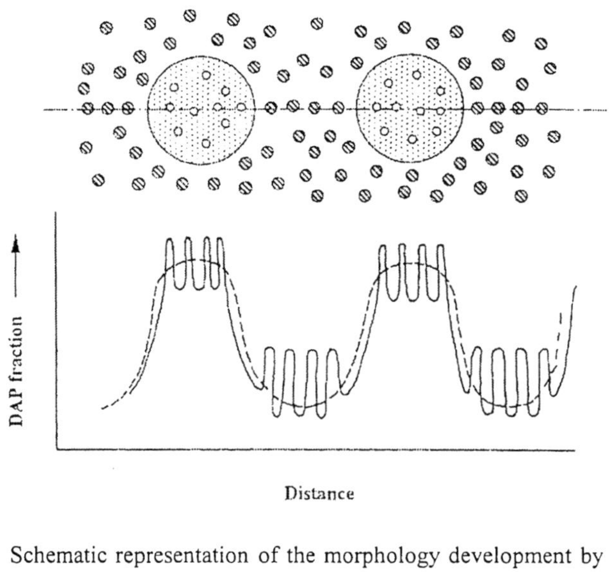

Figure 6. Schematic representation of the morphology development by the two-step spinodal decomposition and the concentration profile by the first spinodal decomposition (broken line) and that by the second one (solid line).

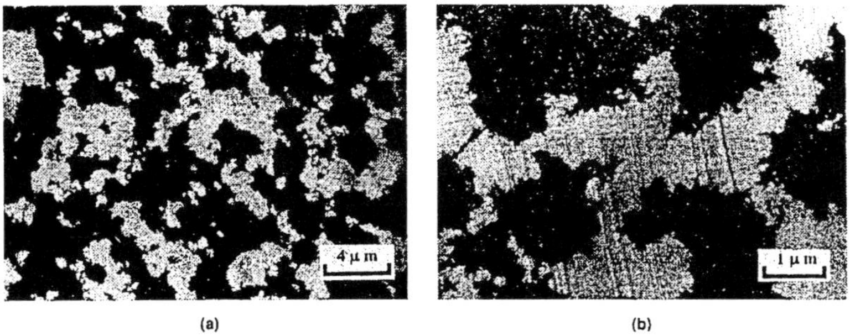

Figure 7. Transmission electron micrographs of the DAP/PPE/PBP 80/20/1.75 system cured at 120°C for 12 h then at 150°C for 5 h: (a) low and (b) high magnification.

morphology developed by 'the first spinodal decomposition' would be arrested by partial crosslinking at a certain stage and then the quench depth further increases without significant change in morphology for a while; finally, 'the second decomposition' would start to yield much smaller domains.

In Figure 7 are shown TEM micrographs of DAP/PPE/PBP 80/20/1.75 by the two-step cure (cured at 120°C for 12 h and then 150°C for 5 h). The morphology is more bicontinuous, compared with Fig. 4. This is expected, because at the low temperature the rate of phase separation is expected to decrease drastically; e.g., following the WLF-type temperature dependence of the chain mobility, while the reaction rate does not decrease so much (slow down to a half by a 10°C depression), and then the structure could be

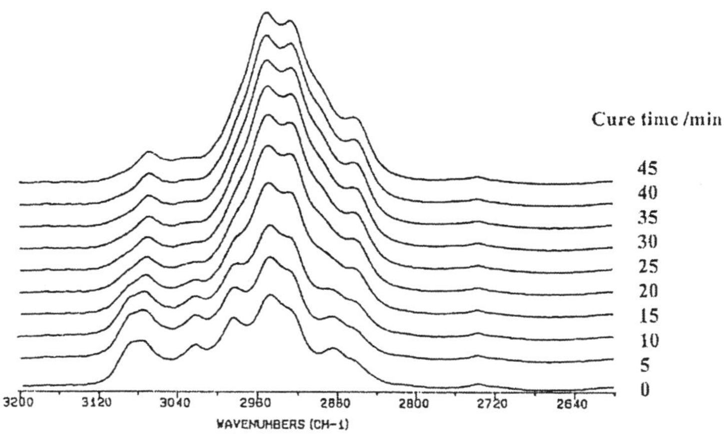

Figure 8. FT-IR spectra of the DAP/PPE/PBP 80/20/1.75 blend cured at 170°C.

fixed at an intermediate stage of spinodal decomposition to provide the bicontinuous structure as shown in Fig. 3b or c. Also in this material, the fine domains of scale 10 nm are seen in both regions (Fig. 7b), implying that the morphology developed by the two-step spinodal decomposition.

Note that the volume ratio of DAP-rich region and PPE-rich region in Fig. 4, 5 and 7 is very different from the charge ratio (80/20 DAP/PPE). For example, the area ratio (~volume ratio) of matrix (dark)/dispersed particles (bright) in Fig. 4b is approximately 70/30. The dark matrix contains bright domains, as discussed above (Fig. 5) and the dark region may be ca. 60%. Then, omitting the PPE domains in the dispersed particles, the dark area in the matrix is counted to be 42% (= 70% × 60%). This is much larger than the charged amount of PPE (20%). The situation is same for Fig. 4a and 7. It implies that the dark matrix is not pure PPE phase but a significant amount of DAP should be incorporated in the dark matrix. The incorporation could be realized by grafting of polyDAP on PPE chains.

Figure 8 shows FT-IR spectra of DAP/PPE/PBP 80/20/1.75 at various cure times. A 3020 cm^{-1} band is assigned to C–H stretching vibration in the vinyl group of DAP. It decreases with cure and eventually disappears, suggesting simply the polymerization of DAP. A 2923 cm^{-1} band is assigned to the stretching vibration of C–H in $-CH_2-$ and methyl groups attached to a phenyl ring. It increases with cure, suggesting the graft reaction of DAP on the methyl group of PPE. The grafting is expected to generate two C–H

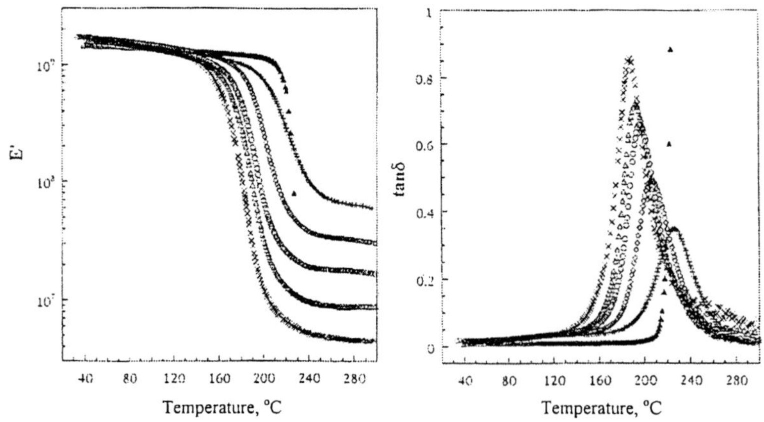

Figure 9. Dynamic mechanical properties of PPE, DAP and their blends. DAP and blends were cured at 120°C for 12 h and 150°C for 5 h and then postcured in vacuum at 180°C for 1 h and at 250°C for 0.5 h: ▲, PPE; +, polyDAP; ◊, DAP/PPE 80/20; O, DAP/PPE 70/30; Δ, DAP/PPE 60/40; ×, DAP/PPE 50/50 (PBP 1.75 phr).

bonds by consuming one C–H bond of the PPE methyl group and one C=C double bond of DAP.

The graft reaction was confirmed by the solvent extraction experiment. After dipping in boiling trichloroethylene for 72 h, the cured blend of DAP/PPE/PBP 80/20/1.75 showed neither weight loss nor any change in shape. The results implies that all PPE chains are connected to polyDAP by chemical bond and/or crosslinked with DAP.

Figure 9 shows the results of dynamic mechanical analysis. For all cured blends, the storage modulus exhibits a rubbery plateau (Fig. 9a), suggesting the nature of three-dimensional network. The blend with the higher PPE content shows the lower plateau modulus, i.e., the lower crosslink density. Since the bulk modulus depends mostly on that of matrix phase, the results imply that even in the blend with the highest PPE content, the PPE-rich matrix is completely crosslinked to form a three dimensional network.

The dynamic loss (tan δ) shows a single peak (single glass transition temperature, T_g) for all cured blends. As has been discussed, the cured blends are the four-phase materials. However, the four-phase nature is not apparent in tan δ–temperature curve. This is simply due to the close T_g values of component polymers (PPE and polyDAP). Peak position of tan δ shifts to lower temperature as PPE content increases. The T_g depression could be caused by dangling chains, i.e., short polyDAP chains dangling from PPE-polyDAP network may play a role of internal plasticizer to depress the T_g.

Figure 10a shows the stress–strain curves of the cured materials. The tensile modulus and the elongation at break increases with increasing PPE content. The area under stress-strain curve and the elongation at break are plotted as a function of PPE content in Fig. 10b, representing the toughening

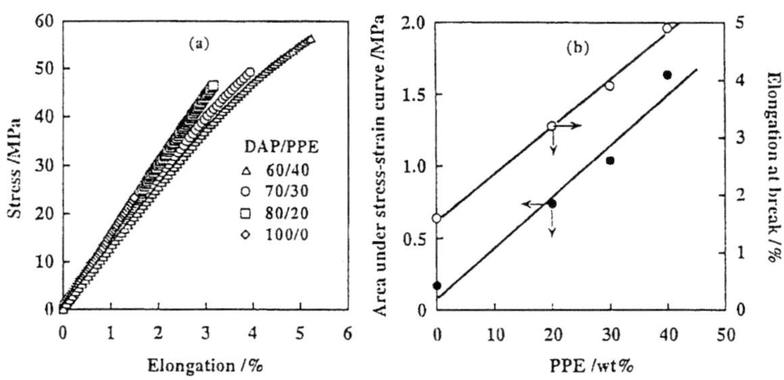

Figure 10. Tensile properties of DAP/PPE blends cured at same condition as in Fig. 9; (a) stress–strain curves and (b) the elongation at break and the area under stress–strain curve vs. PPE content.

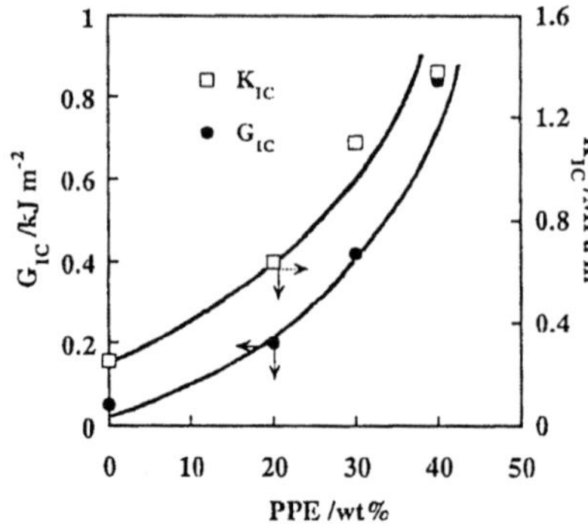

Figure 11. Fracture toughness vs. PPE content: same samples as in Figs 9 and 10.

effect of PPE incorporation. The fracture toughness in terms of the critical stress intensity factor K_{IC} and the critical strain energy release rate G_{IC} is shown as a function of PPE content in Fig. 11. The fracture toughness of DAP/PPE 70/30 blend, for example, is more than ten times higher than that of neat polyDAP. This is surprising, if one remembers that in the thermoset/thermoplastic systems, such as epoxy/PPE and epoxy/PES blends, the toughness can be usually improved in a range of 50%–100% by loading thermoplastic at a 20–30 wt % level [11, 12]. The dramatic improvement of toughness may be mostly caused by the formation of PPE-rich matrix, even for low PPE content system (20–30 wt% level). PPE is intrinsically ductile and its ductility would not be deteriorated so much by the crosslinking. An excellent adhesion between PPE-rich and polyDAP-rich phases could be achieved by the grafting. It may partly contribute to the toughening.

CONCLUSIONS

Thus, the two-step spinodal decomposition in DAP/PPE system driven by cure reaction yielded the cured blend with a unique four-phase morphology having the PPE (minor component)-rich network matrix. The cured material showed excellent chemical resistance, high temperature property and mechanical properties. Such properties are provided by the reactivity and high T_g characters of PPE and it seems to be an interesting constituent polymer for thermoset/thermoplastic blends.

REFERENCES

1. Olabishi O., Robeson L.M., and Shaw M.T., *Polymer–Polymer Miscibility*, Academic Press, New York, 1979.
2. *Dynamics of Ordering in Condensed Matter*, Komura S. and Furukawa H., Eds., Plenum Press, New York, 1988.
3. Inoue T., *Prog. Polym. Sci.*, **20**, 119 (1995).
4. Chen W.J., Kobayashi S., Inoue T., Ohnaga T., and Ougizawa T., *Polymer*, **35**, 4015 (1994).
5. Yamanaka K. and Inoue T., *Polymer*, **30**, 662 (1989).
6. Ougizawa T., Inoue T., and Kammer H.W., *Macromolecules*, **18**, 2089 (1985).
7. Kim B.S., Chiba T., and Inoue T., *Polymer*, **36**, 67 (1995).
8. Yang Y., Fujiwara H., Chiba T., and Inoue T., *Polymer*, **39**, 2745 (1998).
9. Kircher K., *Chemical Reactions in Plastics Processing*, Hanser Press, Munich, 1987.
10. Ohnaga T., Chen W.J., and Inoue T., *Polymer*, **35**, 3774 (1994).
11. Raghava D.R., *J. Polym. Sci., Polym. Phys. Ed.*, **25**, 1017 (1987).
12. Pearson R.A. and Yee A.F., *Polymer*, **34**, 3658 (1993).

Chapter 14

Epoxy-Based PDLCs: Formation, Morphological, and Electrooptical Properties

Lyudmila L. GUR'EVA[1]*, Georgii B. NOSOV[2], Vladimir K. GERASIMOV[3], Anatolii E. CHALYKH[3], Anatolii S. SONIN[2], and Boris A. ROZENBERG[1]

[1]*Institute of Problems of Chemical Physics, Russian Academy of Sciences, Chernogolovka, Moscow Region, 142432 Russia*
[2]*Institute of Organoelement Compounds, Russian Academy of Sciences, 28 Vavilov Street, Moscow, 117813 Russia*
[3]*Institute of Physical Chemistry, Russian Academy of Sciences, 31 Leninskii Prospect, Moscow, 117915 Russia*

*e-mail: gur@icp.ac.ru

ABSTRACT

Kinetics of polymerization and polycondensation of diglycidyl ether of bisphenol A under the action of tertiary and primary amines in the presence of nematic liquid crystals (LC) based on *n*-pentylcyanbiphenyl was studied by isothermal calorimetry and IR spectroscopy. The presence of LC was shown to decelerate these reactions. An explanation of this phenomenon is proposed. The temperature of LC-isotropic phase transition was measured as a function of the reaction conversion. The morphology of obtained polymer dispersed liquid crystals (PDLC) with different nematic concentrations and network densities of polymer matrix was studied by optical and scanning microscopy. Some electrooptical properties of PDLC films were investigated. Transmission and relaxation times depending on high frequency pulse voltage were measured. It was shown that decrease of network density of polymer matrix led to a decrease in transmission and threshold strength of the electric field, to an increase in the pulse relaxation time, and to distortion of the electrooptical response. The data obtained were explained by the difference between the refractive indices of nematics and polymer matrix and a decrease in the matrix rigidity.

INTRODUCTION

Polymeric liquid-crystalline (LC) heterophase materials represent a new wide class, microdispersions of LC in polymer matrices (PDLC). The optical properties of thin layers of these materials can be controlled by the applied electric field. Therefore, PDLC are promising for use in optical displays and indicators [1–3]. The electrooptical properties of PDLC are defined by the optical, electrical, and elastic properties of LC, polymer matrix, and the morphology of PDLC. The morphological parameters of PDLC, such as LC drop size distribution, composition, and volume fractions of the phases, adhesion between LC drops and polymer matrix, are determined by the chemical nature, composition of starting blend, and the reaction kinetics [1–5]. One of the ways for PDLC preparation is polymerization-induced phase separation (PIPS), i.e., polymerization of a monomer in LC solution accompanied by phase separation. Although considerable progress has been achieved recently in PDLC production and application [1–3], the studies on the regularities of PDLC formation with polyacrylic [6–8] and polyepoxy [9–11] matrices are still numerous.

In this work, we investigated formation of PDLC during polymerization and polycondensation of diglycidyl ether of *bis*-phenol A (DGEBA) under the action of tertiary and primary amines in the presence of nematic liquid crystals (LC) based on *n*-pentylcyanobiphenyl (LC-609 and LC-1277). Dimethylbenzylamine (DMBA) and 4,4'-diamine-dicyclohexylmethane (DACGM) were used as tertiary and primary amines, respectively.

The following aspects of the problem were studied in this work: (1) the effect of LC on the cure kinetics; (2) influence of the conversion degree on the nematic–isotropic phase transition; (3) influence of the LC concentration and matrix network density on the morphology and electrooptical properties of resultant PDLC films.

EXPERIMENTAL

Commercially available LC, DGEBA, and phenylglycidyl ether (PGE) were used without further purification. DMBA and DACGM were purified by vacuum distillation. The nematic temperature of LC-609 and LC-1277 ranges from +15 to + 30 and –20 to +62° C, respectively.

Kinetics of DGEBA polymerization and polycondensation was studied by isothermal calorimetry by using a DAK1-1A calorimeter. To calculate the reaction conversion, the heat effect was taken equal to 104.5 kJ/mol [12].

IR spectra of starting and model compounds (benzonitryl and ethanol) were taken (by using drops pressed between KBr and CaF_2 prisms) with a Specord M-80 spectrophotometer.

Solubility of starting reagents in LC was investigated visually and by optical interference [13]. Phase formation during DGEBA polymerization and polycondensation in the presence of LC was observed visually. The reactive blend was exposed (at a preset temperature) for some time until necessary conversion was achieved. Then the blend was cooled down to become turbid, and heated again to measure the temperature of the sample clarification (T_{cl}).

Morphology of PDLC was investigated by using a Boetius optical microscope, a Philips SEM-500 scanning electron microscope, and standard analytical procedures [14].

The glass transition temperature of polyepoxy matrices, LCs and PDLCs, were measured with an UIP-70M thermomechanical instrument.

PDLC films for microscopic and electrooptical investigation were prepared by the PIPS method at 60–70°C with post-curing at 140°C. For electrooptical investigations, PDLC films (about 30 μm thick) were placed between the glass plates coated with transparent ITO electrodes.

To investigate the electrooptical properties of PDLC, we used a special cell for measuring a change in the sample transmission caused by an applied voltage pulse. To reduce the effects of nematics conductivity [15], a high frequency (10 kHz) meander voltage was applied to the cell. The pulse duration was varied within the range 10–20 ms. A He-Ne laser was used as a light source. An FD-24K photodiode was used as a photodetector. The aperture angle of a photomultiplier was below 0.2°. The voltage amplitude, photodiode signal, and their duration were measured with an C8-13 storage oscilloscope. All measurement were carried out at room temperature (19.5 ± 1°C). The measurement accuracy was ±10%.

The refractive index of reactants, polyepoxide matrix, and PDLCs were measured with an IRF-454B Abbe refractometer. The nematics exhibits positive dielectric anisotropy ($\varepsilon_a = \varepsilon_{\|} - \varepsilon_{\perp} \approx 12$) and refractive indices of $n_o \approx 1.52$ and $n_e \approx 1.72$. The refractive index of a polyepoxy matrix n_p is within the range of 1.56–1.58.

RESULTS AND DISCUSSION

Cure Kinetics

Typical kinetic curves for DGEBA polymerization under the action of DMBA in the presence/absence of LC-609 are presented in Fig. 1. The rate of DGEBA polymerization is maximal from the very beginning of the reaction. In the presence of LC, the initial reaction rate decreases sharply; the higher is LC concentration, the lower is the rate (Fig. 2). In the absence of additives, the reaction stops at a 70% conversion because the system undergoes transition to the glassy state. In the presence of LC, conversion of the epoxy groups is complete, although the reaction rate is low. Added LC also reduces the rate of DGEBA polycondensation under the action of DACGM, although to a lesser extent (Fig. 2). It was shown that LC does not react with individual reagents (diepoxide and amines).

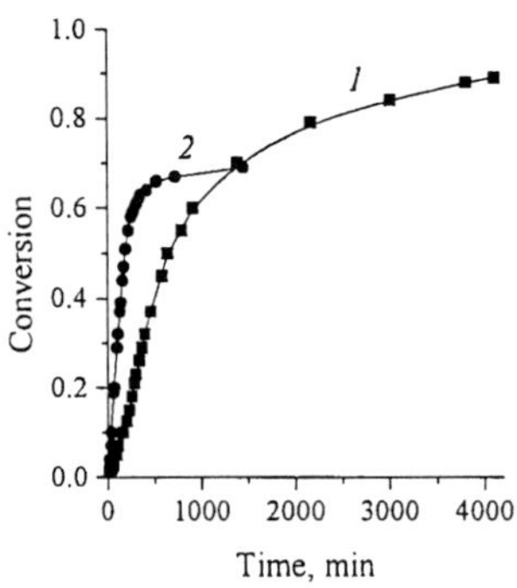

Figure 1. Kinetic curves of DGEBA polymerization under the action of DMBA (3 wt %) at 60°C in the presence of 45 wt % LC-609 (*1*) and without additive (*2*).

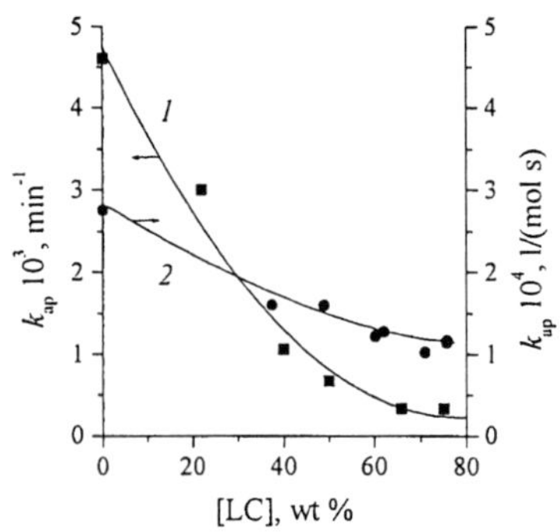

Figure 2. The effective rate constants *vs.* LC-609 concentration (*1*) for DGEBA polymerization under the action of 3 wt % DMBA and (*2*) for DGEBA polycondensation with DACGM at 60°C.

Deceleration of both the polymerization and polycondensation of epoxies may be explained by determining role of proton-donating compounds in the reaction kinetics. These compounds are present in the starting system (as hydroxyl groups in the epoxy monomer or as water impurities), or generated (as hydroxyl groups) in the cure process [16]. Both are reactions of nucleophilic addition, and they require activation of the α-carbon atoms in the epoxy ring through electrophilic assistance of the proton donors, which form activated donor-acceptor complexes with the oxygen atoms of the epoxy rings. The rate of these reactions is proportional to concentrations of donor-acceptor complexes. The presence of other electron donors in the system reduces concentration of proton donors because of the competitive complex formation, and therefore decreases the reaction rate. The nitrile groups of LC

$$\mathrm{ArCN} + \mathrm{ROH} \leftrightarrow \mathrm{ArCN...HOR}$$

are such electron donors, though rather weak. Similar explanation for the cure inhibition in diepoxide–diamine system upon addition of LC was proposed in [11]. However, no special proofs have been presented in that work.

The occurrence of this reaction was experimentally proved by the IR spectra of model compounds, benzonitrile, ethanol, and their mixture. As follows from Fig. 3, a broad strong absorption band of the OH groups in the range 2990–3590 cm^{-1}, typical of intermolecular hydrogen bonds (Fig. 3*a*), transforms into a narrower band at 3140–3590 cm^{-1} (Fig. 3*c*) for the benzonitrile–ethanol mixture. In this case, a considerable decrease in the relative intensity of the nitrile group at 2248 cm^{-1} (Fig. 3*b*) is also observed. These data unambiguously indicate a partial rupture of the intermolecular hydrogen bonds between the ethanol OH groups and formation of ethanol complexes with benzonitrile.

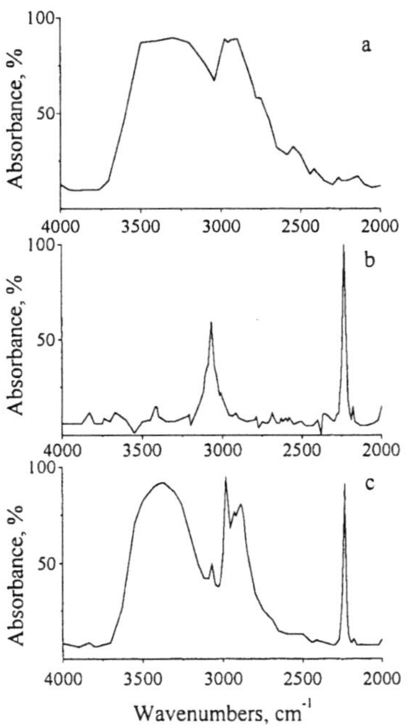

Figure 3. IR spectra of (a) ethanol, (b) benzonitrile, and (c) their mixture.

It should be taken into account that, due to high LC concentration in the reacting system, the concentration of the nitrile groups significantly exceeds that of proton donors, in particular, at the early stage of the reaction. Therefore, binding of the OH-groups (even by such a weak electron donor as the nitrile group conjugated with the aromatic ring) leads to a drastic deceleration of the epoxide cure. This mechanism of the reaction deceleration also explains the autocatalytic character of the reaction in the presence of LC (Fig. 1). Another argument in favor of this mechanism is that the activation energies of DGEBA polymerization with and without LC are virtually identical (11±1 and 12±1 kcal/mol). This suggests that LC does not affect the polymerization mechanism but only reduces the concentration of OH groups in the system and, hence, the concentration of the activated form of the epoxy group that controls the reaction rate [16]. As follows from Fig. 2, the effect of polycondensation inhibition is weaker compared to that of polymerization because, upon polycondensation, one OH group is formed in each act of interaction between the epoxy and amine groups while, upon polymerization, one OH group is formed after addition of 5–6 epoxy groups through the reaction of chain growth [16].

Phase Separation upon PDLC Formation

As follows from the data of interference measurements, the solubility of starting DGEBA–LC systems is virtually unlimited above the temperature of the nematic–isotropic phase transition (T_{n-i}) in LC-609 and LC-1277. For $T < T_{n-i}$, the solubility of LC in DGEBA is limited (Fig. 4). It is seen in Fig. 4, the blends DGEBA–LC-609 and DGEBA–LC-1277 are isotropic at ambient temperatures for the DGEBA concentrations above 6% and 20%, respectively.

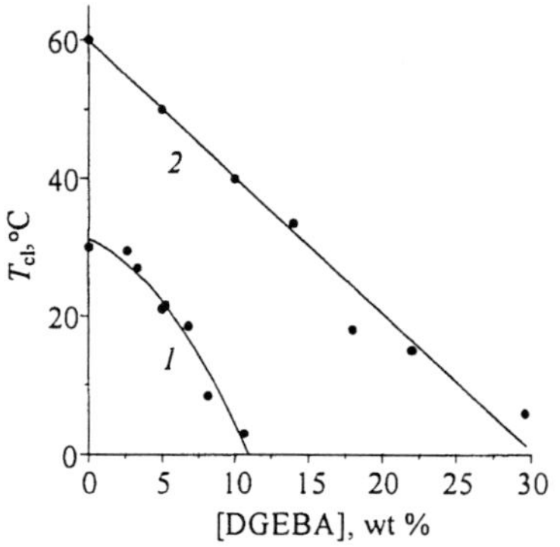

Figure 4. The clarification temperature of DGEBA blends with (1) LC-609 and (2) LC-1277 *vs*. DGEBA concentration.

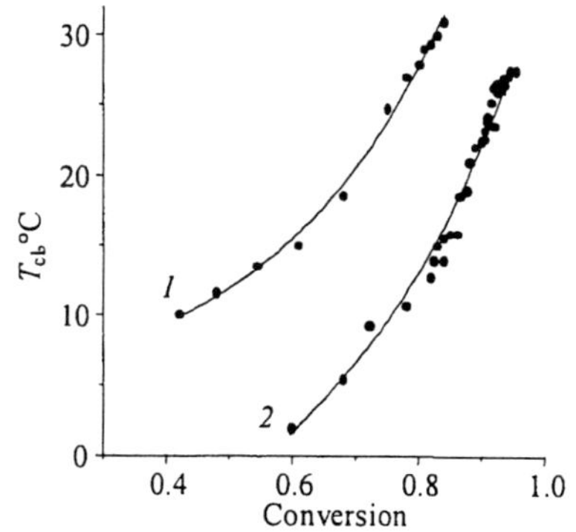

Figure 5. The clarification temperature of the reactive blends *vs*. conversion at (1) DGEBA polycondensation and (2) DGEBA polymerization with 45 wt % LC-609 at 70°C.

As follows from Fig. 5, the clarification temperature of the system, T_{cl}, increases during polymerization and approaches the phase transition temperature for individual LC-609 (T_{n-i} = 30°C). A sample that achieves the maximal conversion becomes anisotropic at room temperature. During polymerization, T_{cl} does not reach T_{n-i} even at the maximum reaction conversion (Fig. 5, curve *2*), whereas for polycondensation T_{cl} reaches 30°C at a conversion degree of 0.84 (Fig. 5, curve *1*). Phase separation at the maximal conversion of polycondensation and polymerization is observed only for the LC concentration above 40%.

In nematics-containing curing systems, phase separation takes place only when the cure temperature, T_c, is below T_{n-i} for pure LC. In this case, phase separation is similar to the isotropic–nematic phase transition in LC. When cure occurs for $T_c > T_{n-i}$, phase separation takes place only upon cooling to some temperature that depends on the reaction conversion. The higher is the conversion, the higher is the transformation temperature and the closer it is to T_{n-i} for individual nematics. Although the observed phase transition is thermal in character, it may be considered, with minor

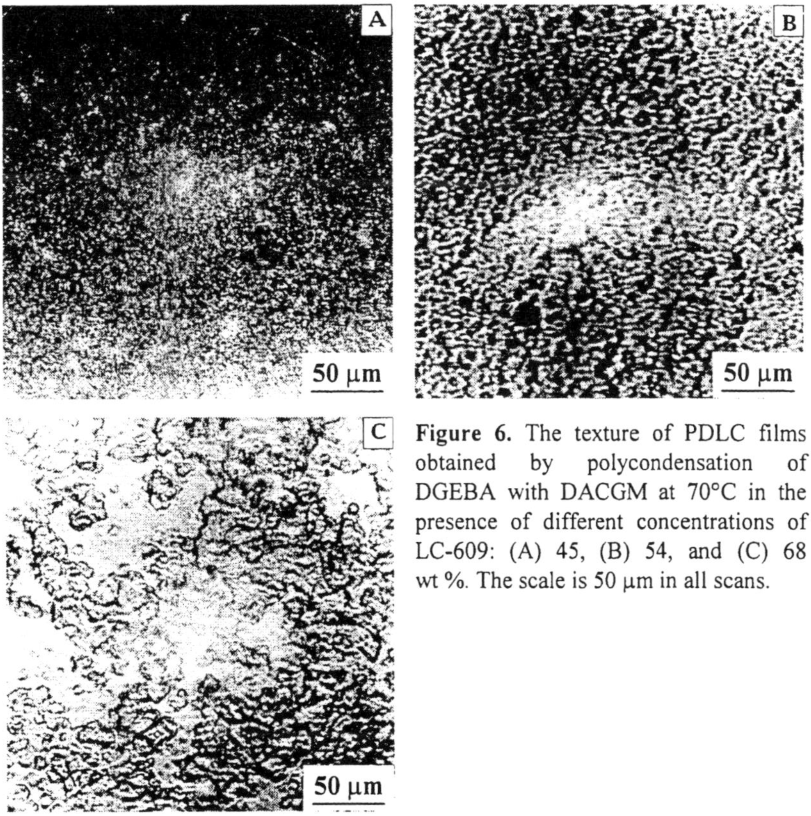

Figure 6. The texture of PDLC films obtained by polycondensation of DGEBA with DACGM at 70°C in the presence of different concentrations of LC-609: (A) 45, (B) 54, and (C) 68 wt %. The scale is 50 μm in all scans.

reservations, as a reaction-induced phase separation. It is the chemical reaction that changes a medium where nematics has been dissolved, and finally, defines the transition temperature. The role of chemical reaction is most evident when it occurs below T_{n-i} for pure nematics.

PDLC Morphology

The texture of PDLC films obtained by polycondensation of DGEBA with DACGM in the presence of LC-609 are shown in Fig. 6.

PDLC films containing 45 wt % LC-609 have a uniform texture with a nematic drop size about 1 μm (Fig. 6*a*). Separate drops (5–10 μm in size) can be seen against the background of fine-grained inhomogeneities 1–2 μm in size (in samples containing 54 wt % LC-609, Fig. 6*b*). PDLC containing 68 wt % LC-609 has a percolation structure with a drop cluster size up to 50 μm (Fig. 6*c*). The PDLC films based on LC-1277 have a similar texture. The resolving power of the optical microscope is not sufficient to characterize the texture of PDLC films obtained by polymerization of DGEBA with DMBA in the presence of LC. This suggests that The LC drop size is very small, less than 1 μm. Therefore, the PDLC morphology is sensitive to LC concentration and to the character of polymer matrix.

The influence of the polymer matrix on the PDLC morphology was studied by using samples with a constant LC concentration and different matrix network density. The matrix network density was varied by changing the epoxy to amine stoichiometric ratio, NH/E. No phase separation was observed with excess DGEBA (with respect to DACGM), while excess DACGM (with respect to DGEBA) resulted in phase separation. It should be noted that introduction of monoamine (aniline) into the starting diepoxide–diamine system did not result in phase separation, while the matrix network density changed. The electron micrographs of PDLC films (with 45 % LC-609 different epoxy–amine matrix network density) are shown in Fig. 7. The size of LC drops increases as the network density decreases. Nematic drops in PDLC have a non-spherical shape for equifunctional contents of the reactive groups, i.e., for the densest network (Fig. 7*a*). However, the drops become more spherical and their surface smoother (Fig. 7*b*, *c*) as the network density decreases. This indicates that, at some certain cooling rate, phase separation occurs in a less rigid matrix (because of decrease in the network density) and becomes more equilibrium.

The size distribution of nematic drops (plotted according to [14]) shows (Fig. 8) that the average size of LC drops ranges between 0.94 μm for PDLC with NH/E = 1 and 6.2 μm for PDLC with NH/E = 2.8. Furthermore, the size distribution of LC drops is unimodal for NH/E < 2 and becomes bimodal only for PDLC with NH/E = 2.8.

As it follows from Fig. 7, the LC phase volume fraction in PDLC is close to that in the initial blend. Two glass-transition temperatures T_g (–60 and –56°C) for PDLC samples are lower than those for pure nematics (–50°C) and polyepoxy matrix (140°C). That means that each phase (polyepoxy matrix and dispersed LC phase) contains a certain amount of dissolved components.

All the samples described above were prepared as films between two glass plates. Electron micrographs of PDLC films of the same composition (prepared on open glass plates when a partial evaporation of the components is possible) were different (Fig. 7*d*). LC drops have nearly spherical shape and are packed in a dense hexagonal lattice. The latter fact needs additional investigation.

Electrooptical properties of PDLC

Although PDLC obtained by polymerization scatter the light, their transmission does not change under the action of electric field up to 300 V.

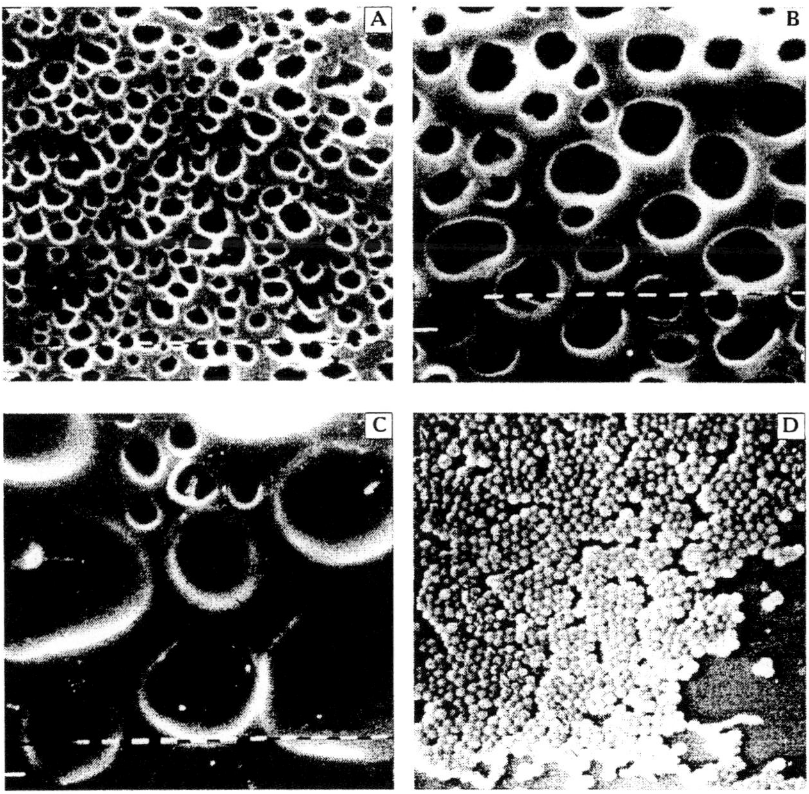

Figure 7. Electron micrographs of PDLC films based on DGEBA + DACGM + LC--609 (45 wt %) system: NH/E = (A) 1, (B) 2.1, (C) 2.8, and (D) 1 (open substrate).

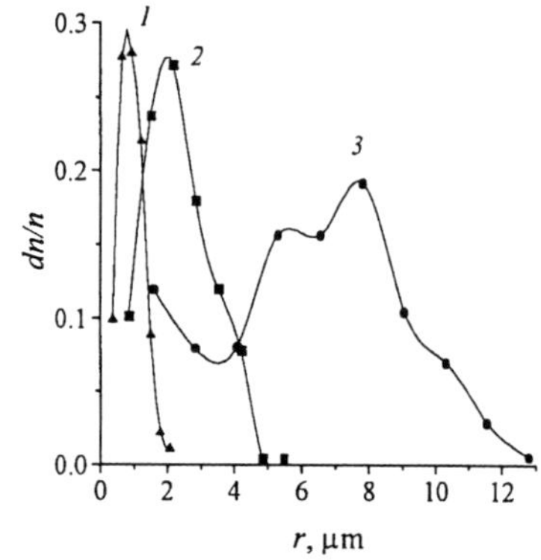

Figure 8. Size distribution of nematic drops in PDLC based on the DGEBA + DACGM + LC-609 (45 wt %) system: NH/E = (*1*) 1, (*2*) 2.1, and (*3*) 2.8.

It is due to high conductivity of the system and to small nematic drop size (< 1 μm).

The electrooptical response to a high-frequency voltage pulse of PDLC films obtained by polycondensation of diamine and diepoxide (NH/E = 1) depends on the LC-609 content. The time dependence of PDLC transmission (44 wt % finely dispersed nematics) has a classical overall view [1]. The transmittance grows from 0.1% in low field to 30% for the maximal one. The response is different for PDLC with a higher LC content and larger drop size. The transmittance of PDLC films (56 wt %) exhibits two minima: one appears during the action of the electric field while the second appears upon switching the field off. The transmittance of the samples changes slightly from 0.5 to 2–3%. Electrooptical response to a high-frequency voltage pulse for 68% PDLC films is similar to the previous one. However, relaxation time is much higher than in the previous case because of larger nematic areas in these PDLC films (Fig. 6*c*).

Although details are not completely clear, this can be explained as follows. Relatively good agreement between refractive indices of nematics, n, and matrix, n_p, for small drops of approximately equal size and shape (Fig. 6a) is due to the drop surface influence, which prevents, for given fields, all nematic molecules to orient along the field and the $n < n_p$ condition to be realized. Here n is an average refraction index of the nematics drop ($n_o < n < n_e$). For large drops (Figs. 6*b*, 6*c*) in the samples containing more than 50% of nematic, the effect of surface is smaller. A nematic structure with $n < n_p$ is realized at relatively low field strengths. Inconsistency of refraction indices of nematics and matrix ($n \rightarrow n_o$) increases with increasing field strength. This results in increased light scattering and, accordingly, in lower transmission. In addition, polydispersity of the nematic phase, in a particular drop size distribution must play an important role. Competitive influence of drop size and shape on the sample transmittance becomes effective in some certain field. A change of transmission with time can be due to this competition, taking into account different values of threshold fields and relaxation times for small and large drops.

One of the objectives of this research was to elucidate the influence of matrix network density on PDLC electrooptical properties. The methods for

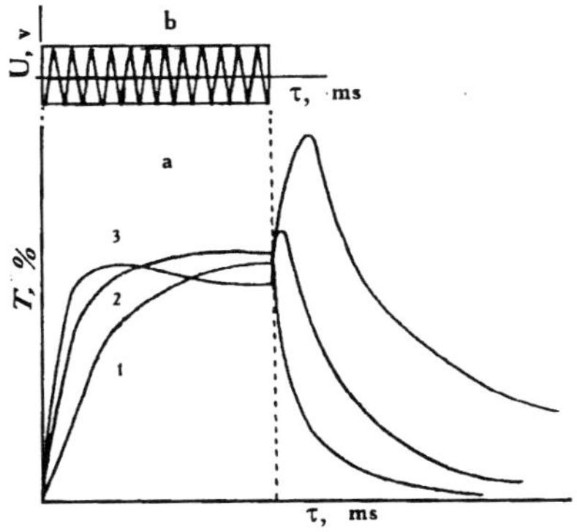

Figure 9. (a) Optical transmission *vs.* time for PDLC film based on DGEBA + DACGM + LC-609 (44 wt %) at (b) applied voltage: U = (1) 50, (2) 70, and (3) 100 V.

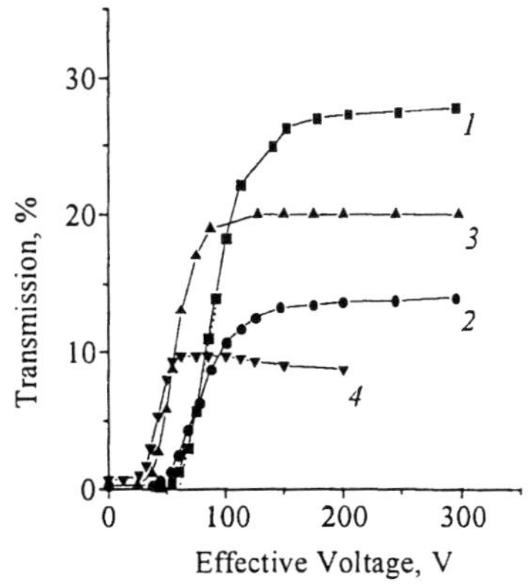

Figure 10. Optical transmission of PDLC films based on DGEBA + DACGM + LC-609 (44 wt %) *vs.* effective voltage: NH/E = (*1*) 1, (*2*) 1.66, (*3*) 2.1, and (*4*) 2.8.

varying the matrix network density and PDLC morphology were described above. A typical electrooptical response of PDLC samples is shown in Fig. 9. The response has a classical shape at small voltages (Fig. 9, curve *1*). But as voltage increases, the transmittance increases during the pulse time, reaches maximum and then decreases a little. An unusual increase in transmission is observed after the pulse is switched off (Fig. 9, curves *2* and *3*). As the pulse duration grows, the transmission maximum after the pulse increases. The maximum becomes most evident in an excess of diamine. For matrix with NH/E = 1, the maximum is very small and is observed only at high U (250–300 V).

Figure 10 shows a typical dependence of maximal transmittance on U for PDLC containing 45% nematic phase in matrices with different network density. For samples with NH/E in the range of 1 to 2 and similar nematic drops (Fig. 8, curves *1* and *2*), these diagrams have a classical overall view (Fig. 10, curves *1*–*3*). Only for samples with NH/E = 2.8 and a drop size several times larger than in the former case (Fig. 8, curve *3*), the plot exhibits a maximum (Fig. 10, curve *4*).

One of the most important electric characteristics of PDLC is the threshold field strength, E_{th}. Its value was calculated from equation $E_{th} = U_{th}/l$, where l is the sample thickness. U_{th} is determined as the intercept of the extrapolated linear portion of the $T(U)$ curve with the level of initial transmittance (Fig. 10). The values obtained are presented in Fig. 11.

It follows that the threshold field strength decreases with increasing amount of excess amine and, hence, with diminishing matrix network

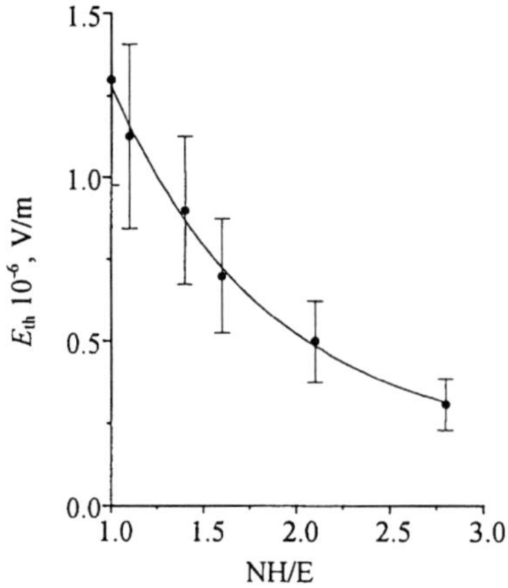

Figure 11. Threshold field strength E_{th} *vs.* NH/E ratio for PDLC films based on the blend DGEBA + DACGM + LC-609 (44 wt %).

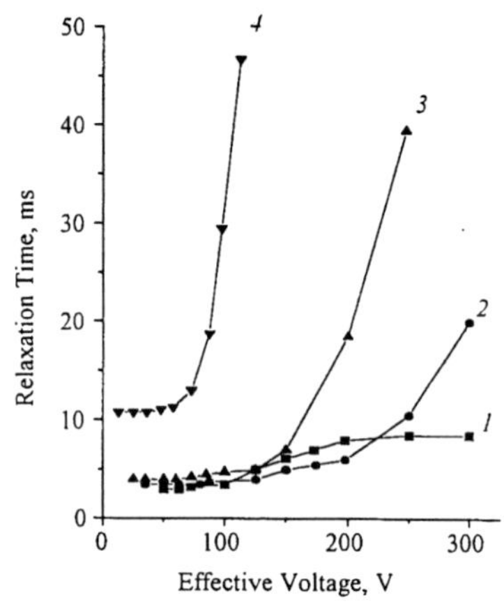

Figure 12. Electrooptical response relaxation time. *vs.* voltage for PDLC films based on DGEBA + DADCGM + LC-609 (44 wt %): NH/E = (*1*) 1, (*2*) 1.66, (*3*) 2.1, and *4*) 2.8.

density. However, a marked scatter in the measured values is due to instrumental errors and different conditions of film preparation. Influence of drop size distribution on E_{th} may also be significant.

Figure 12 shows the dependence of relaxation time, τ_{off}, on U. Relaxation time was determined at the level of $0.1(T_m - T_0)$ of the response, where T_m is the maximum sample transmittance during the pulse, and T_0 is the initial transmittance (see Fig. 9). As follows from Fig. 12 (curves *1–3*), the initial portions of the curves (for the samples with NH/E $\approx$ 1–2) coincide. This means that the relaxation time at small U is virtually independent of the matrix type, provided that nematic drop diameters and shapes and nematic structures are similar in all samples. However, the relaxation time grows drastically with increasing voltage (Fig. 12, curve *4*). For PDLC with NH/E = 2.8, the relaxation time (at U = 150 V) grows by a factor of ten.

The character of the $\tau_{off}(U)$ curves suggests that, at some field strength and matrix network density, a slow-relaxation mechanism becomes operative. The lower is the network density and, consequently, matrix rigidity, the more significant is the orientation of the surface nematic molecules under the action of electric field. At field strength, the boundary conditions on the drop surface may change because of changing the structure of nematic –matrix transition layer. When the field is switched off, the nematic relaxation time must increase with decreasing matrix rigidity.

This model explains the increase in transmittance after the pulse was switched off (Fig. 9), especially with regard to the fact that the refractive index of matrix is higher than that of the nematics oriented along the field. When orientation of the surface layers changes after switching off the field, nematic phase adapts to new boundary conditions, so that the drop structure also changes, and a new refractive index is better matched to that of the matrix. Undoubtedly, further thorough experimental and theoretical elaboration is necessary.

The electrooptical properties of PDLC with LC-1277 are in general similar to those for PDLC with LC-609. But the relaxation time of the former is much higher mainly because of a larger size nematic drops in these systems. Furthermore, when the matrix network density of PDLC was varied upon addition of phenylglycidyl ether, similar electrooptical behavior was observed. To prepare these composites, we used equivalent concentrations of the amine and epoxy groups. In this case, the above effects are more pronounced.

CONCLUSIONS

The influence of cyan-biphenyl based LC on the kinetics of epoxy cure reactions by polymerization and polycondensation mechanisms was investigated. LC additives were shown to decelerate these processes because of interaction between the CN groups in LC and OH groups in the curing system, which decreases the OH concentration that is responsible for the cure rate.

The influence of conversion degree on the nematic–isotropic phase transition during the epoxy cure reaction was studied. In this case, phase separation proceeds by the mechanism of LC isotropic–nematic phase transition. The higher is the conversion degree, the higher is the transition temperature and the closer it is to T_{n-i} for individual LC.

PDLC morphology was shown to be sensitive to LC concentration and to matrix network density. As matrix network density decreases, the phase separation becomes more equilibrium process and LC drops grow with their shapes becoming more spherical and their boundary getting smoother.

The electrooptical properties of PDLC films were studied. The PDLC films obtained by polymerization do not change their transmittance under the action of electric field up to 300 V. Electrooptical response to a high-frequency voltage pulse for PDLC films obtained by polycondensation of equimolar quantities of diamine and diepoxide (NH/E = 1) depend on the LC-609 concentration. This is due to a competitive action of electric field and orientation of nematics. The influence of matrix network density on electrooptical properties of PDLC was elucidated. Unusual increase of transmittance after the pulse termination was observed for PDLC with a low matrix network density. With decreasing matrix network density, the threshold strength decreases while the relaxation time grows.

Acknowledgements. This work was supported by the International Science and Technology Center (grant no. 358-96) and the Russian Foundation for Basic Research (project no. 96-03-32027).

REFERENCES

1. Zharkova G.M. and Sonin A.S. *Zhidkokristallicheskie kompozity* (Liquid Crystalline Composites), Nauka, Novosibirsk, 1994.
2. Montgomery G.P., Smith G.W., and Vaz N.A., *Liquid Crystalline and Mesomorphic Polymers*, Shybaev V.P. and Lam L, Eds., Springer, New York, 1994, Ch. 5.
3. Drzaic P.S. *Liquid Crystalline Dispersions*, World Scientific, Singapore, 1995.
4. Rozenberg B.A. and Sigalov G.M., *Polym. Adv. Technol.*, **7**, 356 (1996).
5. Williams R.J.J.,.Rozenberg B.A, and Pascault J.-P., *Adv. Polym. Sci.*, **128**, 96 (1997).
6. Serbutoviez C., Kloosterboer J.G., Boots H.M.J., and Touwslager F.J., *Macromolecules*, **29**, 7690 (1996).
7. Serbutoviez C., Kloosterboer J.G., Boots H.M.J., Paulissen F.A.M.A., and Touwslager F.J., *Liq. Cryst.*, **22**, 145 (1997).
8. Boots N.M.J., Kloosterboer J.G., Serbutoeviez C., and Touwslager F.J., *Macromolecules*, **29**, 7683 (1996).
9. Vaz N.A., Smith G.W., and Montgomery G.P., *Mol. Cryst. Liq. Cryst.*, **146**, 17 (1987).
10. Smith G.W. and Vaz N.A., *Liq. Cryst.*, **3**, 543 (1988).
11. Siddiqi H.M., Dumon M., Elondou J.P., and Pascault J.-P., *Polymer*, **37**, 4795 (1996).
12. Dzhavadyan E.A., Irzhak V.I., Rozenberg B.A. *Polymer Sci. A.*, **41**, 413 (1999).
13. Malkin A.Ya. and Chalykh A.E. *Diffuziya i vyazkost': metody issledovaniya* (Diffusion and Viscosity: Methods of Investigation), Khimiya, Moscow, 1979.
14. Chalykh A.E., Aliev A.D., and Rubtsov A.E., *Elektronno-zondovyi mikroanaliz v issledovanii polimerov* (Electron-Probe Microanalysis for Investigation of Polymers), Nauka, Moscow, 1990.
15. Generalova E.V., Nosov G.B., and Sonin A.S., *Mol. Mater.*, **3**, 53 (1993).
16. Rozenberg B.A., *Adv. Polym. Sci.*, **75**, 115 (1986).

Chapter 15

Crosslinking Studies in Rigid and Semi-Rigid Polymers

Shawn JENKINS, Karl I. JACOB, and Satish KUMAR*

School of Textile and Fiber Engineering, Georgia Institute of Technology, 801 First Dr., Atlanta GA, 30332-0295 USA

ABSTRACT

Attempts to improve the compressive properties of high-performance polymeric fibers are summarized. Special attention is given to chemical modifications designed for intermolecular crosslinking. Issues pertaining to the evidence of crosslinking and the effects of crosslinking on the fiber structure and morphology are discussed. While uncrosslinked, high-performance polymeric fibers (e.g., PBZT) exhibit a fibrillar morphology, a morphological transition from fibrillar to non-fibrillar is

*e-mail: satish.kumar@textiles.gatech.edu

observed to take place in methyl pendant PBZT fibers on crosslinking. A small degree of crosslinking can result in a significant improvement in compressive strength. However, internal stresses may build-up in the fiber on crosslinking, resulting in low tensile strength. Based on the work to date, it is concluded that intermolecular crosslinking, which does not result in the build-up of stresses within the fiber, will improve the compressive strength of high performance polymeric fibers without a concomitant decrease in tensile strength. An example of a system, in which this may occur, is presented.

INTRODUCTION

In the late 1970s, several rigid-rod polymeric systems, such as poly(*p*-phenylene benzobisoxazole) (PBO) and poly(*p*-phenylene benzobisthiazole) (PBZT) (Fig. 1), were developed out of the United States Air Force Ordered Polymer Research Program. Interest in these systems stemmed from their potential for forming lightweight, heat-resistant, high strength, high modulus fibers by way of ordering the liquid crystalline domains present in solutions of these materials. Prior development and commercialization of the semi-rigid, lyotropic, aromatic polyamide, poly(*p*-phenylene terephthalamide) (PPTA), otherwise known as DuPont's Kevlar™, confirmed that the aforementioned material properties could be attained in a polymeric fiber. Since their development, several liquid crystalline systems, either thermotropic or lyotropic in nature, have been

a)

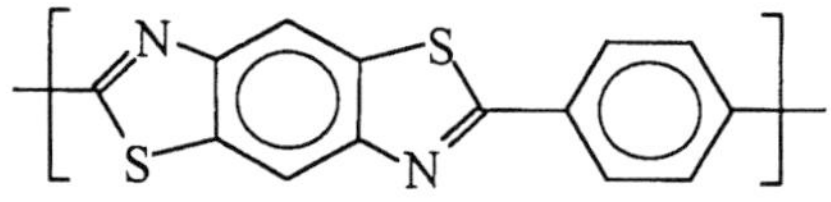

b)

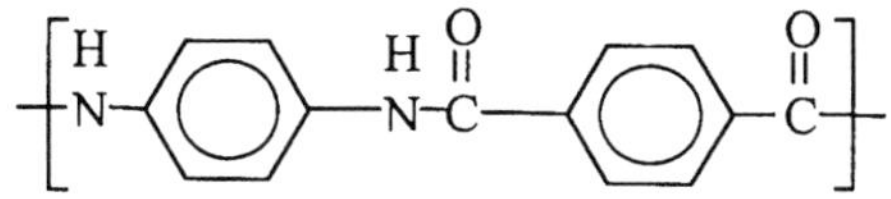

c)

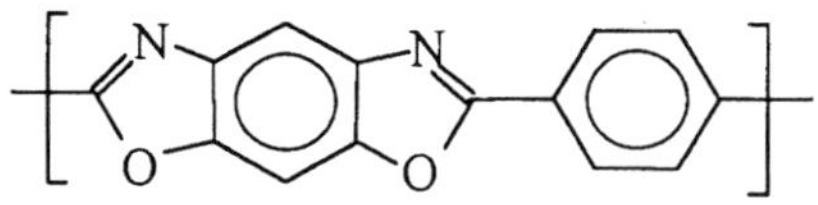

d)

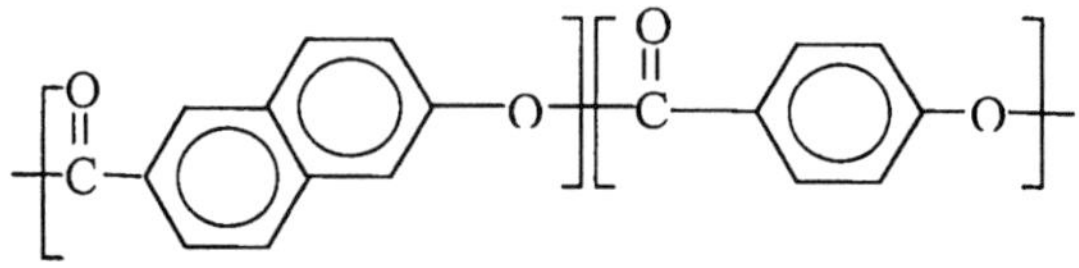

Figure 1. Chemical structures of some rigid and semi-rigid polymers; a) poly{p-phenylene benzobisthiazole} (PBZT), b) poly{p-phenylene terephthalamide} (PPTA), c) poly{p-phenylene benzobisoxazole} (PBO), d) poly(hydroxybenzoic acid –co- hydroxynaphthoic acid) (HBA-co-HNA).

Table 1. Properties of high-performance fibers

	Density, (g/cm^3)	Tensile Strength, (GPa)	Tensile Modulus, (GPa)	Compressive Strength, (GPa)
Kevlar™ 149	1.45	3.5	185	.360
PBZT	1.58	2.6-3.9	200-300	.2-.3
PBO [1]	1.56	5.8	280	.2-.35
Vectran™ [2]	1.41	2.8	65	.18
PIPD [3]	1.70	4.6	325	1.6
P-100 [4]	2.15	2.2	725	.5
T-300 [5]	1.79	3.2	230	2.7
Boron	2.5	3.5	415	5.9
Nicalon™ (SiC)	2.8	2.8	200	3.1

[1] Toyobo technical literature on Zylon™.
[2] Hoechst Celanese technical literature on Vectran™ and Ref. [4].
[3] Akzo Nobel data sheet on PIPD fiber. All other values taken from Ref. [1].
[4] Pitch-based carbon fiber.
[5] PAN-based carbon fiber.

synthesized and investigated. One semi-rigid, thermotropic copolyester of commercial importance is Hoechst Celanese's Vectran™ fiber, which is a copolyester of p-hydroxybenzoic acid and 6-hydroxy-2-naphthoic acid (HBA-co-HNA). Furthermore, in 1998, Toyobo commercialized PBO fiber under the trade name Zylon™.

While rigid-rod polymeric fibers have excellent axial tensile properties, their axial compressive, torsional, and transverse properties are poor. Typically, the axial compressive strengths of these polymers are an order of magnitude lower than their tensile strengths [1]. By comparison, the compressive strength of inorganic fibers can be greater than their tensile strength, with the compressive strength of carbon fibers being intermediate to that of the polymeric and the inorganic fibers (Table 1) [1]. Compressive failure in these fibers is a result of plastic deformation via kink band formation [1, 2, 3]. Fig. 2 shows kink bands formed in some high-performance polymeric fibers as a result of an axial compressive stress.

Several theories have been developed to explain compression failure in high performance polymeric fibers. A model based on the elastic instability of a collection of perfectly oriented and extended chains, interacting laterally via a "matrix" of secondary bonds, has been developed [5, 6]. From this work, a relationship between compressive strength and the transverse or shear modulus, depending on the buckling mode, is suggested. The morphology of semi-rigid and rigid-rod polymeric fibers is known to be fibrillar, with fibril diameters ranging from 0.01μm to several microns [7]. The theory applied to oriented molecules [5] has also been applied to oriented fibrils [8, 9], with the conclusion that the compressive failure is

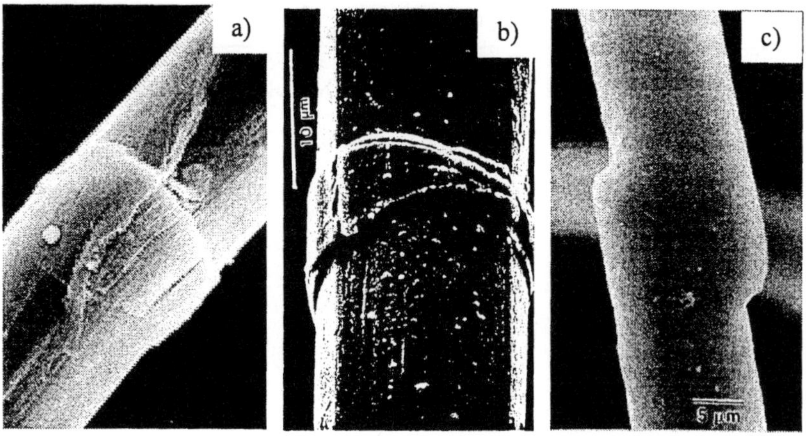

Figure 2. Scanning electron micrographs of kink bands formed in a) Vectran™, b) PBZT and c) Kevlar™ 49 fibers under recoil compression [4].

dictated by fibrillar instability rather than molecular instability. Other theories have been developed that consider the misorientation between fibrils, crystallites, or domains of well-aligned molecules and the direction of the applied load [10–12]. Northolt et al. [12] has incorporated a distribution of domain orientations. In this theory, failure in tension as well as in compression is brought about by irreversible rotation of domains, which is thought to involve the disruption of the relatively weak interchain secondary bonds. A common thread between these theories is that the compressive strength of well-oriented, highly aligned materials is influenced by intermolecular and/or interfibrillar interactions. An abundance of research has been performed in an attempt to improve lateral interactions in rigid and semi-rigid polymers. This paper summarizes these efforts, primarily focusing on the effect of crosslinking on structure, morphology, and properties.

DISCUSSION

Modification of Morphology

The compressive strength of PBO fiber has been reported to improve upon the introduction of interfibrillar entanglements by twisting and drawing the polymer dope during fiber spinning [13]. Attempts have been made to increase the fibril transverse dimension by varying coagulation conditions. The compressive strength of PBO fiber slowly coagulated in acetone was reported to be 300 MPa, compared to 100 MPa for fiber coagulated in water

[14]. Tensile and compressive strengths of PPTA fibers have been reported to decrease [15] if coagulation takes place in ethanol or aqueous solutions of potassium iodide as opposed to water. The compressive strength of heat-treated PBO fibers coagulated in water, steam, methanol or ammonium hydroxide, was reported to be in the 200 to 300 MPa range [16].

Polymer Infiltration

It has been suggested [8, 9] that a matrix material infiltrated within the fiber may provide lateral support to the fibrils. Several matrix materials, crosslinkable or otherwise, have been investigated with this intention. No significant improvement in compressive strength was observed [17] upon exposure of reactive resin-infiltrated PBO fibers to irradiation. A slight increase (~13%) in compressive strength, as well as an increase in tensile strength and elongation to break, was reported [18] when PPTA fibers were infiltrated with thermosetting epoxies, novolac, or bismaleimides. Likewise, the compressive strength of PBO and PBZT fibers infiltrated with silica glass [19] and nylon [20], respectively, have been reported to show only moderate improvements on infiltration.

Chemical Modification

Several chemical modifications to rigid and semi-rigid polymers have been made in hopes of improving compressive strength. Some of these include introducing bulky side-groups [21] and incorporating a small fraction of a multifunctional monomer, from which a "multidimensional" polymer is produced [4, 22, 23]. However, introducing crosslinkable moieties into these polymers has been considered most promising and, as a result, most research efforts have focused on this approach. Figure 3 gives the chemical structures of some modified semi-rigid and rigid-rod polymers, incorporating a variety of crosslinkable groups. Most of these polymers are expected to undergo crosslinking when heated or irradiated with electromagnetic or ionizing radiation. Lack of solubility has generally been used as conclusive evidence of crosslinking in rigid-rod polymers. However, work done on methyl-pendant PBZT (MePBZT) fibers has shown insolubility to be inconclusive in demonstrating the presence of crosslinking [24]. As-spun MePBZT fibers are soluble in methane sulfonic acid. Fibers heat-treated at 450°C were insoluble in methane sulfonic acid; however, no evidence of crosslinking was observed by ^{13}C solid state NMR. Fibers heat-treated between 475 to 550°C do show crosslinking evidence from solid state NMR. Thus, lack of solubility alone cannot be used as the evidence for crosslinking. Much of the pertinent literature reports compressive strength studies on fibers presumably crosslinked under heat-treatment or irradiation. Very limited compressive strength data exists on fibers on which direct evidence of crosslinking is shown.

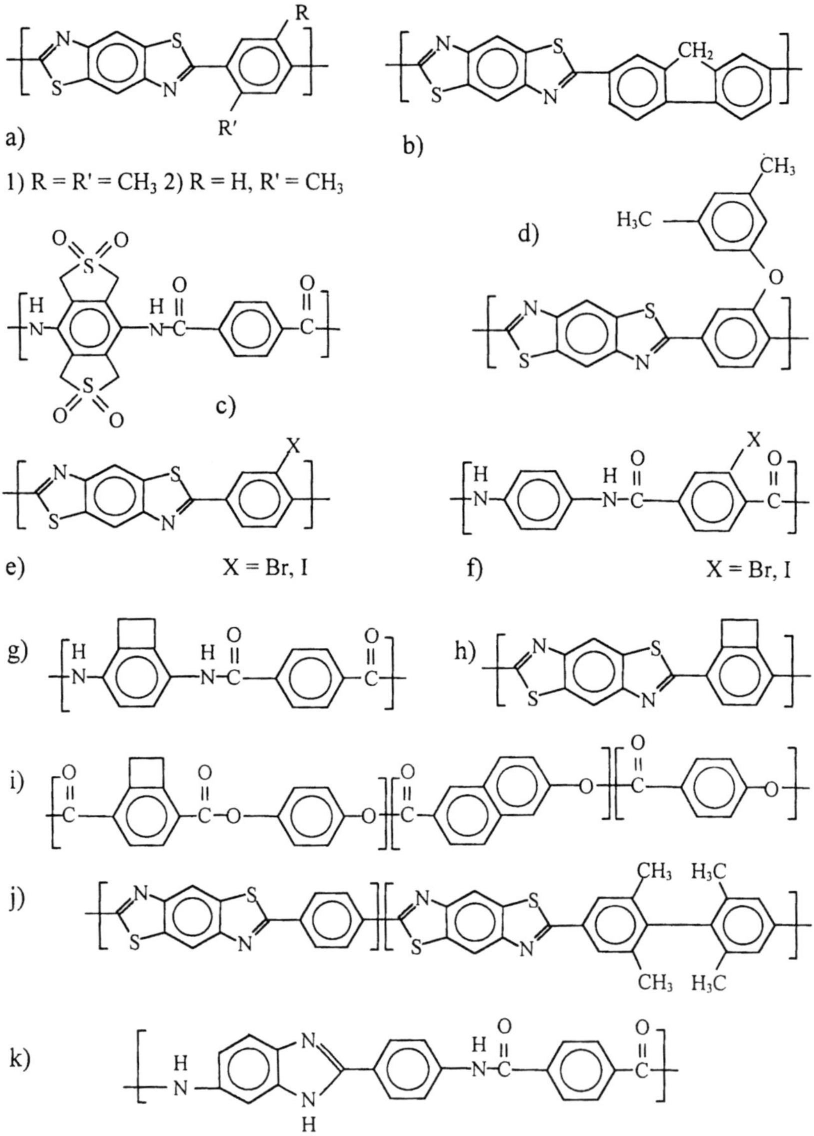

Figure 3. Modified rigid and semi-rigid polymers that incorporate crosslinkable or strongly associating chemical groups: a) methyl/dimethyl PBZT, b) PBZT modified with the fluorene moiety, c) Disulfone-diamine PPTA, d) PBZT with a dimethyl-phenoxy pendant group, e) & f) halogenated PBZT and PPTA, respectively, g)–i) PPTA, PBZT and HBA-co-HNA modified with the benzocyclobutene moiety, j) PBZT containing a tetramethyl-biphenyl moiety, k) poly(amidebenz-imidizole).

l) 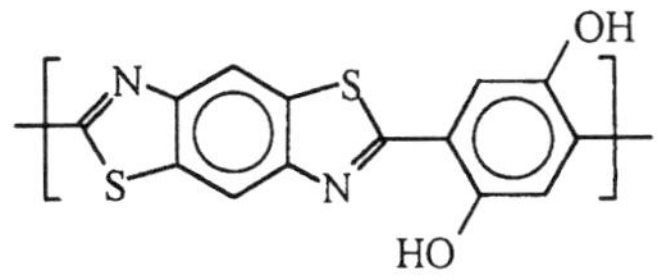 m) 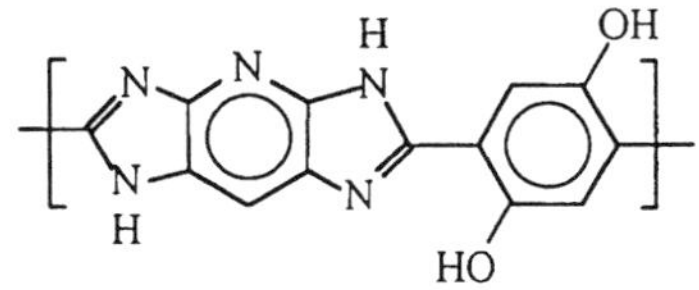

Figure 3 (cont-d). l) di-hydroxy PBZT, m) PIPD. Properties of these polymers are summarized in Table 2.

Initially, labile methyl pendant groups were incorporated on the phenyl ring within the PBZT structure [25]. While the evolution of methane from MePBZT was observed on heating above 450°C [26] and is a by-product of the crosslinking reaction in MePBZT, crosslinking was assumed to take place on account of the insolubility of the heat-treated fiber. Compressive strength of tension heat-treated (at 525°C) MePBZT fiber was found to be 440 MPa, compared to 220 MPa for comparably heat-treated PBZT. The compressive strength of free annealed MePBZT fibers, which exhibited direct evidence of crosslinking and showed a non-fibrillar structure (Fig. 4a), could not be measured due to transverse surface cracks (Fig. 4c). It should be noted that the fibrillar morphology remained in tension heat-treated MePBZT fiber (Fig. 4b). It was reported that the properties of dimethyl PBZT fiber could not be measured due to surface blistering that occurred on heat treatment [23]. A fluorene moiety has been incorporated into PBZT [27], as the reaction of this moiety would not involve the evolution of byproducts. The compressive strength of fluorene-modified PBZT was found to increase marginally over that

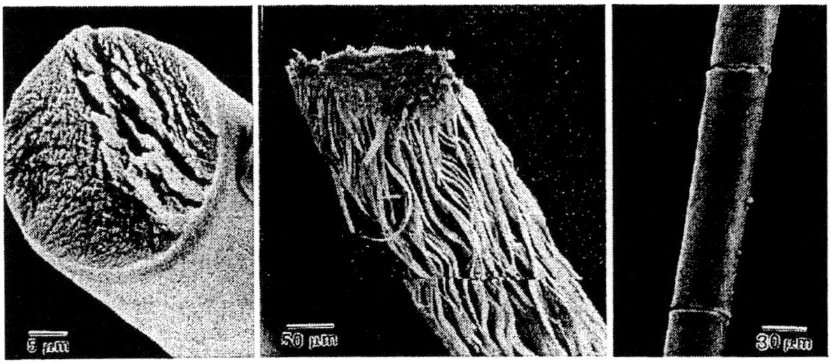

Figure 4. Scanning electron micrographs of heat-treated MePBZT fibers. a) Fiber treated in methane sulfonic acid overnight after free annealing at 525°C. b) Fiber treated in methane sulfonic acid overnight after tension heat-treatment at 530°C, c) Fiber free annealed at 475°C showing surface cracks. Fibers a) and b) were washed in water and dried following treatment in methanesulfonic acid [28].

Table 2. Properties of various polymers shown in Figure 3

Structures (see Fig. 3)	Reference	Crosslinking Method	Evidence of Crosslinking	Compressive Strength (GPa)	Comments
A	17, 24, 25, 29	Thermal Proton radiation Electron radiation	Solubility/ Solid state NMR	0.44-0.52	No systematic correlation between methyl-pendant content and fiber compressive strength was apparent. The compressive strength of free annealed and crosslinked MePBZT homopolymer (exhibiting non-fibrillar morphology) could not be measured due to surface cracks.
B	27	Thermal	Solubility	0.31-.056	
C	33	Thermal	Solid state NMR	0.11-0.26	Properties were found to be time dependent, with the compressive and tensile strengths decreasing more than 14% over one year in a copolymer containing 5 mol% of the disulfone-diamine comonomer. The time dependent behavior was attributed to the formation of radical species on heat-treatment.
D	30, 31	Thermal	Solubility	0.3-0.4	
E	32	Thermal	Solubility/ Halogen Loss	0.24-0.69*	Weak correlation between halogen loss and compressive strength.
F	32	Thermal	Solubility/ Halogen Loss	0.38**	Properties inferior to PPTA control. Decrease in properties on halogen loss attributed to chain degradation.

* Converted from gpd to GPa assuming a density of 1.58 g/cm^3.
** Converted from gpd to GPa assuming a density of 1.45 g/cm^3.

Table 2 (cont-d). Properties of various polymers shown in Figure 3

Structures (see Fig. 3)	Reference	Crosslinking Method	Evidence of Crosslinking	Compressive Strength (GPa)	Comments
G	34	Thermal	Solubility	0.29	Morphology was reported to change from fibrillar to non-fibrillar on heat-treatment. The morphological change was attributed to chain degradation.
H	35	Thermal	Solubility	0.10-0.48	Compressive strength greatly influenced by heat-treatment temperature. Compressive strength of fiber heat-treated at 500°C increased by a factor of 4 as compared to the fiber heat-treated at 450°C.
I	36	Thermal	Solubility	Not reported	Viscosity increase observed on heat-treatment.
J	37, 38	Thermal	Solid state NMR	0.76	Compressive strength measured on crosslinked 25/75 mol % biphenyl-PBZT/PBZT copolymer. Degree of crosslinking 2.5%.
K	10	x-ray	Solubility	0.56	
L	39	N/A	N/A	0.14-0.41	Intra-molecular hydrogen bonding reported.
M	40	N/A	N/A	1.6	Bi-directional, intermolecular hydrogen bonding reported. Highest reported compressive strength for a polymeric fiber.

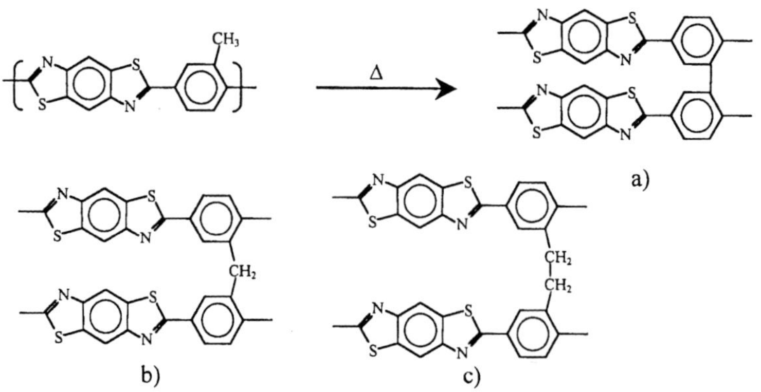

Figure 5. Possible crosslinks formed during heat-treatment of MePBZT: a) Crosslinking via coupling of adjacent phenyl rings, b) Crosslinking via the formation of a methylene bridge, c) Crosslinking via the formation of an ethylene bridge [41].

of unmodified PBZT at a heat-treatment temperature of 600°C. Less of an improvement was noted at lower and higher heat-treatment temperatures.

The observation of a non-fibrillar morphology [29] in MePBZT fiber, suggests the presence of inter- as well as intra-fibrillar crosslinking; however, chain degradation may also be a factor in this change in morphology [34]. As mentioned above, crosslinked, free annealed MePBZT fibers were observed to develop transverse (i.e., perpendicular to the fiber axis) surface cracks (Fig. 4c). Comparatively, tension heat-treated fibers were found to develop a skin/core morphology. Moreover, fibers heat-treated under tension did not develop surface cracks, but the skin was found to dissolve in methane sulfonic acid indicating the absence of crosslinking in the skin. It is believed that the fibrillar core may be crosslinked to some extent, however due to sample limitations, evidence of crosslinking has yet to be obtained in tension annealed samples showing fibrillar morphology.

Possible crosslinks formed in heat-treated MePBZT are shown in Fig. 5. The effects of crosslinking in MePBZT have been modeled using atomistic simulation techniques [41]. Results of this work indicate the development of a significant axial stress upon direct coupling of adjacent phenyl groups (Fig. 5a). Moreover, the same trends were not observed on the introduction of methylene or ethylene type crosslinks (Figures 5b and c). The predicted stress build-up as a function of phenyl-to-phenyl coupling is shown in Fig. 6. In this figure, Cell 1 and Cell 2 represent different spatial arrangements of methyl-pendants on neighboring chains, between which crosslinking may occur. Both unit cells may be present in MePBZT due to the stochastic nature of polymerization. The stress predicted to occur on

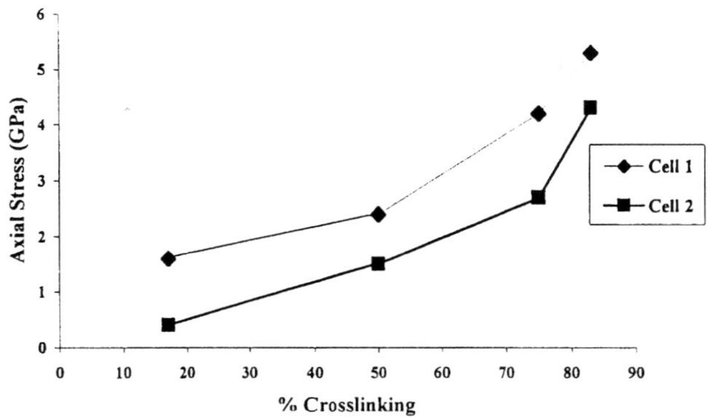

Figure 6. A plot of predicted stress build-up as a function of crosslink density for direct coupling of adjacent phenyl groups (shown in Figure 5a) [41].

crosslinking serves to contract the material along the axial direction and is a possible explanation for the appearance (or growth) of the transverse surface cracks in the skin of the free annealed MePBZT fibers.

A PBZT, with a 2,6-dimethylphenoxy pendant group incorporated, has been synthesized [30] and subsequently spun and tested [31]. The compressive strength of this polymer remained unchanged on heat-treatment. The synthesis and properties of a halogenated PPTA and a halogenated PBZT system [32] have been reported where, on heat-treatment, halogen gas was evolved. The degree of crosslinking was presumed to be a function of halogen loss. The tensile as well as compressive properties of the halogenated PPTA system were found to decrease on heat-treatment. This was attributed to the formation of halogen radicals on heat-treatment resulting in chain cleavage. Both the shear modulus and compressive strength of the halogenated PBZT system were found to increase as a function of crosslink density (i.e., halogen loss). A wide scatter was reported in tensile strength subsequent to heat-treatment, with the properties of some samples being inferior to the control and others being comparable.

A thermally reactive PPTA copolymer has been synthesized [33] and spun [42]; the comonomer being based on a disulfone-diamine. The properties of this copolymer were found to be time dependant and inferior to PPTA. Specifically, the compressive strength of a copolymer containing 5 mol % of the disulfone-diamine comonomer decreased 14% over the period of one year. In addition, the tensile strength and modulus decreased 14 and 36 %, respectively. The time dependant behavior was attributed to the generation of free radicals on reaction, which resulted in chain scission over time. A thermally reactive PPTA copolymer incorporating a comonomer based on a benzocyclobutene derivative of terephthalic acid (i.e., PPTA-co-PPTAXA) has also been investigated [34]. On heat treatment the compressive strength was reported to increase 14%. In a later work on the homopolymer of PPTAXA, it was reported that [43] by varying the heat-treatment temperature and therefore varying the crosslink density, fewer kinks per unit length were initiated after compressive deformation. This was attributed to an increase in the energy of kink formation consequent to crosslinking. This led to the conclusion that intermolecular interactions were of importance to compressive behavior.

The benzocyclobutene derivative of terephthalic acid has also been incorporated into PBO [44], PBZT [45], and HBA-co-HBN [36]. However, mechanical properties are only reported for the modified PBZT [35]. For this system, both a homopolymer and copolymers of PBZT containing the benzocyclobutene derivative were prepared. While the compressive strength of the copolymer was poorer than that for the control PBZT, the homopolymer was reported to have modestly improved compressive strength. A wide variation in compressive strength was observed with heat-treatment temperature.

The effects of electron and proton radiation on dimethyl PBZT fiber has been investigated [45]. Electron irradiation was found to improve compressive strength roughly 40% over that of the unirradiated sample, while proton radiation left the compressive strength unchanged. In the same work, PBO fiber was subjected to proton, electron, and gamma radiation. In the case of PBO fiber, proton radiation was found to increase the compressive strength over 140% while gamma and electron irradiation showed only an 84 and 38% improvement, respectively. Swelling studies were used as a means to detect the presence of crosslinking in these

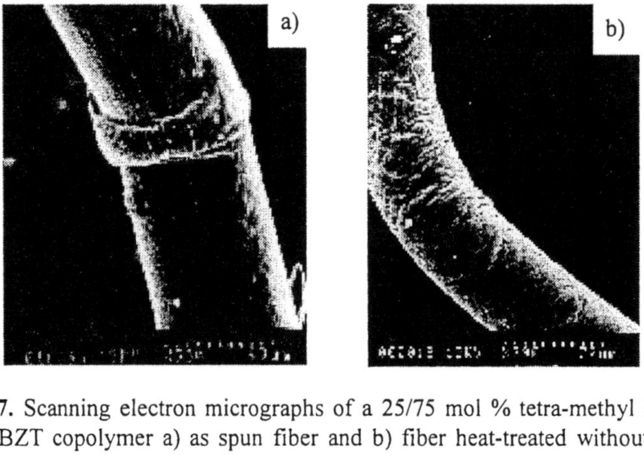

Figure 7. Scanning electron micrographs of a 25/75 mol % tetra-methyl biphenyl PBZT/PBZT copolymer a) as spun fiber and b) fiber heat-treated without tension [46].

materials. Electron radiation dosages as high as 30 Mrad were used [47]. Kozey and Berlin [10] have reported a 17% increase in the compressive strength of a poly(amide benzimidizole) when exposed to 50 Mrad x-ray irradiation dosage.

A system with the potential of forming a three-dimensional, covalently bonded network has recently been reported [37]. This system incorporates a tetramethyl-biphenyl moiety along the main chain of PBZT. The tetramethyl-biphenyl moiety is in a minimum energy conformation when the biphenyls are approximately 90° with respect to one another. This is due to the steric hindrance between methyl pairs on adjacent rings in the biphenyl moiety. The biphenyl group would, thus, be expected to form crosslinks in two dimensions. The compressive strength of the thermally crosslinked 25/75 tetramethyl-biphenyl PBZT/PBZT copolymer is about three times the compressive strength of PBZT [38]. The effect of crosslinking on the compression behavior of tetramethyl-biphenyl PBZT/PBZT copolymer is shown in Fig. 7. Figure 7a shows the development of kink in the as spun fiber; however, at roughly the same bending curvature, a kink was not observed in the crosslinked fiber (Fig. 7b).

The relatively high compressive strength of PPTA, as compared to PBO and PBZT, is attributed to interchain hydrogen-bonding acting in much the same way (albeit much weaker) as covalent crosslinks. Therefore, it is of interest to introduce intermolecular hydrogen bonding capability into rigid-rod polymers. Tan *et al.* [39] incorporated two hydroxyl groups on the phenyl ring in PBZT. In doing so, it was found that the hydroxyl groups had a detrimental effect on the compressive strength. This was attributed to the formation of *intra*-molecular hydrogen bonds, the presence of which has been substantiated in model compounds [39]. Researchers at Akzo Nobel have had significant success in improving the compressive strength through the introduction of bi-directional, intermolecular hydrogen bonds. PIPD, the structure of which is given in Fig. 3m, results in bi-directional hydrogen bonding [40]. This is unlike the planar arrangement of hydrogen bonding found in PPTA [48]. This suggests that the bi-directional hydrogen bonding plays an important role, as PIPD fiber is reported to have a compressive strength of 1.6 GPa, the highest compressive strength value reported for a polymeric fiber to-date.

The potential for crosslinking in a diacetylene-modified PPTA system (refer to Fig. 8 for structures) has been explored using atomistic simulation [49]. Diacetylenes require strict conformance to certain topochemical criteria in order to react in the solid state [50]. However, if these criteria are satisfied, reaction takes place with minimal disruption of molecular packing. This has been illustrated in work done on a diacetylene-modified aliphatic polyamide, where it was found that the diacetylene reaction

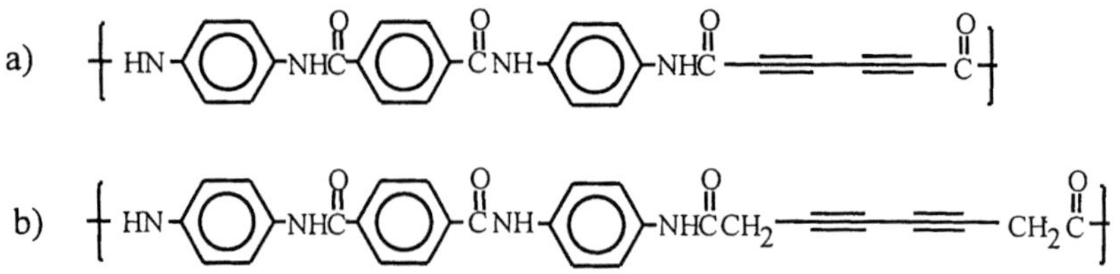

Figure 8. Potentially crosslinkable diacetylene-modified PPTA [49].

occurred predominantly in the crystalline domains and did not disrupt the previously formed hydrogen bonds [49, 51]. Both of the structures shown in Fig. 8 were simulated. Structure 8b incorporates methylene groups flanking the diacetylene moiety, which are thought to be important in allowing adjacent diacetylenes to react [50]. Molecular mechanics simulations predicted [49] structure 8a not to undergo reaction, as adjacent diacetylene groups did not lie within the acceptable topochemical limits. Structure 8b was found to allow for reaction between mutually hydrogen-bonded chains. However, with the addition of thermal vibrations (i.e., molecular dynamics), groups in adjacent planes of hydrogen bonds were found to oscillate within the required topochemical range, also allowing for reaction. Moreover, when crosslinks were incorporated into the structure of Fig. 8b, unit cell parameters and atomic positions were found to be left relatively unperturbed. Thus, this system would seemingly provide lateral cohesion between mutually hydrogen-bonded molecules, consisting of primary and secondary bonds, in addition to possessing the potential for two-dimensional covalent crosslinking (i.e., crosslinking within and across hydrogen bonding planes).

CONCLUDING REMARKS

A significant amount of research work has been carried out over the last decade to improve the axial compressive strength of high performance polymeric fibers. Work reported within the last few years [28, 40, 50] clearly indicates the significant effect of intermolecular interaction or intermolecular crosslinking on compressive strength. A polymeric fiber (PIPD) with a compressive strength of 1.6 GPa has been reported. The high compressive strength of PIPD fiber has been attributed to bi-directional, intermolecular hydrogen bonding. Moreover, a compressive strength of over 700 MPa has been reported for tetramethyl bi-phenyl PBZT at a 2.5% degree of crosslinking. By comparison, the compressive strength of PBZT is 220 MPa.

Achievement of a high degree of intermolecular, covalent crosslinking should result in higher compressive strength, as compared to systems in which only hydrogen bonding is present. However, crosslinking may also make the fiber more brittle. While a significant amount of work has been carried out to achieve intermolecular crosslinking, a number of important issues remain in this area. Some of these are summarized as follows:

a) Conditions under which thermal crosslinking occurs may also lead to polymer degradation [24]. This calls for the development of systems that can be crosslinked below the main-chain degradation temperature.
b) Crosslinking may result in the development of internal stresses, which may result in lower tensile strength and increased brittleness [29, 41]. Structures in which crosslinking will not lead to the development of internal stresses, are desirable. An example of such a structure is given in Fig. 8b.
c) Insolubility is not sufficient evidence for crosslinking in rigid-rod polymers [24]. A more direct means of observing the presence of crosslinking (e.g. spectroscopic techniques) is necessary.
d) It is important that the observation of crosslinking and mechanical property measurements be carried out on fibers treated under similar conditions. Fibers (or bulk polymer) may undergo crosslinking when heat-treated without tension. Fibers heat-treated under tension may crosslink to a lesser degree or not at all.
e) Other crosslinking methods (e.g., via radiation) must be explored in greater detail. Direct evidence of crosslinking has not yet been reported for high performance polymers in which radiation crosslinking has been attempted. Radiation may also result in a different crosslinked structure than achieved in thermal treatment. On exposure to 500 Mrad electron radiation, MePBZT fibers did not exhibit any evidence of crosslinking from ^{13}C solid-state NMR [52]. Thus, radiation crosslinking may have to be carried out at high temperature, enabling the molecules to have sufficient mobility for the crosslinking reaction to occur.

REFERENCES

1. (a) Kumar S. and Helminiak T.E., *Mater. Res. Soc. Symp. Proceedings*, **134**, 363 (1989). (b) Kozey V.V., Jiang H., Mehta V.R., and Kumar S., *J. Mat. Res.*, **10**, 1044 (1995).
2. Greenwood J.H. and Rose P.G., *J. Mater. Sci.*, **9**, 1809 (1974).
3. Allen S.R., *J. Mater. Sci.* **22**, 853 (1987).
4. Bai Y., *M.S. Thesis*, Georgia Institute of Technology, 1998.
5. DeTeresa S.J., Porter R.S., and Farris R.J., *J. Mater. Sci.*, **20**, 1645 (1985).

6. DeTeresa S.J., Porter R.S., and Farris R.J., *J. Mater. Sci.*, **23**, 1886 (1988).
7. Sawyer L. and Jaffe M., *High Performance Polymers*, E. Baer and A. Moet, Eds., Hanser, New York, 1992.
8. Lee C.Y.-C. and Santhosh U., *Polym. Eng. Sci.*, **33**, 907 (1993).
9. Rein D.M. and Cohen Y., *J. Mater. Sci.*, **30**, 3587 (1995).
10. Kozey V.V. and Berlin A.A., *III Japan–USSR Joint Symp. Adv. Comp. Mater.*, Moscow, 1991.
11. Vezie D., *Ph.D. Thesis*, MIT, 1993.
12. Northolt M.G., Baltussen J.J.M., and Schaffers-Korff B., *Polymer*, **36**, 3485 (1995).
13. Kumar S., Hunsaker M., Adams W.W., and Helminiak T.E., USA Patent 5,174,940, 1992.
14. Gillie J.K., Newsham M., Nolan S.J., Jear V.S., and Bubeck R.A., *Am. Phys. Soc. Bull.* **38**, 292 (1993).
15. Rakas M.R. and Farris R.J., *Mater. Res. Soc. Symp. Proceedings*, **134**, 277 (1989).
16. Kumar S., in *Encyclopedia of Composites*, Lee S. M., Ed., VCH, New York, 1991, p. 51.
17. Kovar R.F., Richard J., Druy M., Tripathy S., Thomas E.L., and Anwar A., *Polym. Preprints* (ACS), **35**, 900 (1994).
18. Mathur A. and Netravali A.N., *Textile Res. J.* **66**, 201 (1996).
19. Kovar R.F., Haghighat R.R., and Lusignea R.W., *Mater. Res. Soc. Symp. Proc.*, **134**, 389 (1989).
20. Hwang C.R., Malone M.F., Farris R.J., Martin D.C., and Thomas E.L., *J. Mater. Sci.*, **26**, 2365 (1991).
21. Wang C.S., Burkett J., Bhattacharya S., Chuah H.H., and Arnold F.E., *Polym. Mater. Sci. Eng.*, **650**, 767 (1989).
22. Yang F., *M.S. Thesis*, Georgia Institute of Technology, 1998.
23. Dean D.R., Husband D.M., Dotrong M., Wang C.S., Dotrong M.H., Click W.E., and Evers R.C., *J. Polym. Sci.: Polym. Chem.*, **35**, 3457 (1997).
24. Mehta V.R., Kumar S., Polk M.B., VanderHart D.L., Arnold F.E., and Dang T.D., *J. Polym. Sci.: Polym. Phys.*, **34**, 1881 (1996).
25. Chuah H.H., Tsai T.T., Wei K.H., Wang C-S., and Arnold F.E., *Polym. Mat. Sci. Eng.*, ACS, **60**, 517 (1989).
26. Tsai T.T. and Arnold F.E., *Polym. Preprints*. (ACS), **29**, 324, 1988.
27. Bhattacharya S., Chuah H.H., Dotrong M., Wei K.H., Wang C-S., Vezie D., Day A., and Adams W.W., *Polym. Mater. Sci. Eng.*, ACS, **60**, 512 (1989).
28. Mehta V.R., *Ph.D. Thesis*, Georgia Institute of Technology, 1996.
29. Mehta V.R. and Kumar S., *J. Appl. Polym. Sci.*, submitted.
30. Dotrong M., Dotrong M.H., and Evers R.C., *Polym. Mater. Sci. Eng.*, ACS, **65**, 39 (1991).
31. Santhosh U., Dotrong M.H., Song H.H., and Lee C.Y-C., *Polym. Mater. Sci. Eng.*, ACS, **65**, 40 (1991).
32. Sweeny W., *J. Polym. Sci.: Polym. Chem.*, **30**, 1111 (1992).
33. Rickert C., Neuenschwander P., and Suter U.W., *Macromol. Chem. Phys.*, **195**, 511 (1994).
34. Jiang T., Rigney J., Jones M.G., Markoski L.J., Spilman G.E., Mielewski D.F., and Martin D.C., *Macromolecules*, **28**, 3301 (1995).

35. Dang T.D., Wang C.S., Click W.E., Chuah H.H., Tsai T.T., Husband D.M., and Arnold F.E., *Polymer*, **38**, 621 (1997).
36. Mather P.T., Chaffee K.P., Romo-Uribe A., Spilman G.E., Jiang T., and Martin D.C., *Polymer*, **38**, 6009 (1997).
37. Hu X., Kumar S., Polk M.B., and VanderHart D.L., *J. Polym. Sci.: Polym. Chem.*, **36**, 1407 (1998).
38. Hu X., Kumar S., and Polk M.B., in preparation.
39. Tan L-S., Arnold F.E., Dang T.D., Chuah H.H., and Wei K.H., *Polymer*, **35**, 14 (1994).
40. Klop E.A. and Lammers M., *The Fiber Society Book of Abstracts*, Spring 1997 Joint Conference in Mulhouse, France.
41. Jenkins S., Jacob K.I., and Kumar S., *Polymer*, submitted.
42. Glomm B., Rickert C., Neuenschwander P., and Suter U.W., *Macromol. Chem. Phys.*, **195**, 525 (1994).
43. Jones M.-C.G. and Martin D.C., *J. Mater. Sci.*, **32**, 2291 (1997).
44. Referenced as Martin D.C., unpublished work, 1995 in Ref. 36.
45. Jiang T., Rigney J., Jones M.G., Markoski L.J., Spilman G.E., Mielewski D.F., and Martin D.C., *Macromolecules*, **28**, 3301 (1995).
46. X. Hu, *Ph.D. Thesis*, Georgia Institute of Technology, 1997.
47. Kovar R.F., personal communication.
48. Northolt M.G., *Eur. Polym. J.*, **10**, 799 (1974).
49. Jenkins S., Jacob K.I., and Kumar S., *Polymer*, submitted.
50. Enkelmann V., *Adv. Polym. Sci.*, **63**, 91 (1984).
51. Beckham H.W. and Rubner M.F., *Macromolecules*, **26**, 5192 (1993).
52. Jenkins S., *Ph.D. Thesis*, Georgia Institute of Technology, in progress.

Chapter 16

Modeling Shrinkage Defects in Fiber Composites

Vladimir N. KOROTKOV and Boris A. ROZENBERG*

Institute of Problems of Chemical Physics, Russian Academy of Sciences, Chernogolovka, Moscow Region, 142432 Russia

ABSTRACT

Experimental and theoretical models for formation of shrinkage defects during curing epoxy-amine systems under conditions of restricted shrinkage are analyzed in comparison with recent data obtained for the systems cured by anionic and radical polymerization. The resistance to formation of shrinkage defects was found to

*e-mail: rozen@icp.ac.ru

decrease for the systems cured by polycondensation, anionic, and radical polymerization, respectively.

INTRODUCTION

Since polymer composites comprise two components with different mechanical properties, this inevitably leads to appearance of internal (technological) stresses in these materials. The factors responsible (on the micro- and macroscopic level) for appearance of internal stresses and possible ways for their elimination were reviewed in [1–3].

On the microlevel, residual internal stresses arise owing to nonisothermal conditions for composite formation. Shrinkage stresses caused by chemical reaction are often ignored. This is associated with low elasticity of polymer matrix at this stage. But under some conditions (specified below), the importance of reaction-induced shrinkage stresses may be expected to markedly grow. Apparently, these stresses may result in formation of shrinkage flaws on the microscopic level.

RESULTS AND DISCUSSION

Curing under Conditions of Restricted Shrinkage

Normally the temperature profile in the bulk of polymer composite during its preparation is fairly complicated. For our purposes, it will suffice to consider only three stages: (1) heating up to the curing temperature (will not be considered below since, until that moment, the polymer matrix is liquid), (2) isothermal curing, and (3) subsequent cooling to room temperature. Unless otherwise stated, we will assume that curing proceeds above the glass transition temperature T_g^∞ for a completely cured binder. At the stage (2), binder undergoes transition from viscous to elastic state, while at the stage (3) it is transferred to the glassy state.

Upon cooling and especially curing, the mechanical properties of polymer matrix change drastically. Meanwhile, behavior of most important parameters (e.g., strength) during curing has been studied inadequately. The role of internal stresses caused by temperature and chemical reaction can be illustrated as follows. For most existing polymer networks, the total chemical shrinkage ε_c attains a value of 5–10 %. Upon curing, the volumetric shrinkage of the solid (above the gel point) $\varepsilon_c^s \approx 2$–3 % (accordingly, the linear shrinkage $e_c^s = \varepsilon_c^s/3 \approx 1$ %). Given that fibers do not undergo chemical shrinkage, $\varepsilon_c^s = \Delta\varepsilon_c^s$, where $\Delta\varepsilon_c^s$ is the difference between the volumetric shrinkage of the components (after curing). A typical value for the linear thermal expansion coefficient of polymer matrix is $\alpha_r = 2\cdot10^{-4}$ K^{-1} in the highly elastic state and $\alpha_g = 0.8\cdot10^{-4}$ K^{-1} in the glassy state. Let us assume that the temperature of curing $T_c = 150°C$, glass transition temperature $T_g = 100°C$, and final temperature $T_f = 20°C$. Then the linear temperature shrinkage $e_t = 1.6$ % while the volumetric change is $\varepsilon_t = 3e_t \approx 5\%$. The α value for fiber is negligibly small compared to that for a polymer matrix. Then at the stage of shrinkage, $\Delta\varepsilon_t \approx \varepsilon_t$, while $\Delta\varepsilon_t$ and $\Delta\varepsilon_c^s$ differ only by a factor of two.

Let us consider two limiting cases. Since the fibers are much more rigid than the matrix, then at low degree of filling the microstress is given by a product of the effective elasticity modulus of matrix E and the difference between the linear structural shrinkage of the components. Structural shrinkage is defined as the sum of chemical and temperature-induced shrinkage. More accurate estimates may be performed in terms polydisperse model, but this does not seem necessary for our purposes. For the glassy state during cooling, typical value of $E_g \approx 2$ GPa; during reaction in the highly elastic state, $E_r \approx 50$ MPa. It follows that $\sigma_c \approx E_r\Delta e_c^s \approx 0.5$ MPa. At the stage of cooling, $\sigma_t \approx E_g\alpha_g(T_g - T_f) \approx 10$ MPA which one order of magnitude higher than σ_c. Even with account of the effect of T and extent of conversion on E, this relation remains virtually unaffected.

The situation changes drastically when approaching the theoretical limit of filling degree. For the sake of simplicity, let us consider the limit case of osculating fibers. Due to high rigidity of fibers, the matrix must undergo uniform stretching upon curing, so that the stress will be defined by the modulus of uniform stretching, K. In contrast to classical E, modulus K is known [4] to be a conservative magnitude that changes only by a factor of several times during the process under consideration. For epoxies, the measured [5] values are $K_r = 1.6$ GPa for $T > T_g^\infty$ and $K_g = 3.3$ GPa at room temperature. It follows that, under these conditions, $\sigma_c \approx K_r\Delta\varepsilon_c^s \approx 50$ MPa that is close to $\sigma_t \approx K_g\Delta\varepsilon_t \approx 150$ MPa. Given that the strength in the glassy state may be one order of magnitude greater than that in the highly elastic state [6], we can conclude that the probability for formation of shrinkage flaws is higher at the stage of curing rather than during cooling.

Modeling Defect Formation in Epoxies Cured by Polycondensation

Model studies are prompted by the lack of experimental data on the strength of polymer networks at different stages of curing. The main results obtained in the field will be reviewed below.

We developed a number of experimental techniques for modeling the process of curing under condition of restricted shrinkage typical of composite curing. The use of glass cells makes possible visual observation of appearance and development of shrinkage defects during isothermal curing and subsequent cooling.

For this purpose, elongated tubes (with $L \gg 10D$) were found to be most convenient. Upon attaining the highly elastic state, curing in the middle of the tubes proceeds under conditions close to isochoric ones. In this area, the state of the system is close to uniform stretching: $\sigma \approx -K\varepsilon_c^s$ [8]. Hereinafter, these conditions will be termed quasi-isochoric (isochoric in the solid state). Under these conditions, all of the epoxies studied exhibited formation of cohesion shrinkage defects during isothermal curing.

For short tubes ($L \approx D$), internal stresses are close to uniform biaxial stretching. In this case, $\sigma \approx -E\varepsilon_c^s$, where E is the equilibrium elasticity modulus and ε_c^s is the free chemical shrinkage in the bulk of solid. Under these conditions, no shrinkage defects were observed. The two above situations take place during curing of low- and high-filled composites.

Another situation takes place during curing under conditions of 3D restricted shrinkage, e.g., between 2 sheets separated by a rigid gasket (planar model) [9].

In order to characterize the intensity of defect formation, some parameters were suggested. For $L \geq 10D$, it is possible to measure the mean separation l between shrinkage defects. It is also convenient to use the dimensionless parameter $\bar{l} = l/D$. Since σ decreases with decreasing l (effect of free ends), there exists some L_{min} at which the shrinkage defects still arise. For this reason, the parameter $\bar{L}_{min} = L_{min}/D$ is also useful.

These techniques were used to investigate defect formation in epoxies. The mean separation $\bar{l}$ was found [10] to be independent of D over the range 15 μm–8 mm. This can also be expected to refer to separations of about 1 μm typical of polymer interlayers in fiber-filled composites.

We found three temperature ranges with different outward appearances of cohesion-related shrinkage defects and different $\bar{l}$ and L_{min} [11]. Defect formation is minimal around T_g^{gel}. At higher and lower temperatures, defect formation is relatively intense. The morphology of cohesion-related defects (cracks) was found [12] to depend on the type of relaxation processes during curing. The cracks appear initially at the sample center and then propagate outward. Upon reaching the tube wall, the cracks propagate over the surface in the form of adhesion exfoliation.

Despite the marked difference in the outward appearance of shrinkage defects formed in tubes and between planar sheets, there are some common features in the mode and mechanism of defect formation for each type of relaxation [9].

The above behavior was observed for 10 different epoxies [13]. We suggested a theoretical model [14] of defect formation under the conditions of 3D restricted shrinkage. This model provides adequate explanation for

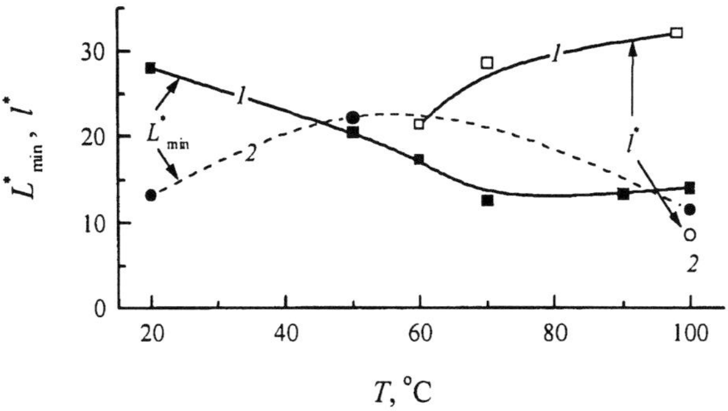

Figure 1. The temperature dependence of l^* and L_{min}^* for epoxies cured via (*1*) anionic polymerization and (*2*) polycondensation. L_{min}^* is the relative length of the smallest sample with a shrinkage flaw; l^* is the average relative distance between cohesive cracks.

the data obtained in model experiments. We also suggested a theoretical model that predicts the probability for formation of shrinkage defects in unidirectional fiber-filled composites as a function of the extent of filling.

Modeling Defect Formation in Epoxies Cured by Anionic and Radical Polymerization

The epoxy systems described above have found wide application as matrices for composite materials. Meanwhile, formation of shrinkage

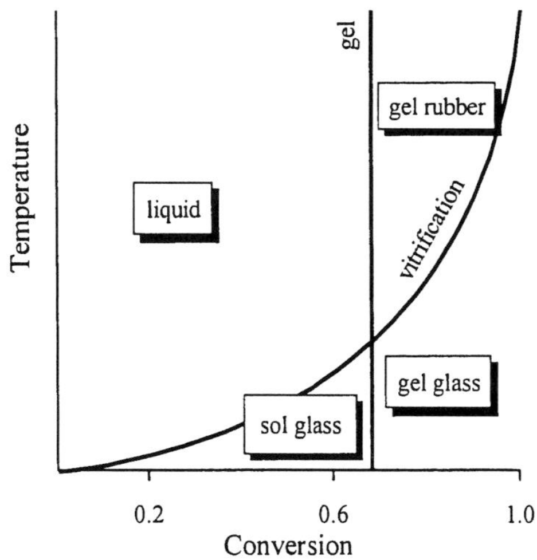

Figure 2. The conversion–temperature–transformation diagram for a typical epoxy cured via polycondensation. The gel point remains unchanged. Shrinkage flaws develop largely in the sol–glass–gel rubber.

defects is of considerable interest also for other polymer binders that are cured by anionic or radical polymerization.

In particular, we compared two chemically close systems differing by the mechanism of chain growth. The DHEBA–bis-methylcyclohexylamine (system **I**) is cured by polycondensation while the DHEBA–tertiary amine (system **II**) is cured by anionic polymerization. In the range 20–100°C, we observed isothermal formation and development of shrinkage flaws in both the systems under the conditions of restricted shrinkage. The temperature dependences of l^* and L_{min}^* are presented in Fig. 1. Because of complexity of defect formation, parameter l^* sometimes could not be determined.

The internal stresses were found to attain a value of 3 MPa for both systems. This can be related to close shrinkage in the solid state. Development of stresses is defined largely by relaxation-related processes (vitrification and gel formation) taking place during curing.

The outward appearance and shrinkage flaw parameters are markedly affected by mutual disposition of T_g^{gel} and T_g^{∞}. Normally, defect formation is minimal around T_g^{gel} [15]. This holds true for system **I**. For system **II**, defect formation is different from that in other related systems, including system **I**. This difference can be explained in terms of the time–temperature–transformation diagrams (Figs. 2, 3).

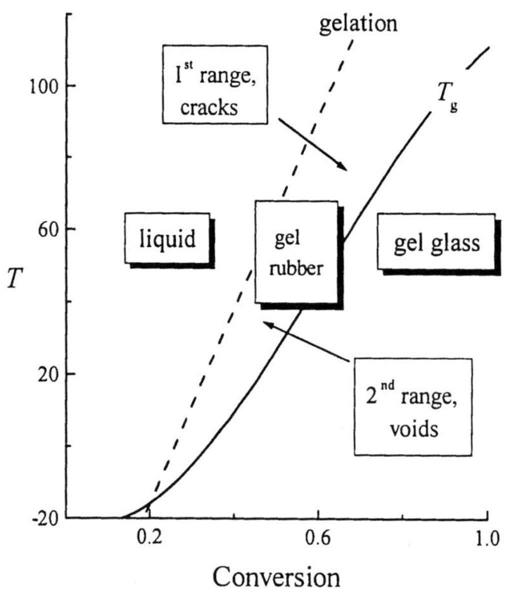

Figure 3. Conversion-temperature transformation diagram for epoxy + tertiary amine system (anionic polymerization). The T_g curve was taken from [15]. Shrinkage flaws develop in gel rubber. Sol glass is absent.

Table 1. Shrinkage flaw formation in different systems

Main factors	Polycondensation, epoxyamines	Anionic polymerization, epoxyamines	Radical polymerization, acrylates
Volume cure shrinkage	5–8%	5–8%	up to 21%
Gel point	0.6–0.7	0.2–0.7	around 0
Temperature dependence of the gel point	no	yes	no
Topological structure of the curing system	homogeneous	homogeneous with topological defects	inhomogeneous
Heterogeneity of curing system	no	no	significant
Adhesion	high	high	low
Order of flaw formation	cohesive → adhesive	cohesive → adhesive	only adhesive; under special conditions: cohesive → adhesive
Resistance to shrinkage flaw formation	high	moderate	low

The diagrams are especially simple in the α–T coordinates. Of key importance is the temperature dependence of separation between the curves corresponding to maxima of the relaxation transitions. For the systems cured by polycondensation, this separation initially drops to zero (slower flaw formation) and then, beginning with T_g^{gel}, gradually grows (increase in the size and amount shrinkage flaws). This leads to appearance of a maximum (at T_g^{gel}) in the temperature dependence of L_{min}*.

For the systems cured by anionic polymerization, the position of the gel point is temperature-dependent. As a result, separation between the maxima of relaxational transitions is only weakly dependent on T (Fig. 2). Accordingly, the parameters of shrinkage flaws change only slightly with increasing temperature.

For system **II**, the T_g^{gel} value is close to the lower T used in our experiments. This implies that the maximum of defect formation is far beyond the limits of Fig. 1, which shows only the descending portion of the temperature dependence of L_{min}*. At higher temperatures, the values of l* and L_{min}* are strongly different, which is indicative of adhesive defect formation. The first cohesive defect formed in the system was found to rapidly transform into the adhesive flaw that propagated all over the entire cylindrical sample. This propagation prevents formation of other cohesive cracks.

For the systems cured by radical polymerization, defect formation proceeds still further differently. Weak adhesion to the cell wall leads to adhesive exfoliation at earlier stages of curing. In case of better adhesion (adhesion layers, special additives), the process of defect formation gets qualitatively closer to that observed for the systems cured by polycondensation. Nevertheless, the rate of defect formation was always higher in case of radically polymerized systems.

The factors affecting the formation of shrinkage flaws in 3D restricted systems are compared in Table 1. It follows that the systems cured by polycondensation are most resistant to shrinkage flaws while those cured by radical polymerization, less resistant. This is consistent with the above experimental data.

High resistance to formation of shrinkage flaws under the conditions of restricted 3D shrinkage is apparently the main factor for wide used of these epoxies as composite matrices.

CONCLUSIONS

It was established that the resistance to crack formation during formation of network polymers formation depends on the mechanism of curing and decrease in the following sequence: polycondensation, anionic, and radical polymerization.

Acknowledgments. This work was supported by the International Science and Technology Center (grant no. 358-96) and the Russian Foundation for Basic Research (project no. 96-03-32027).

REFERENCES

1. Bolotin V.V. and Novichkov Yu.N., *Mekhanika mnogosloinykh konstruktsii* (Mechanics of Multilayer Structures), Mashinostroenie, Moscow, 1980.
2. Kostritskii S.N. and Ryabov V.M., *Mekh. Polim. Kompozit.*, **19**, 701 (1983).
3. Kusznewski V., Waker G., Chate A., and Bledzky A.K., *Mekh. Kompozit. Mater.*, **30**, 579 (1994).
4. Tabor D., *Polymer*, **35**, 2759 (1994).
5. Plepys A.R. and Farris R.J., *Polymer*, **31**, 1932 (1990).

6. Markevich M.A., Irzhak V.I., and Prut E.V., *Crosslinked Epoxies: Proc. IX Discussion Conf.*, Prague, 1987, p. 339.
7. Korotkov V.N., Chekanov Yu.A., and Rozenberg B.A., *J. Mater. Sci. Lett.*, **10**, 896 (1991).
8. Korotkov V.N., Chekanov Yu.A., Smirnov Yu.N., and Zenkov I.D., *Vysokomol. Soedin., Ser. A*, **38**, No. 6, 1025 (1996).
9. Chekanov Ya.A. and Korotkov V.N., *J. Mater. Sci. Lett.*, **15**, No. 24, 2168 (1996).
10. Chekanov Yu.A., Bogdanova L.M., and Korotkov V.N., *Mekh. Kompozit. Mater.*, **31**, 163 (1995).
11. Chekanov Yu.A., Korotkov V.N., Rozenberg B.A., Dzhavadyan E.A., Bogdanova L.M., Chernov Yu.P., and Kulichikhin S.G., *J. Mater. Sci.*, **28**, 3869 (1993).
12. Korotkov V.N., Chekanov Ya.A., Smirnov Yu.N., and Zenkov I.D., *VII European Conf. on Composite Mater.*, London, 1996, Vol. 1, p. 141.
13. Chekanov Yu.A., Korotkov V.N., Rozenberg B.A., Dzhavadyan E.A., and Bogdanova L.M., *Polymer*, **36**, 2013 (1995).
14. Korotkov V.N., *Vysokomol. Soedin., Ser. A*, **39**, 677 (1997).
15. Dzhavadyan E.A., Bogdanova L.M., Irzhak V.I., and Rozenberg B.A., *Vysokomol. Soedin., Ser. A*, **39**, 591 (1997) [Engl. Transl. *Polymer Science A*, **39**, 383 (1997)].

Chapter 17

Analysis of Shrinkage Cracking in Three-Dimensionally Constrained Heterophase Network Polymers

Valentina A. LESNICHAYA and Vladimir N. KOROTKOV*

Institute of Problems of Chemical Physics, Russian Academy of Sciences, Chernogolovka, 142432 Moscow Region, Russia

ABSTRACT

The mechanical phenomena during cure under conditions of omnidirectional constraints of heterophase polymer system, the matrix being in the rubbery state, are considered. It is shown that in heterophase systems, as well as in earlier investigated

*e-mail: korot@icp.ac.ru

homogeneous systems, development of shrinkage defects during isothermal cure takes place. General analysis of the shrinkage cracking is carried out. Most probable mechanisms of the cracking process are revealed. Stress state of the heterophase system on the initial stage of fracture is investigated. Possible directions for further research are discussed.

INTRODUCTION

Heterophase polymeric systems are widely used to achieve high mechanical properties, which cannot be achieved in homophase systems. Heterophase systems with thermosetting matrix and elastomeric dispersed particles represent an important type of polymer materials. Usually the mechanical properties of such systems are studied at room temperature [1]. At the same time, there is practical interest in the study of mechanical phenomena in these systems in the course of their formation.

Several years ago it was experimentally shown [2–3] that shrinkage cracks can be formed and developed during isothermal cure under three-dimensionally constrained conditions (TDCC). Such conditions are realized in polymeric composites with a rigid filler and good adhesion between matrix and filler (in particular, in composites with high fiber content) [4]. If cure conditions are close to isochoric, then stresses due to the chemical shrinkage usually exceed the strength of the curing system in the rubbery state [5], and defects of different morphology such as voids, cracks, or debondings may occur [6]. The shrinkage defects develop at any cure temperature both above and below the glass transition temperature of the completely cured network, T_g^{∞}. However, the type and concentration of the defects significantly depend on the cure temperature [7–8]. Although the above studies concerned homophase systems, they allow us ascertain that under TDCC the shrinkage cracking may also take place in heterophase systems. Hence, operational characteristics of heterophase polymers are determined by mechanical phenomena taking place during cure.

The fracture process in the heterophase systems is complex and usually includes several mechanisms of energy dissipation [9]. As far as we know, the present work is the first to consider the fracture in heterophase systems during cure. Therefore, the main objective of the work is a general analysis of mechanical phenomena under TDCC and revealing main mechanisms of fracture process development in the curing heterophase systems. In the present work, we confine ourselves to consideration of initial stage of shrinkage damage, i.e., development of shrinkage microcracks. One of important goals of the work is also to attract attention to the problem of shrinkage damage and to discuss the possible directions of further studies.

GENERAL ANALYSIS

Mechanical Phenomena in Curing Heterophase Systems

Mechanical phenomena during cure of heterophase systems under TDCC are characterized by a number of peculiarities that differentiate them from the mechanical phenomena under operational conditions:

(1) during formation the polymer is in rubbery state, while under operational conditions the system is in glassy state;

(2) tremendous changes in the physical properties of the polymer simultaneously with development of the stress state take place;

(3) macroscopic hydrostatic stress state takes place;

(4) there is no free surface, which is the main source of defects under operational conditions.

Consider these items in more detail. Four basic relaxation states of the matrix phase are possible during cure in a general case: liquid, rubbery, sol glass and gel glass [9]. The rubbery matrix is most interesting for us, as vitrification of the network systems during synthesis is usually attempted to avoid because it hinders the cure process. In the liquid state, development of shrinkage defects is also possible, however special conditions, which are rarely realized in practice, are necessary for this.

Formation of micron size particles, which are optimal for achievement of high mechanical properties of the heterophase systems, is impossible in the network systems. Therefore, we shall consider further the phase separation occurring before the transition of the matrix through the gel-point. Elastomeric particles are usually used in order to increase the toughness of the systems with thermosetting matrix, so vitrification of the dispersed particles during cure takes place only in special cases (when a thermoplastic with high T_g is used as a modifier). Both liquid and rubbery state of the particles in rubbery matrix may be the case.

Increase in the crosslink density during cure leads to significant changes in physical and mechanical properties. The kinetics of chemical reactions within the dispersed particles can considerably differ from the kinetics in the matrix phase. A reaction in particles can proceed both faster or slower than in the matrix. There may be significant changes of the dispersed phase properties or these properties may remain practically constant from the gel point until full completion of the reaction. Both cases should be considered.

The hydrostatic stress state developing during cure of polymeric systems under TDCC virtually never exists under operational conditions. This results in certain differences in the mechanical phenomena that will be considered further in more detail. However, such differences occur only in the first stage of shrinkage cracking until the origin of a macroscopic crack. The point is, e.g., in simple stretching of a macroscopic sample there is a region of hydrostatic tension near the crack tip. Within the scope of the linear fracture mechanics, there is a hydrostatic singularity in the stresses in this region. This means that the stress increases infinitely as it approaches the crack tip. Thus, direct contribution of the uniform stresses that are parallel to the crack plane is insignificant in comparison with the hydrostatic singularity in the region. Therefore, after appearance of a macroscopic crack, the peculiarity of the stress state realized during cure under TDCC disappears.

Absence of the free surface in the sample cured under TDCC leads to absence of the surface defects. This increases the role of microcrack initiation.

The process of shrinkage cracking may be divided into different stages. These are connected with change in the characteristic length of the shrinkage defects. The average diameter of the dispersed particles is a natural scale in the heterophase system. Defects (voids, cracks, and debondings) with characteristic size less than the average size of dispersed particles will be referred to as microdefects. The following stages of shrinkage damage may be distinguished:

(1) development of the shrinkage stresses until the first microdefect appear;

(2) arising of microdefects;

(3) development of cracks exceeding the particle size;

(4) development of a macrocrack.

In the present work, we will restrict ourselves to consideration of stages 1–3, only.

Influence of Dispersed Phase on Stress State in Heterophase Systems before Appearance of Microdefects

A spherically symmetrical two-phase solid is considered as a simple mechanical model of the heterophase system. The inner sphere corresponds to a dispersed particle, and the outer layer corresponds to the matrix phase. Mechanical interaction between particles is neglected in the model. This is admissible when the dispersed phase content is not too high.

TDCC are rarely realized before the gel point of the matrix. Further, as well as in the case of homogeneous systems [5, 10], we shall consider so-called quasi-isochoric conditions. These conditions are realized during the cures of polymer systems in long rigid tubes, which are convenient in experimental studies, or in composites with high fiber content. In this case, it is possible to neglect the viscous stresses in the liquid state and to consider isochoric cure in the solid state. Thus, it is supposed that there are no stresses in the incipient gel:

$$\sigma_{ij}\left(t_{gel}\right)=0 \ . \tag{1}$$

Neglecting the inelastic strain, we shall write down the total strain as a sum of the elastic component ε^e and the cure (shrinkage) component ε^c:

$$\varepsilon_{ij} = \varepsilon_{ij}^{e} + \varepsilon_{ij}^{c} . \tag{2}$$

The shrinkage strain depends linearly on the conversion. As the gel-point is taken as the initial moment,

$$\varepsilon_{ij}^{c} = \varepsilon_{c}^{\infty} \left(\alpha - \alpha_{gel}\right)\delta_{ij} , \tag{3}$$

where α is the conversion, ε_c^∞ is the chemical shrinkage at ultimate conversion, and δ_{ij} is the Kroneker symbol.

On the outer surface of the heterogeneous spherical solid, the isochoric condition of cure was set,

$$u_r^m = 0 , \tag{4}$$

where u^{m} is the displacement in the matrix (outer) layer and subscript r refers to the radial direction.

On the surface of the inner spherical particle, the conditions of continuity of the normal components of displacement and stress were used:

$$u_r^m = u_r^d , \ \sigma_r^m = \sigma_r^d . \tag{5}$$

The Hooke law was used as a constitutive equation. The bulk modulus K only slightly changes during cure, therefore, it was taken to be constant. Within this approach the hydrostatic stress state of a single-phase system at the end of cure process is described by the following expression:

$$s = -K\varepsilon_c^\infty \left(\alpha_\infty - \alpha_{gel}\right) , \tag{6}$$

where ε_c^∞ and α_∞ are volume shrinkage and conversion at the end of cure.

The equilibrium shear modulus undergoes enormous changes during cure. For polycondensation systems there are theoretical [11] and experimental [12] data showing that the following scaling law,

$$G \sim \left(\alpha - \alpha_{\mathrm{gel}}\right)^{\lambda} . \tag{7}$$

can be applied to describe the changes in equilibrium shear modulus within all the range from the gel-point up to ultimate cure. Exponent λ is about 8/3. Equation (7) can be written in the form

$$G = G^{\infty}\left(\frac{\alpha - \alpha_{\text{gel}}}{\alpha_{\infty} - \alpha_{\text{gel}}}\right)^{\lambda}. \tag{8}$$

Here G^{∞} is the shear modulus at ultimate cure.

The analytical solution of the elastic problem for the heterophase model has been derived by a conventional procedure [13]. The hydrostatic stresses in the matrix s_{m} and in the dispersed phase s_{d} are

$$\begin{aligned} s_m &= -(1+3\varphi C)K_m\varepsilon_c^{\infty}(\alpha - \alpha_{\text{gel}}), \\ s_d &= (3(1-\varphi)C - 1)\,K_d\varepsilon_c^{\infty}(\alpha - \alpha_{\text{gel}}), \end{aligned} \tag{9}$$

where

$$C = \frac{K_d - K_m}{3[K_m(\varphi + k) + K_d(1+\varphi)]},$$
$$k = 2(1 - 2\nu_m)/(1 + \nu_m).$$

K_m is the bulk modulus of the matrix, K_d is the bulk modulus of the dispersed particle, ν_m is the Poisson ratio of the matrix, φ is the dispersed phase volume fraction.

Numerical analysis of the equation obtained shows that the stress state in the matrix is hydrostatic, except for a layer near the particle surface. In the particle, the stress state is also hydrostatic. The hydrostatic stresses in the matrix s_{m} and in the dispersed phase s_{d} are constant. The following values of the curing system parameters, which are characteristic for epoxy matrices, were taken: $K_m = 2$ GPa, $G^{\infty} = 20$ MPa, $\varepsilon_c^{\infty} = -8\%$, $\alpha_{\text{gel}} = 0.6$. Development of the curing stresses was investigated in the system at different ratios of the cure degree of the matrix and the dispersed phase and at various values of the dispersed phase volume fraction.

The ratio of hydrostatic stresses at ultimate cure in both phases to the hydrostatic stresses in the homophase system is presented in Table 1. Volume fraction of the dispersed phase is 0.3.

Various ratios of the bulk modulus of the particle to the modulus of the matrix were considered. Possible ratio of the bulk moduli of the phases in rubbery state is usually less than 2. Therefore, the influence of the dispersed phase on the stress state in the system is insignificant. This means that consideration of the next level of approximation, when mechanical interaction between particles is taken into account, is not of special interest. However, if a particle is debonded from the matrix, the situation essentially changes, and the shrinkage stresses sharply decrease even at small volume fraction of pores.

Table 1. Shrinkage hydrostatic stresses in the heterophase system

K_d / K_m	1	2	0.5
s_m / s	1	1.17	0.77
s_d / s	1	1.17	0.76

If the volume fraction of particles is not very high, e.g., if $\varphi < 0.3$, the results obtained are valid within the scope of the linear theory of elasticity not only for the considered spherical model, but also for the case when fine spherical particles are uniformly distributed in the matrix.

Influence of Microdefects on Stress State in Heterophase System

It was established earlier [2, 5, 10] that during cure under TDCC significant shrinkage stresses appear in homophase systems in rubbery state. This results in shrinkage defects arising. As was shown above, the shrinkage stresses in heterophase system are close to the stresses in homogeneous system, and one may expect that shrinkage defects will arise, too. As the stress states inside the particles and the matrix are close to each other, the possibility of occurrence of microdefects in both phases depends on the mechanical properties of the phases.

The failure of homophase network polymers in the rubbery state under hydrostatic tension proceeds through inflation of pores. It can be stated that this mode of failure significantly differs from the failure, e.g., at uniaxial stretching. In the former case, the failure always occurs in the bulk of the material. This feature is very rarely observed at mechanical testing [14]. In the first approximation, the critical size of the hydrostatic stresses, s^*, at which loss of stability of the material and pore growth occur, is proportional to the equilibrium shear modulus. Therefore, within this approximation, it is possible to assume that the ratio of shear modulus of the dispersed and matrix phases defines in which phase first microdefects will arise.

For network polymer near the gel point the equilibrium shear modulus is close to zero. Then the surface energy of the pore should be also taken into consideration along with the elastic energy of the inflating pore. A general criterion of pore inflation can be written in the following form [15]:

$$s^* = \frac{5G}{2} + \frac{2\gamma}{Gr^*}, \tag{9}$$

where γ is the surface tension, r^* is the pore radius at the moment of breaking of the periphery layer and transition to the crack. r^* depends on the distribution of defects nuclei (bubbles, defects of the network structure, and so on), and it can hardly be exactly determined experimentally. At small values of the shear modulus, the second term becomes essential and the uncertainty in the value of r^* can hinder correct analysis.

Both the variants—occurrence of first microdefects in the matrix or in the particles—may take place. In the former case (at $s_m^* < s_d^*$), the process of microdefect development does not differ from the investigated earlier homophase systems [3–8] due to weak influence of the dispersed phase on the stress state. In the latter, this stage has significant peculiarities and is of a special interest.

The first shrinkage defect in homophase systems occurs just after the gel-point. In heterophase systems with soft particles (at $s_m^* > s_d^*$) the first break of particles takes place also after the gel-point. There will be consecutive failures of particles. The microcrack will not penetrate from the particle into the matrix if the latter is glassy or if the adhesive strength of the phase boundary is low. Then a pore actually arises on the place of the particle. This results in reduction in shrinkage stresses in the system in comparison with the case of monolithic material with the same elastic characteristics.

The stress state in the heterophase system with broken dispersed particles may be estimated with the help of the single-particle model considered above. The well-known polydisperse model of a composite gives the same results.

Let us introduce the effective bulk modulus of the heterophase solid, K_{ef}. The shrinkage stress in the system under TDCC is determined by the effective modulus:

$$s_m = -K_{ef}\varepsilon^{\infty}\left(\alpha - \alpha_{gel}\right). \tag{10}$$

The dependence of the dimensionless effective modulus $K_{ef}^0 = K_{ef}/K_m$ on the volume fraction of pores φ_p at different conversions is shown in Fig. 1. This dependence may not be described by linear additive relations since the modulus drops down extremely fast with increase of the volume fraction of pores. Moreover, the effective modulus vanishes as α approaches α_{gel}, while K_m is constant.

The shrinkage stress versus conversion at different volume fraction of pores is plotted in Fig. 2. It should be noted that the bulk modulus of the matrix is taken to be constant, and the volume fraction of pores for each of the curves does not vary. However, values of the shrinkage stresses vary rather strongly. Near the gel point, where the equilibrium shear modulus of the matrix is negligible, the effective bulk modulus, as it was shown above, also decrease down to zero. This leads to nonlinear dependence of the shrinkage stress against conversion. In the homophase system cured under TDCC, the stress is approximately proportional to the conversion after the gel point, equation 6.

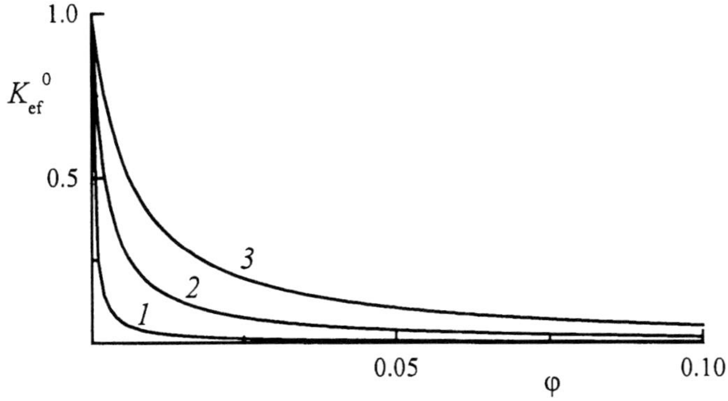

Figure 1. Dimensionless effective modulus versus void fraction: *1*) $\alpha = 0.7$; *2*) $\alpha = 0.8$; *3*) $\alpha = 0.9$.

The shrinkage stresses versus conversion at different volume fractions of dispersed particles are drawn in Fig. 3. The particles are cohesive broken under hydrostatic stresses in accordance with equation (9). The first particle breaks at $\alpha^o = 0.60$, actually just after the gel point. Then consecutive failures of all particles occur. The shrinkage stresses are constant, $s = s(\alpha^o)$, up to the conversion α^*, when all dispersed particles have been broken. For the three particle volume fractions under consideration, $\alpha^* = 0.67$, 0.71, and 0.74. The situation near the gel point is shown in Fig. 4. After α^* the shrinkage stresses grow sharply; nevertheless, in the matrix they will be below s^* during the whole cure process already for $\varphi = 0.01$, curve *1*.

Thus, if during cure after the gel-point the dispersed phase is less strong than the matrix and there is no penetration of microdefects from the

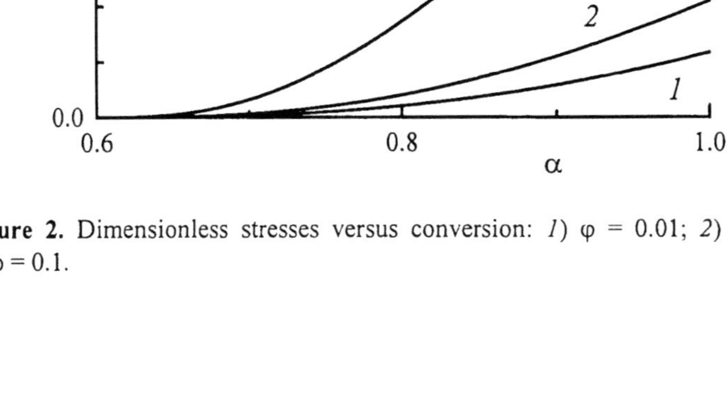

Figure 2. Dimensionless stresses versus conversion: *1*) $\varphi = 0.01$; *2*) $\varphi = 0.05$; *3*) $\varphi = 0.1$.

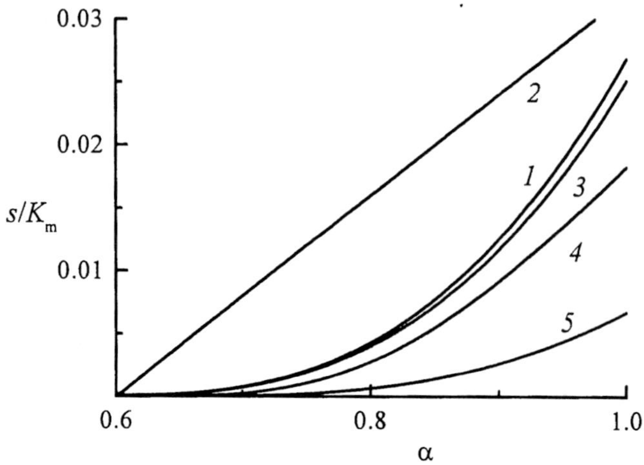

Figure 3. Shrinkage stresses versus conversion at different pore content: *1*) s^*; *2*) $\varphi = 0$; *3*) $\varphi = 0.1$; *4*) $\varphi = 0.05$; *5*) $\varphi = 0.01$.

particles into the matrix, then, even at a very small volume fraction of the dispersed phase, significant decrease in the shrinkage stresses takes place. A certain part of the particles breaks, but shrinkage defects do not arise absolutely in the matrix.

Occurrence of Cracks Exceeding the Particle Size

The above scenario of the fracture process is only true for the case, when the break of a dispersed particle does not lead to development of defects in

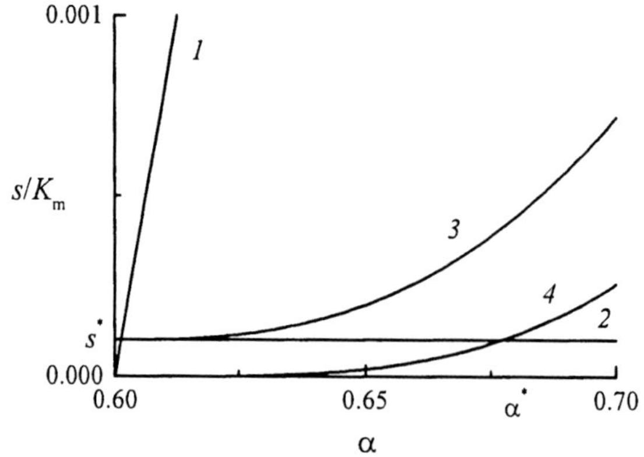

Figure 4. Shrinkage stresses versus conversion near the gel-point: *1*) modulus K_m; *2*) $s^* = 2\gamma/r^*$; *3*) $s^* = 5/2\ G + 2\gamma/r^*$; *4*) modulus K_{ef}.

the surrounding matrix. In this case, the shrinkage defects in the matrix can only arise at a very low volume fraction of the dispersed phase.

However, the curing heterophase system is usually in the rubbery state, and its properties can be close to the properties of the particles. Therefore, the microcrack can spread into the matrix from the particle and then it can develop into a catastrophic crack. In this case, the dispersed particles do not increase the toughness of the system. At the same time, these promote the fracture of material.

CONCLUSIONS

The analysis carried out allows classifying basic events during cure of the heterophase systems (see Table 2).

The direct influence of the heterophase inclusions on the shrinkage stress state during cure is insignificant. This is because the stress state during cure is determined first of all by the bulk modulus, which slightly depends on the composition of polymer system. The modulus also changes slightly during cure until vitrification occurs. As a good approximation at a moderate volume fraction of the dispersed phase, the influence of the dispersed particles on the stress state of the system can be neglected until the first microdefects arise.

If the strength of the particles is higher than the strength of the matrix, then the first microdefects appear in the matrix, soon after the gel point, as if the system was homophase.

If the strength of the matrix is higher, then the first microdefects arise inside the particles. Then two scenarios are possible. In the first case, the microcracks freely propagate into the matrix. In this case dispersed particles initiate shrinkage cracking. Additional study is required to clarify the difference between shrinkage cracking in heterophase and homophase systems near the gel point.

In the second case, penetration of microdefects from the particles into the matrix does not take place. This can result from weak adhesion between the particles and the matrix or high strength of the matrix. Then the broken particles, which have zero effective bulk modulus, decrease the shrinkage stress in the system.

The analysis carried out shows that very small concentration of broken particles could prevent shrinkage cracking in the matrix.

The experimental data available now show that in heterophase systems cracking occurs in the matrix. This testifies that the behavior of the investigated heterophase systems is similar to that of homophase ones. It is not clear yet how general this case is.

Table 2. Development of shrinkage cracking in heterophase systems cured under TDCC

Stage of shrinkage cracking	Condition of realization	Details of kinetics	Mechanical phenomena
1. Development of shrinkage stresses until the first microcrack	$\varphi \le 0.3$		Stress state is close to the state of homophase system
	$\varphi > 0.3$		Stress state depends on φ and the mechanical properties of both phases
2. Arising of microdefects			
a) in particles	$s_m^* > s_d^*$	Reaction rate in the matrix is higher	Constancy of the stress level
b) in matrix	$s_m^* < s_d^*$	Reaction rate in the particles is higher	This stage is similar to cracking in the homophase system
3. Development of cracks			
a) penetration from particles into matrix	High strength of phase boundary	Reaction rate in the matrix is higher, particles are grafted to the matrix	Shrinkage cracking is close to the cracking in the homophase system, particles are the nuclei of the shrinkage cracks
b) in the matrix	$s_m^* < s_d^*$	Reaction rate in the particles is higher	Shrinkage cracking is close to the cracking in homophase system
c) in the matrix after breaking of all particles	$s_m^* < s_d^*$, $\varphi < \varphi^*$	Reaction rate in the matrix is higher	Stress level is significantly lower than in homophase system
4. Development of a macrocrack	$\varphi < \varphi^*$		The fracture process is getting close to the fracture of the heterophase system under external tension

The following problems deserve the greatest attention in the further studies. First, it is necessary to carry out experimental studies of the process of shrinkage cracking. They should include observations of the cure kinetics and changes in the relaxation state not only in the matrix, but also in the particles. Second, theoretical study of the conditions of microcrack penetration from the particles into the matrix is essential. These conditions determine the role of the dispersed phase during cure: do the particles prevent or facilitate shrinkage cracking in the matrix. At last, the problem of the development of macrocrack exceeding the particle sizes has not been analyzed at all. As the crack grows, the stress state gets closer to the ordinary case of simple stretching of a heterophase system, but, contrary to the case of normal operational conditions, the matrix under cure is in the rubbery state.

Acknowledgments. This work was supported by the International Science and Technology Center (grant no. 358-96) and the Russian Foundation for Basic Research (project no. 96-03-32027).

REFERENCES

1. Kozey V.V. and Rozenberg B.A., *Vysokomol. Soedin., Ser. A*, **34**, 3 (1992).
2. Plepys A.R. and Farris R.J., *Polymer*, **31**, 1932 (1990).
3. Korotkov V.N., Chekanov Yu.A., and Rozenberg B.A., *J. Mater. Sci. Lett.*, **10**, 896 (1991).
4. Korotkov V.N. and Rozenberg B.A., *Mech. Comp. Mater.*, **34**, 264 (1998).
5. Korotkov V.N., Chekanov Yu.A., Smirnov Yu.N., and Zenkov I.D., *Polym. Sci., Ser. A*, **38**, 1025 (1996).
6. Chekanov Yu.A. and Korotkov V.N., *J. Mater. Sci. Lett.*, **15**, 2168 (1996).
7. Chekanov Yu.A., Korotkov V.N., Rozenberg B.A., Dzhavadyan E.A., Bogdanova L.M., Chernov Yu.P., and Kulichikhin S.G., *J. Mater. Sci.*, **28**, 3869 (1993).
8. Chekanov Yu.A., Korotkov V.N., Rozenberg B.A., Dzhavadyan E.A., and Bogdanova L.M., *Polymer*, **36**, 2013 (1995).
9. Gillham J.K.and Enns J.B., *Trends in Polym. Sci.*, **2**, 406 (1994).
10. Korotkov V.N., *Polym. Sci., Ser. A*, **39**, 464 (1997).
11. Martin J.E., Adolf D., and Wilcoxon J.P., *Phys. Rev. A*, **39**, 1325 (1989).
12. Adolf D. and Martin J.E., *Macromolecules*, **23**, 3700 (1990).
13. Bucknell C.B. *Toughened Plastics*, Applied Science Publishers, London, 1977. Ch. 5.
14. Lindsey G.H., *J. Appl. Phys.*, **38**, 4843 (1967).
15. Gent A.N. and Tompkins D.A., *J. Appl. Phys.*, **40**, 2520 (1969).

Chapter 18

Modeling Polymer Film Formation by Spin Coating

Aleksei K. ALEKSEEV, Sergei M. BATURIN,
Georgii A. PAVLOV*, and Anatolii A. SHIRYAEV

Institute of Problems of Chemical Physics, Russian Academy of Sciences, Chernogolovka, Moscow Region, 142432 Russia

ABSTRACT

Formation of polymer films under the action of mass forces has been modeled. The instability of a non-Newtonian liquid front edge at the initial stage of flow over disc is examined. The factors that determine the final shape of a surface of polymer coating are discussed.

*e-mail: pavlov@icp.ac.ru

INTRODUCTION

A promising way for manufacturing hard-magnetic discs is the coating of disc base with ferroliquid lacquer (FL) by using spin coating. After attaining a desired thickness, the disc is placed in a magnetic field for providing the magnetic anisotropy of coating. The coating is then solidified, polished, washed, and lubricated. The coating is formed due to spreading ferroliquid lacquer under the action of the centrifugal and Coriolis forces. The lacquer is initially located near hard disc rotation axis. During FL motion on the rotating disc, the front FL edge is as a rule subjected to finite disturbances. The drain of magnetic carriers (located in the bulk of initial FL) occurs in parallel with their generation due to possible roughness of the base surface. Partial loss of solvent from multicomponent FL due to vaporization and other processes may be expected to occur as well. The above factors (typical of centrifuging technology) give rise to both the magnetic and geometric nonuniformity of the coating.

In this work, we modeled the initial and major stages of viscous spreading of a non-Newtonian liquid over a rotating disc.

RESULTS AND DISCUSSION

Formulation of the Problem

The FL flow over rotating base of a magnetic disc is accompanied by different physical processes that markedly affect the final quality of the magnetic support. These processes are (*a*) solvent vaporization, (*b*) cooling of the FL upper layers due to solvent vaporization, (*c*) solvent diffusion toward FL upper layers, (*d*) polymer and solvent sedimentation on ferroparticles, and new phase nuclei, and (*e*) instability in the FL front edge. Below, we will discuss features of the above processes.

Vaporization of solvent from the FL surface is governed [1] by following equation:

$$\dot{m} = (RT_s / M_p)^{-1/2}(P_{hp} - P_{p,s}), \tag{1}$$

where R is gas constant, M_p is a solvent mass, T_s is a surface temperature; P_{hp} and $P_{p,s}$ are the saturation pressure and solvent vapor pressure, respectively. For typical spreading conditions, (1) gives $\dot{m} \approx 1\ \text{g}\cdot\text{cm}^{-2}\,\text{s}^{-1}$.

Investigation of the temperature variation in the vicinity of the film free surface is important for elucidating the mechanism of FL solidification.

Qualitatively, the temperature profile and related time of thermal conductivity may be determined from the equation of heat conduction [2]:

$$\rho C_p \frac{\partial T}{\partial t} = \lambda \frac{\partial^2 T}{\partial z^2}, \quad t_2 = \rho C_p \frac{H^2}{\lambda}, \tag{2}$$

where ρ, C_p, λ, H are density, specific heat, thermal conductivity, and thickness of the FL. For typical values of physical constants in (2), t_2 attains a value of about 1 s.

The solvent diffusion to FL upper layers can be estimated from the Fick equation. For typical values of physical constants in (2) we find that t_3 values range around 1 s.

After discussion of the 'surface' effects, we estimate reference values for the 'bulk' effects. It is expedient to introduce such a delimitation, despite the conventionality of these notions for thin films.

The sedimentation of polymer and solvent on ferroparticles prevent their agglomeration due to steric repulsion of long polymer molecules. Sedimentation in a model polymer–solvent system was investigated in [3]. According to [3], the weight ratio of polymer to solvent sedimented at ferroparticles does not exceed 1 : 10, while formation of a polymer and solvent layer about 20 nm thick occurs in time t_4 (about 1 min).

We estimate also, according to [4], the time t_5 required for formation of a new phase from a low-molecular fraction in the approximation of still spherical non-interacting particles of new phase that grow due to polymer diffusion from surrounding FL. We found that $t_5 \approx 3$ min.

Therefore, the process parameters as determined from analysis of processes (*a*)–(*e*) ($t_{\text{flow}} \propto \omega^{-1}$) show that the flow process can be divided into two stages. During the first stage ($t \approx t_{\text{flow}}$) when FL achieves the disk edge, the film front edge can be expected to be disturbed, no significant change in the physicochemical properties of FL takes place, so that the front edge stability problem may be considered in a time that is independent of film properties. Apparently, physicochemical processes (*a*)–(*e*) should be taken into account in modeling the second (longer) stage of magnetic coating formation, when the film thickness decreases to several microns due to loss of FL upper layers from a rotating disc.

Instability of Front Film Edge

The investigation of front edge stability at certain volume of fluid spreading over surfaces of different shapes is of interest both from the academic and practical points of view since front instability often leads to nonuniform coatings. The surface stability in case of Newtonian liquid flow was investigated in [5]. Most coatings are formed by non-Newtonian liquids such as polymer solutions, multicomponent polymer suspensions, various varnishes, etc. Below, we will discuss the flow of non-Newtonian liquid over simple surfaces: inclined plane and rotating disc.

The motion of an incompressible liquid film over an inclined surface with velocity u when inertial members are neglected and under the conditions when all variables depend only on the coordinates along surface x and across surface z in lubrication approximation [6] is governed by the following equation (for Newtonian liquid $n = 1$):

$$\rho g \sin\alpha + \sigma \frac{\partial}{\partial x}\frac{\partial^2 h}{\partial x^2} + \eta \frac{\partial}{\partial z}\left(\frac{\partial u}{\partial z}\right)^n = 0 , \tag{3}$$

where ρ, σ, h, and g are the density, surface tension coefficient, film thickness and acceleration due to gravity; α is the inclination angle. The viscous shear stress tensor for non-Newtonian liquid is chosen in a form [7]:

$$\tau' = 2\eta S^{n-1} D , \tag{4}$$

where S is the double contraction of deformation rate $D = (\nabla \mathbf{v} + \nabla \mathbf{v}^{\mathrm{T}}) / 2$, η and n are the parameters of non-Newtonian liquid. Because we aim to analyze the shape and stability of film surface, we substitute the averaged over film thickness (V is the velocity averaged over z) result of summation of equation (3) into the continuity equation

$$\frac{dh}{dt} + \nabla hV = 0 . \tag{5}$$

After these transformations, we find the equation for quasi-stationary surface profile in the dimensionless form [8, 9]:

$$\left(1 - b^{2+1/n}\right)\left(f_0 - 1\right)\left(b - 1\right)^{-1} - 1 + \left(1 + \frac{\partial^3 f_0}{\partial \xi^3}\right)^{1/n} f_0^{2+1/n} = 0 \tag{6}$$

$$\xi = x/l, \ f_0 = h/H_N , \ b = H_c/H_N .$$

Transition from the normal to precursion film [10] takes place at H_c ($b << 1$), equation (6) is written in the vicinity of spreading film front in the region of capillary order length ($\sim[\sigma/\rho g]^{1/2}$) in a coordinate system attached to a moving front; H_N is the film thickness at the front, l is the characteristic length that is defined when equation (9) is made dimensionless (see below). In (6), time is a parameter since $b = b(t)$.

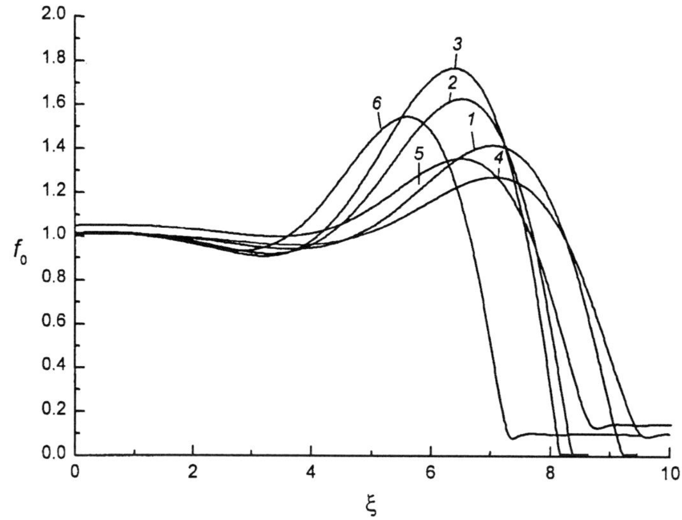

Figure 1. The shape of quasi-stationary front edge for different ξ: (*1*) $n = 0.5$, $b = 0.01$; (*2*) $n = 1.0$, $b = 0.01$; (*3*) $n = 1.2$, $b = 0.01$; (*4*) $n = 0.5$, $b = 0.1$; (*5*) $n = 1.0$, $b = 0.1$; (*6*) $n = 2.5$, $b = 0.1$.

Equation (6) is an ordinary differential equation with following boundary conditions:

$$f_0 \to 1 \text{ for } \xi \to -\infty; \quad f_0 \to b \text{ for } \xi \to \infty. \tag{7}$$

Since (6) does not directly contain ξ, there is no need for a third boundary condition, and the solution depends on b. The characteristic shape of f_0 is presented in Fig. 1. The front edge shape is defined by the character a 'crest' whose height depends on the thickness of the precursion film and the physicochemical properties of spreading liquid [8, 9] (cf. [11]).

We investigate the stability of linear solution in the vicinity of spreading film front. For the purpose, we introduced the following variables $\xi = (x - x_f)/l$, $s = y/l$, $f = h/H_N$ in Eq. (5) rewritten in appropriate coordinate system. We consider $f = f_0(\xi) + f_1(\xi, s, t)$ and $f_1/f_0 << 1$. We pose $\tau = l/u_0$, $u_0 = \rho g H_N^2 \sin\alpha[n/(2n+1)]/\mu$, $\beta = (\rho g \sin\alpha/\eta)^{1/n} H_N^{1+1/n} u_0^{-1}$, $v = u u_0^{-1}\beta^{-1}$, $\mu = \text{const}$; note that $\beta = 1$ for Newtonian liquid. After substituting the sum $f_0 + f_1$ in V and linearization, we find the equation for investigating the linear stability of spreading non-Newtonian liquid front edge. The present equation obtained from (6) for f_0 (at $b \to 0$) and suitable for qualitative analysis of spreading liquid front edge should be solved with the following disturbance at the 'boundary': $\xi_b = A(s)B(t)$, $A(s) = \cos(qs)$, where $-\pi/2 < qs < \pi/2$ forms the so-called finger that grows for $dB/dt > 0$, $B(t) = B_0 \exp(\theta\tau)$; in this case, f_1 can be presented in the form:

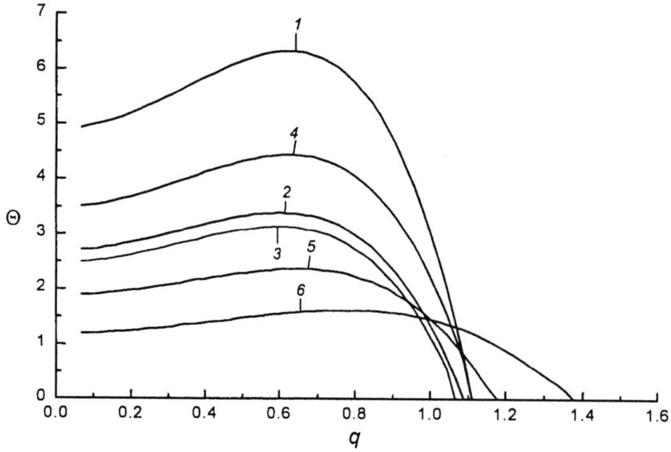

Figure 2. Maximum eigenvalue Θ vs. wave vector q ($\beta \approx 1$) : (*1*) $n = 0.5$, $b = 0.01$; (*2*) $n = 1.0$, $b = 0.01$; (*3*) $n = 1.2$, $b = 0.01$; (*4*) $n = 0.5$, $b = 0.1$; (*5*) $n = 1.0$, $b = 0.1$; (*6*) $n = 2.5$, $b = 0.1$.

$$f_1(\xi, s, \tau) = A(s)G(\xi)\exp(\theta\tau). \tag{8}$$

Boundary conditions for f_1 are as follows: when $\xi \to -\infty$, $\xi_b f_1 \to 0$, $f_{1\xi} \to 0$. In order to analyze the linear stability, we determined the spectrum of the linear differential equation of the fourth order:

$$\theta G = -\frac{\partial}{\partial \xi}\left\{\left[\frac{\partial}{\partial \xi}(G_{\xi\xi} - q^2 G) f_0^{n+2} / n + \left(1 + \frac{1}{n}\right)G + (1-\beta)G\right]\right\}\beta -$$

$$-\left(f_0^{2+1/n}\right)\left(q^4 G - q^2 G_{\xi\xi}\right)\beta. \tag{9}$$

Here the f_0 function is simplified by a step function, which reduces our analysis to investigation of the respective algebraic equation and to matching conditions for G at points f_0 of abrupt change in the step function. Thus determined the $\Theta(q)$ function is presented Fig. 2. Naturally, this function is determined by two physical parameters: n and b.

Typically the curve exhibits a maximum corresponding to growing disturbances that determine the shape film edge at the initial stage of the disturbance development. Similar analysis for the case of non-Newtonian liquid flow on the rotating disc when rz component of viscous shear stress tensor in cylindrical coordinates is written in a form reported in [12], also demonstrate instability of film front edge practically under the same parameters.

Therefore, the instability of front shape is demonstrated by using Eqs. (6), (9) that determine the shape and stability of non-Newtonian liquid film spreading over surfaces with τ' given by (4) in the vicinity of front edge.

It is necessary to account the nature of the spreading liquid by experimentally checking the prediction of our theoretical analysis. Non-Newtonian liquid may, in general, correspond to a Reiner–Rivlin liquid, exhibit viscoelastic properties, etc. The flow of such liquid films is determined by equations that differ from equations (6), (9).

Formation of Polymer Coating on a Rotating Disc

Now we will consider the second stage of coating flow. During the second stage (markedly longer compared to the first one), the film thickness is decreasing to a value of several micrometers. Hence, all the pertinent physicochemical processes of film flow are considered, *viz.*, solvent evaporation from liquid film surface, cooling of coating upper layers due to solvent evaporation, solvent diffusion to upper layers. The comparison of parameters (see above) demonstrated the need in proper account of these physicochemical processes in the numerical modeling of the second stage, during which the film properties abruptly change with time. The main attention is paid to investigation of film thickness uniformity along radius during flow of coating that is, in general, a non-Newtonian liquid. It is demonstrated [12] that free surface initial nonuniformities decrease during the flow of Newtonian liquid on a rotating disc, so that the flow becomes asymptotically uniform along the radius. During the flow of non-Newtonian liquid with rheological properties (4) (or Bingham-type liquid, see below), the resultant film thickness turns out nonuniform along the radius.

Spatially two-dimensional nonstationary flow model in the cylindrical coordinate system (r, z, φ) with velocity components $\mathbf{v}$ (u, v, w) and symmetrical with respect to φ was chosen for numerical investigation of non-Newtonian liquid flow on the disc rotating with frequency ω. The Newtonian liquid flow was also subjected to numerical testing. We have written the system of conservation equations that corresponds to the FL flow on the disc surface. We assumed FL to be a two-component liquid (solvent + polymer with ferroparticles) capable of laminar flow. Let the gas pressure over the surface be zero (solvent vapors are blown away due to base disc rotation), FL density $\rho = \rho_1 + \rho_2$, $c = \rho_2/\rho$, where ρ_2 is the solvent density.

$$\frac{\partial \rho}{\partial t} + \frac{1}{\rho}\frac{\partial (r\rho u)}{\partial r} + \frac{\partial (\rho v)}{\partial z} = 0, \tag{10}$$

$$\frac{\partial(\rho u)}{\partial t}+\frac{1}{r}\frac{\partial(r\rho u^2)}{\partial r}+\frac{\partial(\rho uv)}{\partial z}-\rho\omega^2 r=-\frac{\partial P}{\partial r}+\frac{\partial}{\partial z}\left(\eta\frac{\partial u}{\partial z}-\frac{\partial v}{\partial r}\right)+\alpha_r, \quad (11)$$

$$\frac{\partial(\rho v)}{\partial t}+\frac{1}{r}\frac{\partial(r\rho uv)}{\partial r}+\frac{\partial(\rho v^2)}{\partial z}=-\frac{\partial D}{\partial z}+\frac{1}{r}\frac{\partial}{\partial r}\left(r\eta\frac{\partial v}{\partial r}-\frac{\partial u}{\partial z}\right)+\rho g+\alpha_z, \quad (12)$$

$$\rho\frac{\partial T}{\partial t}+\bar{V}\,\nabla T=\frac{1}{\rho C_p}\lambda\Delta T, \quad (13)$$

$$\rho\frac{\partial c}{\partial t}+\bar{V}\nabla c=D\Delta c. \quad (14)$$

In (10)–(14), η is the viscosity coefficient depending, in general, on the velocity gradient and other hydrodynamic variables (temperature, solvent concentration, etc.). The acceleration α in equations for (u, v) is associated with the surface tension and can be assessed from the surface profile $H(r,t)$. It is necessary to the supplement two-dimensional nonstationary system of equations written in a motionless coordinate system by an equation that governs variation of film thickness with time [13]:

$$\frac{dH}{dt}=v(H,t) \quad (15)$$

at the boundary conditions:

$$\text{at } z=H\text{: } \tau'_{\varphi z}=\eta\left(\frac{\partial u}{\partial z}-\frac{\partial v}{\partial r}\right)=0,\ \frac{\partial v}{\partial z}=0,\ -\lambda\frac{\partial T}{\partial z}=L\dot{m},\ -\rho D\frac{\partial c}{\partial z}=\dot{m};$$

$$\text{at } z=0\text{: } u, v=0;\ T=T_s,\ \frac{\partial c}{\partial z}=0.$$

At a free surface, the conditions for the velocity stem from the absence of solvent vapors over the film surface; the boundary conditions for temperature and concentrations arise from the mass/heat balance at the surface, L is the heat of vaporization, $\dot{m}$ is determined by (1). The symmetry relative to the rotation axis is assumed together with the condition $u = 0$; at the outflow boundary, the extrapolation condition $\partial f/\partial r = 0$ is posed. The initial conditions were assumed to be uniform: thickness, temperature, and concentration are independent of radius, the velocities are zero.

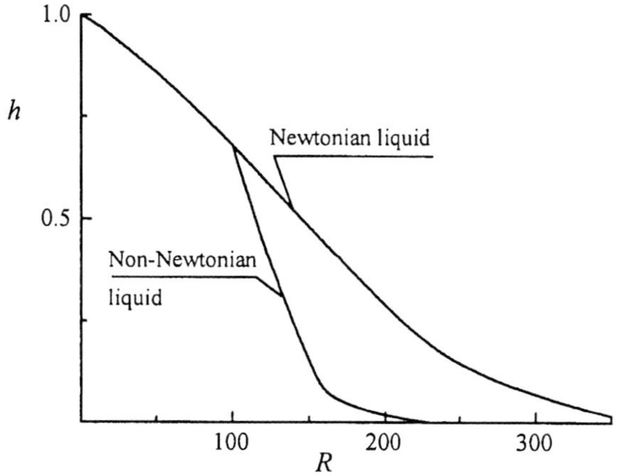

Figure 3. Film thickness h as a function of radius R.

The deviations of rheological behavior of polymer solution from Newtonian on modeling the second stage of disc coating were as follows: (a) as given by Eq. (4) and (b) as in the Bingham liquid. The equations obtained were studied in their dimensionless form.

Numerical experiments with the second-stage model were performed in two modes: space evolution and time evolution [13]. The space evolution problem corresponds to constant fluid inflow in the vicinity of axis and outflow at disk edge. The flow field was calculated for proceeding along the radius. At each new step, the flow parameters were determined by the solution to nonstationary equations (found by quasi-relaxation technique). The calculation field contained about 100 nodes in transverse direction and about 250–900 steps in the radial direction. The results obtained for the Newtonian and non-Newtonian liquids are compared in Fig. 3.

The time evolution problem corresponds to the absence of fluid inflow from outside (which corresponds to our experimental conditions). At every

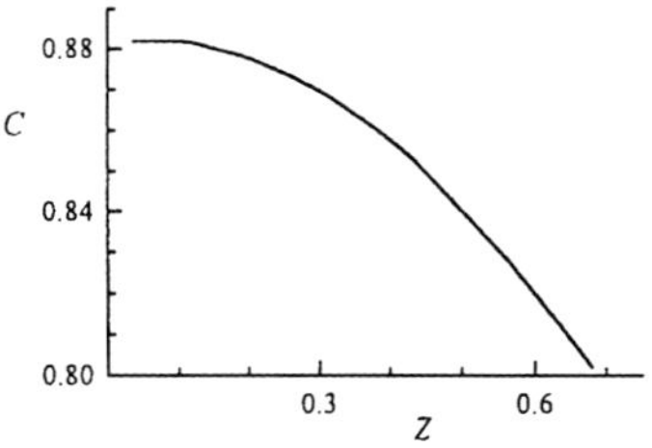

Figure 4. Typical form of concentration profile.

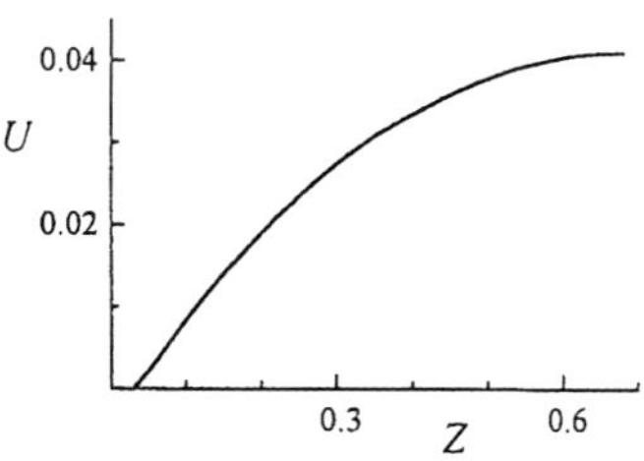

Figure 5. Typical form of velocity profile.

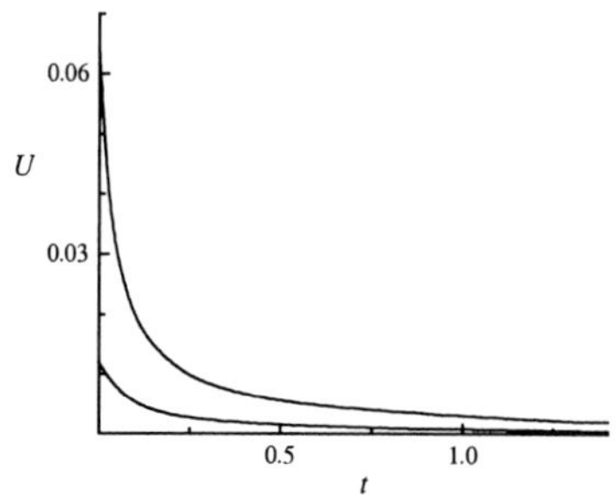

Figure 6. Time evolution of velocity for different points on radius

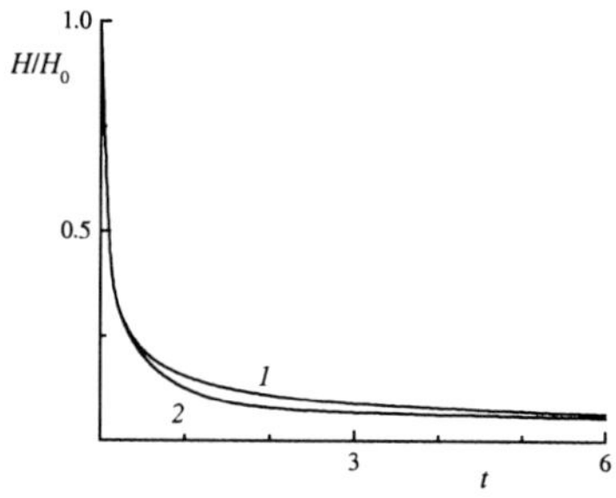

Figure 7. Time dependence of thickness in the presence/absence of evaporation: 1) without evaporation; 2) evaporation about 0.003 kg/m^2 sec.

new step, the flow field is determined from a solution to nonstationary equations. The computational field contains up to 50 nodes in cross-section and about 50–100 nodes in radial direction.

The testing was performed by using the data taken from [12, 15, 16] (see Figs. 4–8), expression for α being similar to that reported in [17]. The values used in our calculations were close to those in [15]; ω = 50 rps, H_0 = 0.1–2 mm, $T_W \approx$ 300 K. Figure 4 presents a typical concentration profile in cross-section, while Fig. 5 represents a standard velocity profile ($t \approx 1$ s, $H_0 \approx 1$ mm).

The evolution of velocity at two points on the radius (R_2/R_1 = 2) is presented in Fig. 6. Figure 7 presents the evolution of film thickness for the Newtonian fluid in the presence/absence of vaporization. The time dependence of film thickness is shown in Fig. 8 for three different values of initial thickness.

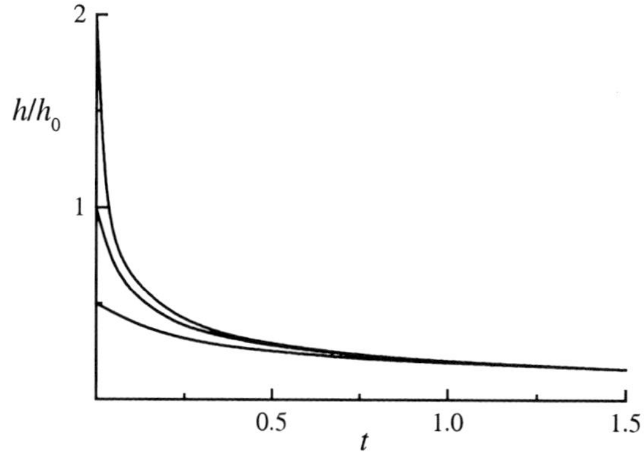

Figure 8. Time dependence of film thickness.

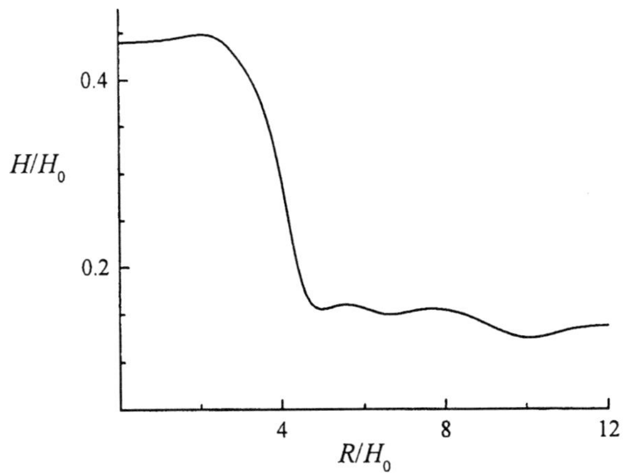

Figure 9. Film thickness at a viscosity drop at some stress.

The non-Newtonian properties (dependence of the effective viscosity on the shear rate) can cause the radial nonuniformity of deposited film. Figure 9 presents the $H(R)$ function for a model non-Newtonian (Bingham-type) liquid with $\mu = \mu_0$, $\partial U/\partial z < \gamma_0$, $\mu = 0.1\mu_0$, $\partial U/\partial z > \gamma_0$. These data show a marked dependence of thickness on the radius.

Numerical experiments allow us to conclude that the dependence of the rheological behavior on temperature (or solvent concentration) can be used to improve the surface uniformity.

CONCLUSIONS

Modeling two qualitatively different stages of polymer solution flow on a rotating disc is performed. The quasi-stationary shape and stability of front edge for a non-Newtonian fluid at the first stage of the process were studied. In modeling the second stage, special attention was paid to the effects caused by a two-dimensional character of the flow. Using complex calculations of two-dimensional nonstationary flow of a non-Newtonian liquid, we explored the impact of rheological properties of liquid on the surface profile. Factors that are responsible for uniformity of resultant surface were found to be the dependence of the rheological behavior on the shear rate, temperature, solvent concentration, and angular velocity ω.

Acknowledgments. This work was supported by the International Science and Technology Center (grant no. 358-96).

REFERENCES

1. Kutepov A.M., Polyanin A.D., Zapryanov Z.L., Vyasmin A.V., and Kazenin D.A., *Khimicheskaya gidrodinamika* (Chemical Hydrodynamics), Kvant, Moscow, 1997.
2. Landau L.D. and Lifshits E.M., *Gidrodinamika* (Hydrodynamics), Nauka, Moscow, 1988.
3. Inoue H., Fukke H., and Katsumoto M., *IEEE Trans. Magn.*, **26**, 75 (1990).
4. Lifshits E.M. and Pitaevsky L.P., *Fizicheskaya kinetika* (Physical Kinetics), Nauka, Moscow, 1979.
5. Hoking L.M., *J. Fluid. Mech.*, **211**, 373 (1990).
6. Fennelop T.K. and Waldman G.D., *AIAA J.*, **10**, 177 (1972).
7. Yakhno O.M. and Dubovitskii V.F., *Osnovy teorii polimerov* (Fundamentals of Polymer Theory), Naukova Dumka, Kiev, 1976.
8. Baturin S.M. and Pavlov G.A., *Pis'ma Zh. Tekh. Fiz.*, **22**, 80 (1996).
9. Baturin S.M., Pavlov G.A., and Shiryaev A.A., in *Strongly Coupled Coulomb Systems*, Kalman G.J., Rommel J.M., and Blagoev K., Plenum Press, New York, 1998, p. 429.
10. de Gennes P.-J., *Usp. Fiz. Nauk*, **151**, 619 (1987).
11. Spaid M.A. and Homsey G.M., *Phys. Fluids*, **8**, 460 (1996).
12. Acrivos A., Shan M.J., and Petersen E.E., *J. Appl. Phys.*, **31**, 963 (1960).
13. Belotserkovsky O.M., Guschin V.Y., and Konshin V.N., *Zh. Vychisl. Matem. Matem. Fiz.*, **27**, 594 (1987).
14. Lawrence C.J., *Phys. Fluids*, **31**, 2786 (1988).
15. Ohara T., Matsumoto Y., and Ohashi H., *Phys. Fluids A*, **1**, 1949 (1989).
16. Borkar A.V., Tsamopolous J.A., Gupta S.A., and Gupta R.K., *Phys. Fluids*, **6**, 3539 (1994).
17. Matsumoto Y., Ohara T., Teruya I., and Ohashi H., *JSME Int. J., Ser. II*, **32**, 52 (1989).

Chapter 19

Thermal Decomposition of Thin Polyepoxide Films

Lev P. SMIRNOV*

Institute of Problems of Chemical Physics, Russian Academy of Sciences, Chernogolovka, Moscow Region, 142432 Russia

ABSTRACT

Properties of polymers depend on both the structure of their molecular subsystem and their non-equilibrium excessive free volume V_{NEF}. The data of densitometry and NMR spectroscopy for unannealed epoxy films demonstrate the presence of three types of defects in highly crosslinked polymers: *n*-defects of approximately atomic

*e-mail: lsmirn@icp.ac.ru

size, their clusters, and microvoids of nanometric size. The polymeric chains involved in the structure of n-defects and their complexes are stressed. The polymeric chains in a vicinity of a microvoid are unstressed and exhibit increasing molecular mobility for $T < T_g$. A physical model for the formation of the V_{NEF} elements is suggested. It involves (1) a stage of formation n-defects in chemical reaction, (2) stepwise formation of the complexes of n-defects, and (3) a stage of microvoids formation by separation of non-equilibrium (excess) free volume in an individual phase. The duration and mechanism of annealing are associated with the size and molecular mobility of the V_{NEF} structural elements. The n-defects, as the smallest ones, are rapidly annealed, at least partially, during heating a sample up to experimental temperature. The annealing the clusters of n-defects and microvoids occurs much slower and includes a step of their disintegration to form mobile n-defects, which then relax by diffusion. At high temperatures (for $T >> T_g$), the annealing of n-defects is accompanied by decomposition of the polymer network, the polymer chains involved in the structure of n-defects and their clusters exhibiting the fastest thermal decomposition.

INTRODUCTION

For modern polymeric materials, filling is one of major ways for variation in their mechanical properties over a wide range. These materials essentially are heterogeneous systems with highly developed interphase boundaries. Therefore, their properties are determined by surface phenomena and the structure of the surface and interfacial layers. Highly crosslinked polymers are often used as composite binders for polymeric materials and composite systems. Useful properties of highly crosslinked polymers in their glassy state are attained predominantly upon cure when chemical reaction of network formation is accompanied by relaxation of the physical structure of reactive system. The available experimental data show that polymers vitrified under these conditions undergo special relaxation and exhibit the properties that cannot be achieved by other methods [1–5].

As is known, these properties (in particular, thermal stability) are associated (to some extent) with various structural defects existing both in the molecular subsystem and free-volume subsystem. In order to rationalize the relaxation behavior of highly crosslinked polymers in glassy state and, hence, to predict a change in some other properties, the available information about overall imperfectness of materials as characterized by an overall value of the excessive free volume seems insufficient [1]. Phenomenology of macroscopic shrinkage defect formation during synthesis of highly crosslinked polymers has been studied inadequately [6]. Crosslinked polymers of this type were shown to have some size distribution of microdefects, so that only a small part of the excessive free volume (related to the largest microdefects) was found to affect the molecular mobility [7]. For highly crosslinked polymers, the structure and properties of the surface layer were shown to markedly differ from those of polymer in the bulk [8]. However, many problems related to the structure formation in highly crosslinked polymers during chemical reaction (in particular, the structure of their excessive free volume) still remain open to discussion.

This work is focused on analysis of the experimental results of thermal degradation [9] and excessive free volume [10] reported for unannealed films of poly(epoxyamines) (PEA) that is known to have a relatively large non-equilibrium excessive free volume V_{NEF} [11]. The PEA films were obtained by condensation of epoxy resin ED-20 with *m*-phenylenediamine.

RESULTS AND DISCUSSION

Thermal Decomposition of Highly Crosslinked Epoxy Polymers

Durability of composite polymeric materials depends essentially on the thermal stability of their binders. Thermal degradations of unfilled and filled epoxy polymers were studied by gravimetric method using an automated vacuum thermobalance [9, 12].

Figure 1 shows that the kinetics of composite decomposition markedly differs from that of pure polymer, and depends on the type of the filler used. This difference can hardly be explained by chemical reaction between

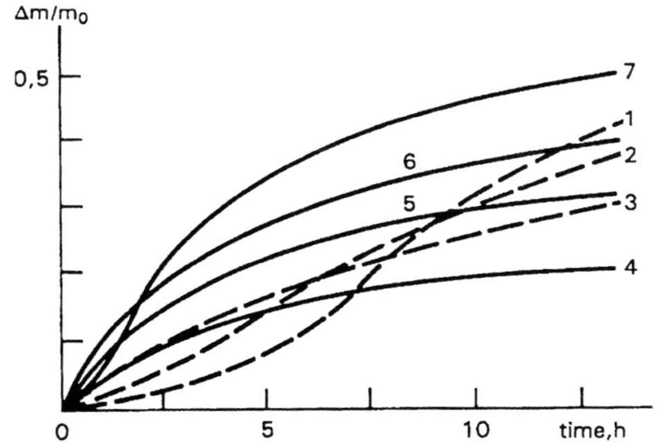

Figure 1. The kinetics of relative weight loss, $\Delta m/m_0$, during thermal degradation of highly crosslinked PEA at 543 K (according to [12]): (1) unfilled sample, 1–2-mm particles; (2) unfilled sample, 0.2-mm film; (3) unfilled sample, 0.09–0.16-mm particles; (4) polymer filled with boracic fiber; (5) polymer filled with organic fiber; (6) polymer filled with glass fiber; (7) polymer filled with carbon fiber.

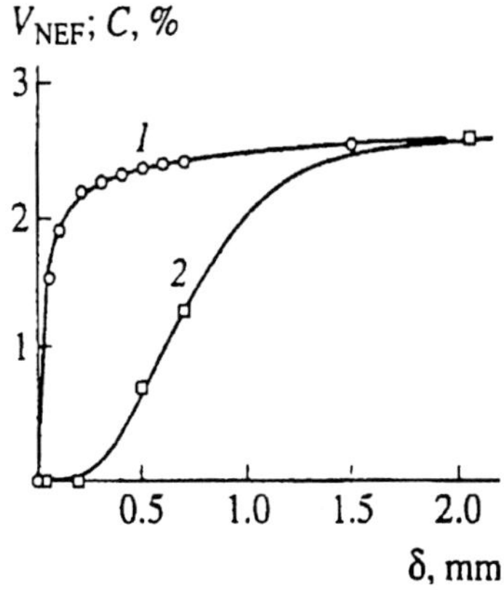

Figure 2. (*1*) The fraction of non-equilibrium excessive free volume V_{NEF} (as taken from [11]) and (*2*) relative concentration of the mobile component *C* (taken from [10]) vs. the film thickness δ.

binder and filler because fillers, being chemically inert, also increase the degradation rate. The effect is similar to 'mechanical activation' of degradation of unfilled polymer. Preliminary mechanical processing of unfilled polymer (fragmentation into pieces, grinding) was shown to increase the degradation rate. Figure 1 shows that degradation rate grows with decreasing particle size, i.e., with increasing extent of mechanical processing. According to modern concepts, tensile stress increases the rate of polymer degradation. Mechanically activated degradation was found to affect the network structure in the same manner as thermal degradation [13]. We assume that an increase in the rate of composite degradation can be associated with residual thermoelastic stresses. The magnitude of thermo-elastic stresses obviously depends on the type of filler, in particular, on the strength of adhesion between filler and binder, and on the difference between their thermal expansion coefficients.

Residual shrinkage stress of composite binders should also be taken into account. Formation of crosslinked polymers is usually accompanied by marked shrinkage. Crosslinked polymers with a high V_{NEF} are formed under conditions of composite production. In some cases, even micropores and cracks are formed. To find out the role of shrinkage stresses and defects in detail, thermal decomposition was investigated for epoxy films with thickness δ ranging between 0.02 and 2.06 mm [9]. The films were obtained under conditions simulating the state of binder in the composite. The films were shown to exhibit high V_{NEF} values [11] (Fig. 2) and no macrodefects.

Figure 3 shows that the kinetics of thermal decomposition depends on the film thickness. This dependence is more complicated than that of excessive free volume. The parameters of many physical and chemical processes (including molecular mobility and reactivity) are known to depend on the total value of V_{NEF}, more exactly, on distribution of excessive free volume over its elements. However, the structural elements of V_{NEF} ranging in their size between the *n*-defect clusters (approximately 10^{-6}–10^{-5} cm) to smallest macroscopic defects (about 10^{-2} cm) have been studied inadequately. Evidently, the structure of V_{NEF} in the PEA films should be investigated in more detail.

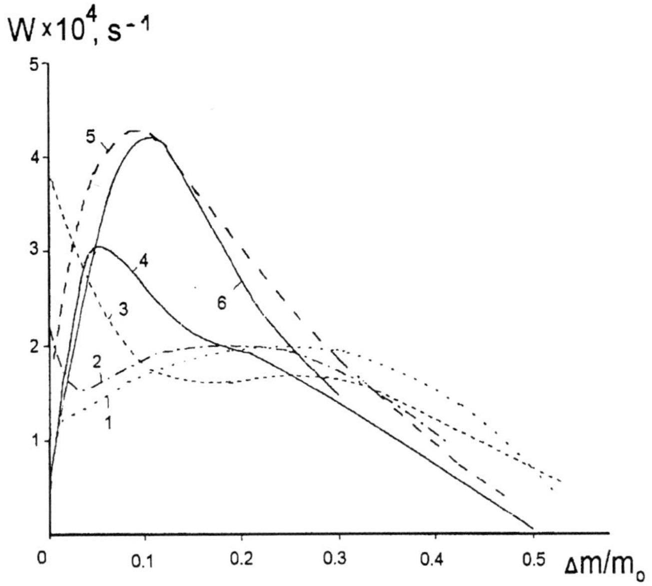

Figure 3. Decomposition rate W vs. $\Delta m/m_0$ at 533 K (taken from [9]). The film thickness δ = (1) 0.04, (2) 0.1, (3) 0.16, (4) 0.50, (5) 0.70, and (6) 2.05 mm.

NMR Study of Unannealed Poly(epoxyamine) Films

To solve this problem, the parameters of NMR spectra of the PEA films were measured in the range between room temperature and T_g [10, 14]. At low temperatures, the NMR spectrum represents a broad Gaussian line typical of amorphous material in its glassy state. A narrow peak appears at 323 K for the films above 0.2 mm thick. This peak corresponds to molecular mobility of a liquid or an amorphous polymer above T_g. Concentration of mobile proton[†], C, increases from 0 to 2.7% with increasing film thickness δ. The nature of this peak is not clear. In [14], it was attributed to impurities. However, the films were prepared under the same conditions and from the same mixture of reagents. Hence, impurities or sol fraction cannot be responsible for appearance of a mobile component in the NMR spectra. The PEA films density was shown to depend on δ [11]. Therefore, the mobile NMR component may be related to the V_{NEF} value, the latter being directly associated with the density decrement. The narrow line observed in the NMR spectra of a glassy polymer was also related to the concentration of free-volume defects [15, 16].

[†]Concentration of a mobile component C was determined as the ratio of the surface area under the peak to the total surface area of the spectrum.

If it is assumed that V_{NEF} affects the molecular mobility of a glassy system [1, 15, 16], we may also admit that protons in the mobile component belong to a polymeric segment in the vicinity of free-volume elements. A characteristic size of these elements must be at least equal to the size of the kinetic unit. Judging from the T_2 values (about 100 μs), the motion under consideration is a segmental one, and its characteristic size is about 2 nm. This estimate of the least size of the considered V_{NEF} element agrees with the data [7] on the presence of elliptic voids in 2.5-mm thick PEA samples. The principal axis of the voids is 10–15 nm long, and their volume fraction (0.02–0.04) coincides with the concentration of mobile protons (2.7%) in 2-mm PEA films (Fig. 2). For convenience, these defects will be referred to below as microvoids. Estimates show that a microvoid volume is above 10^3 atomic volumes. These results well agree with the data on the size of defects in and microheterogeneous character of glassy polymers [15, 16].

Figure 2 shows the V_{NEF} value calculated from the data of [11] in comparison with the concentration of a mobile component as assessed from the NMR spectra. For $\delta > 0.7$ mm, the concentration of a mobile component is maximal and agrees with the V_{NEF} value. This indicates that microvoids are the dominating defects. The films with $\delta \leq 0.2$ mm have a somewhat lower but still significant V_{NEF}. However, they exhibit no mobile component in the NMR spectrum. This implies that, though no microvoids favoring the appearance of segmental mobility are available in films for $T < T_g$, some defects of smaller size are still present in films with $\delta < 0.2$ mm. Samples with $0.2 < \delta < 0.7$ mm apparently contain both large and small defects. The size distribution of microdefects and dependence of their concentration on δ are also supported by the shape of free induction decay measured at 450 K.

Therefore, the above data confirm the presence of three types of microdefects in freshly prepared epoxy films: (a) small *n*-defects of atomic dimensions, (b) clusters of *n*-defects, and (c) microvoids with the dimensions of the kinetic segment.

Physical Model for Formation of Shrinkage Microdefects

The above data suggest the following physical model for formation of shrinkage microdefects during network polymer synthesis. This model takes into account the available data on the structure the surface layer of highly crosslinked polymers [8].

Appearance of non-equilibrium small-size defects can be attributed to chemical reaction of network formation:

$$\sim\text{A} + \text{B}\sim \longrightarrow \sim\text{P}\sim + V,$$

where ~A and B~ are the functional groups that react to form a crosslink ~P~, and V is the average volume change in an elementary event (defect of a minimum size, 'hole' [1]). A hole in PEA must have atomic dimensions. In the initial liquid, the hole dissipates rapidly, and spreads over the volume. As a result, a collection of atoms, including several coordination spheres, has a loosened state (*n*-defect). Because of rapid relaxation via the local and diffusion mechanisms [1, 3], non-equilibrium excessive free volume never reaches high values at low conversion η.

Microvoids are much larger than *n*-defects, and, therefore, cannot immediately appear in the reaction. We believe that a reduction in the molecular mobility of reacting system (especially pronounced after passing the gel point η_g) leads to a marked growth in the concentration of *n*-defects. The resultant supersaturation favors separation of non-equilibrium excessive free volume into an individual phase, predominantly in the central layers of the sample. These central layers are characterized by maximum supersaturation, and they have no factors that determine the structural features of the surface layer. At the first step, clusters V_i (each composed of 10–20 *n*-defects [1]) form by a stepwise mechanism:

$$V_{i-k} + V_k \rightarrow V_i \quad (k,\ i-k \geq 1).$$

The loosened state of the polymer, caused by formation of *n*-defects and their clusters, is associated mostly with the sample internal energy, in other words, with development of local structural stresses.

On reaching a critical size, the cluster of *n*-defects transforms into a microvoid M_j:

$$V_j \rightarrow M_j \quad (j = j_{cr}).$$

This transformation is apparently 'triggered' by the rupture of a few overstressed chemical bonds. Formation of microvoids is a thermodynamically favorable process, resulting in the decrease of internal energy (release of local structural stresses) and growth of entropy [7].

Structure of a microvoid in the native state can be represented as a nanoparticle comprising a loosened polymer layer around a free space (microvoid). Macrochains in this layer are unstressed and, hence, possess a significant molecular mobility even for $T < T_g$. On the supramolecular and topological levels, this layer has a less-ordered structure as compared to the surface layer (due to different conditions of their formation). The local concentration of *n*-defects in the vicinity of a microvoid is determined by temperature.

Microvoids may absorb smaller ones and attach the 'holes' formed in reaction. As a result, microvoids may grow and reach, under certain conditions, macroscopic dimensions [6]. Formation and growth of micro- and macrodefects significantly slows down in the region of glass transition because of a drop in the system molecular mobility for $\eta \geq \eta_v$ (η_v is the conversion corresponding to the moment of concentration vitrification). If the chemical reaction proceeds upon attaining η_v, the growth of structural stresses leads to the rupture of constrained chemical bonds, and formation of microcracks [6].

Physical Model for Annealing Shrinkage Defects

The *n*-defects, their clusters, and microvoids are essentially non-equilibrium species that can be annealed. There are two mechanisms for annealing of defects [1]: local and diffusion-assisted. The duration and mechanism of annealing the structural elements of V_{NEF} are associated with their size and molecular mobility: the larger the defect size, the longer is the time of its annealing. Being the smallest, *n*-defects are rapidly annealed, at least partially, via the local mechanism during sample heating up to annealing temperature, so that no migration of *n*-defects to the sample surface takes place.

Annealing of *n*-defect clusters proceeds via the diffusion mechanism [1]. Though this defect is not mobile (because of its size), at sufficiently high temperature, local equilibrium between *n*-defect clusters and *n*-defects in its vicinity is shifted. Because of the cluster dissociation, the concentration of mobile *n*-defects increases, i.e. annealing of the clusters includes their disintegration with formation of mobile *n*-defects:

$$V_i \leftrightarrow V_{i-1} + V.$$

This is followed by migration of *n*-defects to the sample surface or microvoids, provided that the annealing temperature is not too high.

We believe that the mechanism of the microvoid annealing is similar, in some respects, to that of annealing the *n*-defect clusters in polymers or vacancy pores in crystals. At sufficiently high temperature, the local equilibrium between a microvoid and *n*-defects arranged in its vicinity is shifted:

$$M_i \rightarrow M_{i-1} + V.$$

The concentration of mobile *n*-defects increases during the microvoid disintegration. Then mobile *n*-defects migrate to the sample surface, if the annealing temperature is not too high. At high temperature ($T >> T_g$), annealing the shrinkage defects is accompanied by decomposition of a network with polymeric chains that are involved in the *n*-defects, their clusters having the highest rate of decomposition.

Physical Model for Thermal Decomposition of Unannealed Poly(epoxyamine) Films

Thermal decomposition of unannealed polymers occurs simultaneously with annealing of their defects.

This model explains the kinetic curves at the initial steps of thermal degradation (Fig. 3). It should be taken into account that the chemical bonds of polymer chains involved in *n*-defects and their clusters are loaded by tensile stresses which promote an increase in the decomposition rate.

Indeed, the thinnest films have a small initial rate of decomposition because their initial *n*-defects are annealed during sample heating and, hence, do not affect thermal destruction. The samples with $\delta = 0.1$–0.3 mm exhibit the greatest initial rate of thermal decomposition because they have mainly the *n*-defect clusters which do not disappear during sample heating. Their decomposition rate decreases during the initial stage, as the amount of *n*-defect clusters (having a high decomposition rate of stressed polymeric segments) decreases with increasing η.

The thickest samples decompose with the least initial rate, because they have only microvoids, with polymeric segments having no structural stresses and, hence, a low rate of decomposition. The destruction rate of these samples increases during reaction, as the concentration of *n*-defects and their total content grow during disintegration of microvoids. An increase in the rate of thermal degradation gradually diminishes because the size and amount of microvoids reduce with increasing extent of thermal decomposition.

CONCLUSIONS

Thermal destruction of polymers is normally used to determine their durability. The above results illustrate the applicability of the technique of thermal destruction to elucidating the structure of highly crosslinked polyepoxides and its evolution during cure and annealing. Similar approach was also applied for other crosslinked and linear amorphous polymers (polyurethane, polydimethacrylate, and polystyrene) [12].

Acknowledgements. This work was supported by the International Science and Technology Center (grant no. 358-96). The author is grateful to B. A. Rozenberg and V. I. Irzhak for fruitful discussions.

REFERENCES

1. Rostiashvili V.G., Irzhak V.I., and Rozenberg B.A., *Steklovanie polimerov* (Glass Transition in Polymers), Khimiya, Leningrad, 1987.
2. Irzhak V.I., Rozenberg B.A., and Enikolopyan N.S., *Setchatye polimery: Sintez, struktura i svoistva* (Network Polymers: Synthesis, Structure and Properties), Nauka, Moscow, 1979.
3. Rozenberg B.A. and Irzhak V.I., *Prog. Colloid Polym. Sci.*, **90**, 194 (1992).
4. Irzhak V.I. and Rozenberg B.A., *Vysokomol. Soedin., Ser. A*, **27**, 1795 (1985).
5. Berlin A.A., Korolev G.V., Kefeli T.Ya., and Sivergin Yu.M., *Akrilovye oligomery i materialy na ikh osnove* (Acrylic Oligomers and Related Materials), Khimiya, Moscow, 1983.
6. Korotkov V.N., Chekanov Yu.A., and Rozenberg B.A., *Polym. Sci., Ser. A*, **36**, 564 (1994).
7. Gupta V.B. and Brahatheeswaran C., *Polymer*, **32**, 1875 (1991).
8. Grishchenko A.E. and Cherkasov A.N., *Usp. Fiz. Nauk*, **167**, 269 (1997).
9. Volkova N.N., Bogdanova L.M., Irzhak V.I., Summanen E.V., Rozenberg B.A., and Smirnov L.P., *Abstr. All-Union Conf. on Phys. and Technology of Thin-Film Polym. Systems*, Tashkent, 1991, p. 89.
10. Bogdanova L.M., Volkova N.N, Lankin A.V., and Smirnov L.P., *Polym. Sci., Ser. A*, **40**, 160 (1998).
11. Bogdanova L.M., Ponomareva T.I., Irzhak V.I., and Rozenberg B.A., *Vysokomol. Soedin., Ser. A*, **26**, 1400 (1984).
12. Smirnov L.P. and Volkova N.N., *Prog. Colloid Polym. Sci.*, **90**, 222 (1992).
13. Sandakov G.I., Smirnov L.P., Sosikov A.I., Summanen K.T., and Volkova N.N., *Prog. Colloid Polym. Sci.*, **90**, 235 (1992).
14. Ermolaev K.V., Dubovitskii V.A., Volkova N.N., and Erofeev L.N., *Abstr. III Int. Conf. on Magnetic Resonance Spectroscopy*, Würzburg, Germany, 1995, p. 147.
15. Lee K.L., Inglefield P.T., Jones A.A., Bendler J.T., and English A.D., *Macromolecules*, **21**, 2940 (1988).
16. Lee K.L., Jones A.A., Inglefield P.T., and English A.D., *Macromolecules*, **22**, 4298 (1989).

Section V

NEW APPROACHES TO POLYMER NETWORKS CHARACTERIZATION

Chapter 20

Characterization of Molecular Weight Distribution for Linear and Network Polymers in the Bulk

Vadim I. IRZHAK*

Institute of Problems of Chemical Physics, Russian Academy of Sciences, Chernogolovka, Moscow Region, 142432 Russia

ABSTRACT

Current approaches to characterization of MWD without using solutions have been analyzed with special emphasis on the non-isothermal method based on

*e-mail: irzhak@icp.ac.ru

thermomechanical analysis developed at the title Institute. Some aspects of the approach are discussed, such as choosing sufficient loading and plotting the calibration curves. This approach was found applicable to network polymers.

INTRODUCTION

Molecular weight (MW) and molecular weight distribution (MWD) are important structural characteristics of polymers. There are many methods for MW determination, but only some of them can be used for MWD evaluation. The most common and widely spread method is the size exclusive chromatography (SEC) [1]. Similar to all available methods, this one is based on analysis of the properties of dilute polymer solutions. Therefore, it is impossible to characterize MWD of many insoluble polymer systems (for example, network polymers). There are some destructive methods [2] but they are not widely practiced due to their high labor consumption.

In last 20 years, the attempts have been made to use the relaxation properties of polymers in the bulk. This approach is based on the concept that there exists some relation between the polymer relaxation spectrum and MWD.

At the Institute of Problems of Chemical Physics, a series of studies was performed [3, 4] to demonstrate the possibility of MWD evaluation for both linear and crosslinked chains of network polymers by thermomechanical analysis (TMA). This technique will be described in this chapter. Since the problem of MWD determination in the bulk is novel and studied inadequately, we will also discuss some other relaxational approaches to the problem.

ISOTHERMAL METHODS

Analysis of Flow Curves

Evaluation of MWD via the analysis of flow curves is based on the dependence of the viscosity of monodisperse polymers on the shear rate $\dot{\gamma}$ [5]:

$$\eta(\dot{\gamma}) = \begin{cases} \eta_0 \\ \tau_s / \dot{\gamma} \end{cases} \text{at} \begin{cases} \tau \le \tau_s \\ \tau > \tau_s \end{cases} .$$

Here η_0 is the ultimate value of the Newtonian viscosity, and τ_s is the stressvalue at flow failure. Taking into account that $\eta_0(M) = kM^a$, where

M is the polymer chain MW and a is a constant usually equal to 3.4, the following relationship was obtained in [5]:

$$\eta(\dot{\gamma}) = \left[\int_0^{M(\dot{\gamma})} (kM^a)^{1/a} f(M)\,dM + \left(\frac{\tau_s}{\dot{\gamma}}\right)^{1/a} \int_{M(\dot{\gamma})}^{\infty} f(M)dM \right]^a,$$

where $M(\dot{\gamma}) = (\tau_s / k\dot{\gamma})^{1/a}$.

This approach allowed solution of the inverse problem of MWD estimation by analyzing the flow curve for the case when $f(M)$ is a unimodal function [6]; in particular, there were normal logarithmic distributions and the Bizli one. The character of the flow curves was found to be very sensitive to the parameters of distribution [6].

A similar relation was obtained in [7]:

$$\sum_{i=m+1}^{\infty} \omega_i = \frac{(\eta_{m+1}/k)^{1/a} - (\eta_m/k)^{1/a}}{M_{m+1} - M_m},$$

where ω_i and M_i are the weight fraction and MW of the i-th fraction.

The relaxation spectra of monodisperse polymer and the same fraction in polydisperse polymer were assumed [7] to be identical.

The data assessed from analysis of blends with bimodal distribution [8] and for one of commercially available polystyrenes were found consistent with the SEC data.

Dynamic Measurements

The attempts to use the data of dynamic measurement, i.e., the frequency dependence of $G'(\omega)$ and $G''(\omega)$, for MWD characterization were made about 30 years ago [9–11]. This approach is based on the theoretical or empirical relation between the relaxation spectra and MW of polymers [12–14]. The Rouse model as the simplest theory was used by Ninomiya [15–17]. Later [18], a more complex model by Doi–Edwards was used [19], which took into account the factor of high concentration and strong interchain interaction.

The available data obtained for monodisperse polymers and their binary mixtures shows [20–27] the following.

(1) The $G'(\omega)$ curves in the logarithmic coordinates are similar for different MW in a high-frequency range (high elasticity plateau) and differ in a low-frequency range; the curves can be made coincident by shifting the frequency by a factor proportional to M^a, where $a = 3.4$–3.7 [12–14, 28, 29].

(2) The $G''(\omega)$ curves have peaks whose position depends on MW, similar to the low-frequency tails of $G'(\omega)$;

(3) In case of binary mixtures, similar relationships hold true, but the low--frequency tails of $G'(\omega)$ and the peaks of the $G''(\omega)$ curves are shifted: the high-molecular ones are shifted to low-molecular peaks, and vice versa.

The latter is a result of interchain interaction, i.e., the effect of coupling [24, 30–35].

Another feature of $G'(\omega)$ and $G''(\omega)$ curves is that the level of the high elasticity plateau in the first case and the peak value in the second case depend on component concentrations. There are some steps on the $G'(\omega)$ curves; the $G''(\omega)$ curves are split, and the relative peak heights also depend on relative concentration of components.

However, the data reported on relationships between these values are contradictory. Different formulas have been suggested in the literature. In [36], for example, a linear function was proposed:

$$G'(t) \approx \varphi_1 G_1'(t/\lambda_1) + \varphi_2 G_2'(t/\lambda_2),$$

where φ is a weight fraction and λ_i is some empirical coefficient characterizing the effect of coupling (see also [15–17]).

This formula is empirical, but it is based on the theoretical concept of the so-called reptation model for diffusion of polymer chain in its simplest form [19]: the chain motion is supposed to occur inside a 'tube' and does not depend on environment. At the same time, the quadratic and cubic relations were found [12, 37–40], and explanations were given based on more complicated reptation [41–44]. The existing rules of mixing are summarized in paper [36].

A method of MWD characterization based on quadratic dependence of $G'(\omega)$ on concentration was described in [24]. The main idea is as follows: each polymer fraction is characterized by a frequency corresponding to its MW. At lower frequencies, we can ignore the contribution from this fraction to the relaxation modulus. Contributions from fractions with higher MW are proportional to their concentrations, w_i, according to the law:

$$w_i = \left(G'(\omega_i)/G_N^0\right)^{1/2}.$$

Comparison of the data obtained in this way with the SEC data show [22] that, for binary mixtures, the results obtained by the dynamic method are more contradictory than those based on the analysis of flow curve.

When the frequency dependence of viscosity was used, a better consistency was achieved [8]. This result is likely to be due to the fact that the viscosity is a function of both the real and imaginary parts of the complex modulus [12, 13]:

$$\eta(\omega) = \left[\left([G']^2 + [G'']^2\right)/\omega\right]^{1/2}. \tag{1}$$

A comparison of different approaches allows us to conclude [8] that equation (1) leads to the best results: the data assessed from the $G'(\omega)$ curves are slightly overestimated while those from the $G''(\omega)$ curves are slightly underestimated.

However, this research deals with binary mixtures. No data have been reported for more complicated systems with arbitrary MWD. In [45], dynamic characteristics are used for evaluation of the polydispersity index, $\gamma = M_w / M_n$. This value is determined as the point of intersection of the $G'(\omega)$ and $G''(\omega)$ curves. However, this is only an assumption. No data confirming this concept have been reported so far.

Method of Spin–Spin Relaxation (Pulsed NMR)

The above approaches are applicable only to linear polymers. Obviously, the former cannot be used for analysis of network polymers because stationary flow is impossible in this case. Hence, more promising seems to be the method of pulsed NMR. Development of this method was possible due to fundamental studies [46–48] on spin–spin relaxation in linear and network polymers.

The relationship between the profile of the free induction decay curve, $I(\omega)$, and MWD function $P(N)$ is given by [46]:

$$I(\omega) \approx \int P(N)dN \frac{d\omega_{loc}}{d\cos\theta}.$$

Its time dependence has the form:

$$G(t) = \int_1^N \int_0^{\pi/2} g(N,\theta,t)\, P(N)\, d\cos\theta\, dN, \tag{2}$$

where

$$g(N,\theta,t) \approx \exp\left\{-b_{02}^2(3\cos^2\theta - 1)^2 t^2 \ln N\, N^{-2}\right\}, \text{ and } b_{02} = \omega_{loc}.$$

The free induction decay is characterized by a long component with its initial part containing information about MWD, its time-dependent part being proportional to an average length of the chain between the entanglements and inversely proportional to time. Therefore, the MWD of macromolecules with a length more than that of the chain between entanglements N_0 cannot be evaluated. Comparison with the experiment shows [48] that $N_0 \approx 10^5$ segments.

Equation (2) proved to be useful for analysis of network polymers [49–52]. The integral of $G(t)$ was found to have the following value:

$$F(t) = \int_0^t \tau\, G(\tau)\, d\tau \approx -0.03 < N^2 > + < N > t \frac{\pi}{2\sqrt{3}}.$$

This expression can be used to calculate the polydispersity index γ_c for network chains:

$$\gamma_c = \frac{\overline{M}_{wc}}{\overline{M}_{nc}} = \frac{< N^2 >}{< N >^2}.$$

Therefore, MWD of both the network chains and linear polymers can be assessed from the curves of free induction decay.

NONISOTHERMAL METHOD

Background of the Method: Thermomechanical Analysis

Application of thermomechanical analysis (TMA) to determination of MWD is based on a relationship between a temperature spread of high elasticity plateau for linear polymers ($\Delta T = T_g - T_f$, where T_f and T_g are the temperatures of flow and glass transition) and their MW. This relation was established by Kargin and Slonimskii about 50 years ago [53]: it turned out to be similar to the WLF equation. Further studies have shown that the shape of TMA curves is very sensitive to polydispersity of polymer. It is due to the fact that T_f depends on MWD.

The direct problem, i.e., determination of T_f (depending on the polymer chemical structure and MW) in the course of TMA, was discussed in [54]. However, the authors of [54] were pessimistic about the possibility of using this method for the determination of MWD.

The theoretical background of the approach based on the model of physical network was proposed in [55–57].

The classical theory of network elasticity taking into account the temperature dependence of the concentration of physical crosslinks gives the most obvious relation between a value of the relaxation modulus for a linear polymer and its MWD. The elastic state can be related to the network (gel) state, so that transition to flow can be considered as the gel–sol transition. This concept was discussed in [55].

According to [58–60], the equilibrium concentration of physical crosslinks g can be expressed as:

$$g = n(D\exp D - \exp D + 1)/(\exp D - 1)^2 ,$$

where n is the most probable concentration of physical crosslinks in the system, $D = \Delta E / kT$, ΔE is the free energy of crosslink formation, including the free energy of deformation of the chains joined to the crosslink.

Dividing the number of network crosslinks by the number of chains N, we obtain the index of crosslinking with respect to the physical crosslinks. Hence, a criterion for transition to flow can be written as

$$n / N = M \frac{D\exp D - \exp D + 1}{(\exp D - 1)^2} = 1 .$$

As far as destruction of the network occurs at a sufficiently large value of parameter D, we obtain:

$$\ln M \approx D. \tag{3}$$

Taking into account that T_f and T_g coincide for $M = M_s$, where M_s is the MW of the segment, from (3) it follows that

$$\ln M = \ln M_s + \frac{\Delta E}{RT_g}\left(1 - \frac{T_f}{T_g}\right). \tag{4}$$

The factor at the brackets in the right-hand side of (4) appeared [54, 55] to be constant. Therefore, equation (4) can be taken as a basis for plotting the calibration curve M as a function of T [55].

The study of bi- and trimodal polymer blends [3, 55] shows that the temperature of transition of polymer fraction to sol is equal to that of unimodal polymer with an identical MW. This fact allows [3, 55] us to relate the polymer system modulus (or compliance, $I(T)$) to the MWD function:

$$I(T) = \varepsilon / \sigma = I_0 / [1 - \varphi(T)], \tag{5}$$

where

$$\varphi(T) = \int_0^\infty \omega(M) dM \int_{-\infty}^{\Delta T} \delta(t) dt,$$

$\omega(M)$ is the weight function of MWD, $\Delta T = T - T_f(M)$, and $\delta(t)$ is the Dirac delta function with the integral equal 0 for negative values of the argument and 1 for $\Delta T > 0$.

Therefore, Eq. (5) gives the relation between the temperature dependence of the modulus in the region of the elasticity plateau and the MWD of a linear polymer, i.e., $\varphi(T)$ is actually the integral function of MWD expressed in terms of temperature rather than the MW of the polymer fraction.

As mentioned above, there is some relation between the character of network deformation and MWD of interknot chains [4]. This implies that the gel–sol approach cannot be applied to network polymers. Another theoretical idea based on a concept of relaxation of chain conformations in polymer [61] was proposed in [56].

Formal thermomechanical behavior of linear and network polymers was not different, though for the latter the factor of time was taken into account (it is evident that, for the former, the factor of time is negligible).

If the mechanical model of Kelvin and Voigt is taken as a background for describing both the linear and network polymers, one can obtain the following relations between $\varphi(T)$ and deformation (or compliance) of the polymers:

linear polymer: $$\varphi(T) = (\varepsilon - \varepsilon_0) / \varepsilon ; \tag{6}$$

network polymer: $$\varphi(T) = \frac{\varepsilon - \varepsilon_0}{\varepsilon} \frac{\varepsilon_{eq}}{\varepsilon_{eq} - \varepsilon_0}. \tag{7}$$

Here

$$\varphi(T) = \sum_{j=1}^{\infty} \omega_j \delta_j(t, T),$$

$$\delta_j(t, T) \cong \delta(t - \theta_j(T)) = \begin{cases} 0, & t < \theta_j(T) \\ 1, & t > \theta_j(T) \end{cases},$$

$\theta_j(T)$ is the relaxation time for a given chain conformation; ε, ε_{eq} and ε_0 are the current deformation (at reference temperature), equilibrium deformation (in case of network polymers), and deformation of polymer with infinitely high MW, respectively. The ε_0 value refers to that part of the temperature dependence of the modulus that does not depend on the MW and MWD of the polymer.

Practical Realization of the Method

The TMA technique can be implemented for different ways of loading: in standard devices, e.g., UIP-70M (Russia), penetration and compression are being used.

The magnitude of deformation is usually obtained as a difference between the stock displacement curve at a given method of loading and the dilatometric curve. In the case of compression, the real deformation is proportional to this value. In case of penetration of a spherical stock, the value of the stock displacement relates to the material modulus E via the Hertz law [62]:

$$E = \frac{3(1-\mu^2)P}{4r^{1/2}h^{3/2}},$$

where μ is the Poisson coefficient, P is the load, and r is the stock radius.

As a result, deformation ε determined as a load to modulus ratio is proportional to the stock displacement h to power 3/2, i.e., $\varepsilon \propto h^{3/2}$. In this case, equations (6) and (7) can be rewritten in the form:

$$\varphi(T) = \frac{h_T^{3/2} - h_0^{3/2}}{h_T^{3/2}}, \tag{6a}$$

$$\varphi(T) = \frac{h_T^{3/2} - h_0^{3/2}}{h_T^{3/2}} \frac{h_p^{3/2}}{h_p^{3/2} - h_0^{3/2}}. \tag{7a}$$

Assessment of the $\varphi(T)$ function from the data of TMA is illustrated in Fig. 1.

The main problem in practical application of TMA is elucidating a relation between the temperature and MW, i.e., transition from the $\varphi(T)$ function to the $\varphi(M)$ function. As we mentioned, such a transition is possible only under the assumption that the relaxation of the chains with a given MW does not depend on whether the polymer system contains only these chains or they are a part of the system consisting of many fractions with different MW. This holds true for both the linear polymers and network chains. Therefore, the task is seemingly simple: the dependence of the extension of the high-elasticity plateau $\Delta T = T_g - T_f$ on MW has to be found. However, finding this relation faces some experimental difficulties. One is determination of ε_0, the deformation of a polymer with infinite MW. This value and its temperature dependence can be found by using model samples with narrow MWD and very high MW within a wide temperature range of the high-elasticity plateau. Such work can encounter such a problem as the thermal stability of polymer system. But usually such a temperature dependence is rather weak.

Another difficulty is due to the fact that the temperature of transition from the rubbery state to flow depends strongly on MWD. For polymers with a very narrow MWD, this transition is well pronounced [64], so that the transition temperature can be found easily. Therefore, calibrating plot (4) can be obtained using such polymers as a reference [3, 55]. In other cases [3], relationships similar to Kargin–Slonimsky equation were suitable:

for polar polymers: $$\log M = 1.6 + \frac{20(T_f - T_g)}{200 + T_f - T_g}$$

for nonpolar polymers: $$\log M = 2.0 + \frac{10(T_f - T_g)}{100 + T_f - T_g}.$$

In the latter case the calibrating graphs were obtained as follows. The ultimate values of MW from differential MWD curves (data of SEC

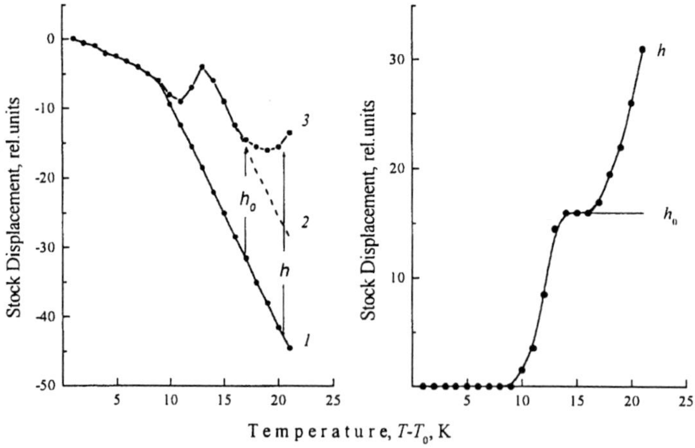

Figure 1. Assessment of the function φ(T) from the TMA data for a linear polymer: (*1*) dilatometric curve, (*2*) stock displacement for a hypothetical polymer with infinitely high MW, and (*3*) stock displacement.

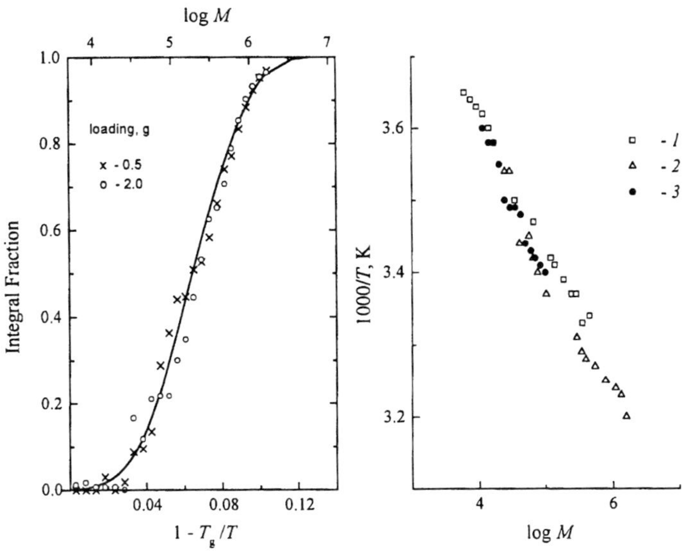

Figure 2. Correlation between TMA and SEC data for polypropylene [63, 64]. Left: direct comparison of φ(*T*) obtained by TMA (point) and φ(*M*) obtained by SEC (curve) methods. 1 and 2 correspond to different loading on the stock. Right: *T*–*M* correlation curve for polypropylene of MW, 10^3: (*1*) 52, (*2*) 189, and (*3*) 330.

analysis) were compared with temperatures of the beginning and the end of the high elasticity plateau. Errors of determination of these values are the highest. Therefore, such approach is considered to be rather incorrect. A more correct solution of the task was proposed in [63, 64].

The integral MWD obtained from SEC data should be compared with integral curve φ(*t*) obtained from TMA. Coordinates φ(*T*) vs. $(1 - T_g/T)$ could be used. Comparison of MW and T values corresponding to the same values of the integral fraction allows a relation between them to be found, i.e., the calibrating graph to be obtained. Since the methodology is based on a total set of points, the errors will be inconsiderable. The graph obtained for amorphous polypropylene (Fig. 2) is shown as an example.

Table 1. Parameters of molecular weight distribution for linear polybutadiene and respective network chains

Polymer	$\overline{M}_n$	$\overline{M}_w$	$\overline{M}_w / \overline{M}_n$
Linear	1820	2920	1.60
Crosslinked 1	1620	2760	1.64
Crosslinked 2	2000	2770	1.39
Crosslinked 3	1930	3070	1.59

As one can see, there is a good accordance between the data obtained by TMA and SEC methods. Besides, our data show that correlation (4) is apparently correct.

The examples of successful application of the described approach to determination of MWD of linear polymers and network chains are given in [3, 4, 55, 56, 62, 63]. Table 1 presents an example of MWD curves obtained by SEC and TMA for linear polybutadiene and network formed by its crosslinking via the terminal groups under different conditions [56].

This means that the TMA method can be efficient for evaluation of MWD of both linear and network polymers.

CONCLUSIONS

Existing methods of MWD determination are based on analysis of dilute solutions. As far as many systems can not be dissolved altogether (e.g., polymer networks), the problem of finding approaches that could be applied to bulk polymers is of current importance. This chapter shows that appropriate techniques are still lacking. Probably, this is due to difficulties of taking into account the effect of coupling. The approach based on NMR cannot be applied to wide experimental and technological practice due to its technical and theoretical complexity. On the contrary, the methodology based on TMA is promising due to its technical simplicity and clear interpretation. In our opinion, it is not difficult to design a proper device. We also believe that there are no principal obstacles for application of this approach to polymer blends and filled polymer systems. But this work demands hard additional efforts.

Acknowledgments. This work was supported by the International Science and Technology Center (grant no. 358-96).

REFERENCES

1. Belen'kii V.G. and Vilenchik L.Z., *Khromatografiya polimerov* (Chromatography of Polymers), Khimiya, Leningrad, 1978.
2. Berlin Al.Al. and Volfson S.A, *Kineticheskii metod v sinteze polymerov* (Kinetic Method in Polymer Synthesis), Khimiya, Moscow, 1973.
3. Russ. Patent 1 763 952, 1993.
4. Russ. Patent 2 023 255, 1994.
5. Malkin A.Ya. and Teishev A.E., *Vysokomol. Soedin., Ser. A*, **29**, 2230 (1987).
6. Malkin A.Ya. and Teishev A.E., *Vysokomol. Soedin., Ser. A*, **30**, 175 (1988).
7. Bersted B.H. and Slee J.D., *J. Appl. Polym. Sci.*, **21**, 10, 2631 (1977).

8. Tuminello W.H. and Cudre-Mauroux N., *Polym. Eng. Sci.*, **31**, 1496 (1991).
9. Ninomiya K. and Fujita H., *J. Colloid Sci.*, **12**, 204 (1957).
10. Ninomiya K. and Fujita H., *J. Polym. Sci.*, **12**, 233 (1957).
11. Ninomiya K. and Fujita H., *J. Phys. Chem.*, **61**, 814 (1957).
12. Ferry J.D., *Viscoelastic Properties of Polymers*, Wiley, New York, 1980.
13. Vinogradov G.V. and Malkin A.Ya., *Reologiya polimerov* (Rheology of Polymers), Khimiya, Moscow, 1977.
14. Frenkel' S.Ya. and Bartenev G.M., *Fizika polimerov* (Polymer Physics), Khimiya, Leningrad, 1990.
15. Ninomiya K., *J. Colloid Sci.*, **14**, 49 (1959).
16. Ninomiya K., *J. Colloid Sci.*, **17**, 759 (1962).
17. Ninomiya K. and Ferry J.D., *J. Colloid Sci.*, **18**, 421 (1963).
18. Wu S., *Polym. Mater. Sci.*, **50**, 43 (1984).
19. Doi M. and Edvards S.F., *The Theory of Polymer Dynamics*, Clarendon Press, Oxford (UK), 1986.
20. Onogi S. and Masuda T., *Kobunshi*, **17**, 640 (1968) (cited in Ferry J.D., *Viscoelastic Properties of Polymers*, Wiley, New York, 1970).
21. Vinogradov G.V., *Mekh. Polym.*, **1**, 160 (1975).
22. Tuminello W.H., *Polym. Eng. Sci.*, **26**, 1339 (1986).
23. Rubinstein M. and Colby R.H., *J. Chem. Phys.*, **89**, 5291 (1988).
24. Ylitalo C.M., Kornfeld J.H.A., and Fuller G.G., *Macromolecules*, **24**, 749, (1991).
25. Jackson J.K. and Winter H.H., *Macromolecules*, **28**, 3146 (1995).
26. Hayes C., Bokobza L., Boue F., Mendes E., and Monnerie L., *Macromolecules*, **29**, 5036 (1996).
27. Fuchs K., Friedrich C., and Weese J., *Macromolecules*, **29**, 5893 (1996).
28. Fodor J.S. and Hill D.A., *Macromolecules*, **28**, 1271 (1995).
29. Adachi K. and Kotaka T., *Macromolecules*, **17**, 120 (1984).
30. Merrill W.W., Tirrell M., Jarry J.-P., and Monnerie L., *Macromolecules*, **22**, 896 (1989).
31. Doi M., Rearson D., Kornfeld J., and Fuller G.G., *Macromolecules*, **22**, 1488 (1989).
32. Watanabe H., Kotaka T., and Tirrell M., *Macromolecules*, **24**, 201 (1991).
33. Seidel U., Stadler R., and Fuller G.G., *Macromolecules*, **27**, 2066 (1994).
34. Ylitalo C.M., Zawada J.A., Fuller G.G., Abetz V., and Stadler R., *Polymer*, **33**, 2949 (1992).
35. Seidel U., Stadler R., and Fuller G.G., *Macromolecules*, **28**, 3739 (1995).
36. Fujimoto T., Dzaki M., and Nagasawa M. *J. Polym. Sci., Part A2*, **6**, 129 (1968).
37. Kurata M., Osaki K., Einaga Y., and Sugie T. *J. Polym. Sci., Part B: Polym. Phys.*, **12**, 849 (1974).
38. Graessley W.W., *Faraday Symp. Chem. Soc.*, **18**, 7 (1983).
39. Graessley W.W., in *Physical Properties of Polymers*, Mark J.E., Eisenberg A., Graessley W.W., Mandelkern L., and Koenig J.E., Eds., Amer. Chem. Soc., Washington, 1984, p. 97.
40. Tuminello W.H., *Polym. Eng. Sci.*, **29**, 645 (1989).
41. Graessley W. and Struglinski M., *Macromolecules*, **19**, 1754 (1986).
42. Rubinstein M., Hefland E., and Pearson D., *Macromolecules*, **20**, 822 (1987).

43. Watanabe H. and Tirrell M., *Macromolecules*, **22**, 927 (1989).
44. Des Cloiseau J., *Macromolecules*, **23**, 4678 (1990).
45. Bafna S.S., *J. Appl.Polym. Sci.*, **63**, 111 (1997).
46. Marchenkov V.V. and Khitrin A.K., *Khim. Fiz.*, **3**, 1399 (1984).
47. Karnaukh G.E., Lundin A.A., Provotorov B.N., and Summanen K.T., *Zh. Eksp. Teor. Fiz.*, **91**, 2229 (1987).
48. Ivanova E.I. and Provotorov B.N., *Zh. Eksp. Teor. Fiz.*, **107**, 473 (1995).
49. Kulagina T.P., Marchenkov V.V., and Provotorov B.N., *Vysokomol. Soedin, Ser. B*, **30**, 23 (1988).
50. Kulagina T.P., Marchenkov V.V., and Provotorov B.N., *Vysokomol. Soedin, Ser. A*, **31**, 381 (1989).
51. Kulagina T.P., Litvinov V.M. and Summanen K.T., *J. Polym. Sci., Part B: Polym. Phys.*, **31**, 241 (1993).
52. Kulagina T.P., *Doctoral (Phys.-Math.) Dissertation*, Chernogolovka: Inst. of Problems of Chem. Phys., 1995.
53. Kargin V.A. and Slonimskii G.L., *Dokl. Akad. Nauk SSSR.*, **62**, 239 (1948).
54. Matveev Yu.A. and Askadskii A.A., *Vysokomol. Soedin., Ser. A*, **35**, 50 (1993).
55. Ol'khov Yu.A., Baturin S.M., and Irzhak V.I., *Polymer Science, Ser. A*, **38**, 544 (1997).
56. Irzhak T.F., Varyukhin S.E., Ol'khov Yu.A., Baturin S.M., and Irzhak V.I., *Polymer Science, Ser. A*, **39**, 459 (1997).
57. Irzhak V.I., Korolev G.V., and Solov'ev M.E., *Usp. Khim.*, **66**, 179 (1997).
58. Solov'ev M.E, Raukhvarger A.B., and Irzhak V.I., *Vysokomol. Soedin. Ser. B.*, **28**, 106 (1986).
59. Solov'ev M.E., Raukhvarger A.B., Makhonina L.I., Korolev G.V., and Irzhak V.I., *Vysokomol. Soedin. Ser. B*, **31**, 485 (1989).
60. Solov'ev M.E., *Doctoral (Phys.-Math.) Dissertation*, Moscow: Inst. of Chem. Phys., 1994.
61. Irzhak V.I., Varyukhin S.E., and Irzhak T.F., *Progr. Colloid Polym. Sci.*, **102**, 4 (1996).
62. Timoshenko S. and Goodier J.N., *Theory of Elasticity*, McGraw-Hill, New York, 1968.
63. Ol'khov Yu.A. and Irzhak V.I., *Polym. Sci., Ser. B*, **40**, 357 (1998).
64. Al'yanova E.E., Bravaya N.M., Ponomareva T.I., Nedorezova P.M., Tsvetkova V.I., and Irzhak V.I., *Polym. Sci., Ser. B*, **40**, 345 (1998).

Chapter 21

Microphase Separation in Epoxies as Studied by Photoactive Probe Technique

Vladimir F. RAZUMOV*, Sergei B. BRICHKIN, Aleksandr V. VERETENNIKOV, Lyudmila L. GUR'EVA, Lyudmila M. BOGDANOVA, and Boris A. ROZENBERG

Institute of Problems of Chemical Physics, Russian Academy of Sciences, Chernogolovka, Moscow Region, 142432 Russia

ABSTRACT

The technique of photoactive probes was found suitable for investigating microphase separation and structural rearrangement induced by curing reaction in multicomponent epoxy systems.

---

*e-mail: razumov@icp.ac.ru

INTRODUCTION

Photoactive probes whose absorption, emission, and photochemical properties depend on the state of microenvironment are being widely used in investigating microheterogeneous systems [1], polymer matrices [2], and biopolymers [3]. Variation in the parameters of a probe molecule may be expected to reflect the kinetics of physicochemical transformations in a matrix.

As is known, some photochemical reactions may proceed in glassy media over a wide temperature range. Among these are the reactions of *cis–trans*-photoizomerization and photorearrangement. This can be exemplified by photoizomerization of stilbene and other diarylethylenes (DAEs) in polymer matrices [2].

Above the glass transition temperature T_g, the kinetics of isomerization obeys a simple exponential law. Below T_g, the kinetics is multiexponential, which is indicative of different reactivity of molecules. This implies that impurity molecules are located at different sites with different free volume and microscopic viscosity [2]. Numerous studies on photochemical behavior of DAEs show a marked effect of environment on the photochemistry of these molecules. This gives grounds for the hope that these molecules can be used as probes that can provide information about the structure and microscopic characteristics of disordered solids.

For appropriately selected additives, their fluorescence may made to depend on pH, microviscosity, polarity, and other factors [1]. During phase separation in polymer blends, photoactive probes undergo redistribution between a solid matrix and finely dispersed oligomeric phase: as a rule, probe is preconcentrated in the liquid phase [4]. In view of this, we can make use of the effect of local concentration on the luminescence properties of the probe. For instance, we could also add two probes into a system to be studied e.g., electron-donating (D) and electron-accepting (A). In case of the dipole–dipole interaction between D and A, the probability for non-radiative transfer of electronic excitation energy is given by the expression [5]:

$$\omega(R) = \frac{1}{\tau_D} \frac{R_F^6}{R^6}. \tag{1}$$

Here R_F is the Förster radius that is given by:

$$R_F^6 = \frac{8.8 \cdot 10^{-25} \chi^2 \varphi_D}{n^4} \int_0^\infty F_D(\nu) \varepsilon_A(\nu) \frac{d\nu}{\nu^4}, \tag{2}$$

where τ_D is the fluorescence lifetime for donor, φ_D is the quantum yield of donor fluorescence, n is the refractive index of medium, ν is the wave number, and χ^2 is the factor of orientation (2/3 for random distribution of D and A). Therefore, the rate of energy transfer depends on separation R between D and A and overlapping integral for the spectra of donor luminescence $F_D(\nu)$ and acceptor absorption $\varepsilon_A(\nu)$.

Microphase separation during polymerization of epoxies markedly affects the optical parameters of polymer films, which also influences the absorption and emission properties of the entire system.

In this work, we investigated the relationship between cure-induced microphase separation, on one side, and photochemical properties of DAEs and luminescence parameters of luminophores, on the other.

EXPERIMENTAL

In our experiments, we used *cis*- and *trans*-1,2-(1-naphthyl)-ethylene (1N1N) and 1,2-(2-naphthyl)-ethylene (2N2N) as DNEs and anthracene (An) and *trans*-1-(1-naphthyl)-2-(9-anthryl)-ethylene (AnE) as a donor–acceptor pair (where An = D and AnE = A).

The starting reactive mixtures comprised (*a*) diglycidyl ester of bisphenol A (DGEBA), (*b*) 3 wt % dimethylbenzylamine (DMBA) as a curing agent, and (*c*) 1–5 wt % poly(propylene glycol)-*bis*-toluylene-diurethaneethylene acrylate) (PBUA) as an oligomeric additive. The mixtures also contained 1N1N or 2N2N (10^{-2}–10^{-3} M) as photochemical probes or An [$(2.8–3.1)\times10^{-3}$ M] + AnE (10^{-3} M) as a donor–acceptor pair. Higher concentration of donor (An) ensured predominant excitation of D. For the An–AnE pair, the value of R_F as calculated from (2) is 4.9 nm. Therefore, [AnE] was selected so that, upon phase separation (and localization of AnE largely in the liquid phase), the mean D–A separation would be below R_F.

The samples for room temperature (RT) measurements were prepared as follows. A weighed amount of photoactive probe was dissolved in DGEBA, after which aliquot amounts of PBUA and then DMBA were added to the transparent solution. The blends were outgassed in vacuum and then placed between two glass plates (separated by a spacer) pretreated with a 5% solution of dimethyldichlorosilane in toluene (in order to preclude adhesion). The cure schedule was as follows: 7 h at 70°C and 5 h at 150°C.

Fluorescence measurements were carried out at RT. With specially designed thermostatically controlled cells, we could monitor the absorption and luminescence spectra at temperatures up to 100°C. Fluorescence spectra were taken with an Elumin spectrophofluorimeter while absorption spectra,

Table 1. Quantum yields of isomerization (φ_{tc}, φ_{ct}) and photocyclization (φ_{DHP}) in different epoxy matrices and at two λ_{ex}. Starting isomer is indicated in parentheses

Matrix	$\lambda_{ex} = 313$ nm	$\lambda_{ex} = 365$ nm
Epoxy	$\varphi_{DHP} = 0.03$, $\varphi_{ct} = 0.05$ (*cis*-1N1N)	$\varphi_{DHP} = 0.18$, $\varphi_{ct} = 0.05$ (*cis*-1N1N)
	$\varphi_{ct} = 0.015$, $\varphi_{tc} = 0.03$ (*cis*-2N2N)	$\varphi_{ct} = 0.03$, $\varphi_{tc} = 0.017$ (*cis*-2N2N)
	$\varphi_{tc} = 0$ (*trans*-2N2N)	$\varphi_{ct} = 0.12$, $\varphi_{tc} = 0.001$ (*trans*-2N2N)
Epoxy+ PBUA	$\varphi_{DHP} = 0.04$, $\varphi_{ct} = 0.04$ (*cis*-1N1N)	$\varphi_{DHP} = 0.33$, $\varphi_{ct} = 0.07$ (*cis*-1N1N)
Epoxy+ PBUA, annealed	$\varphi_{DHP} = 0.02$, $\varphi_{ct} = 0.01$ (*cis*-1N1N)	$\varphi_{DHP} = 0.36$, $\varphi_{ct} = 0.02$ (*cis*-1N1N)

with a Specord-M40 spectrophotometer. Light scattering was detected (at an angle of 90° and $\lambda = 546$ nm) with a TOP-1 apparatus.

RESULTS AND DISCUSSION

Photochemical Probes

In illuminated liquid solutions of DAEs, some dynamic equilibrium is established between *trans*-DAE, *cis*-DAE, and some product of intramolecular photocyclization of *cis*-DAE (dihydroproduct DHP). This equilibrium can be upset in the presence of atmospheric oxygen or some other oxidizers (I_2, $FeCl_3$) that promote irreversible transformation of DHP to appropriate condensed aromatic compounds. With increasing viscosity of the environment, the quantum yields of photocyclization, φ_{DHP}, and *trans–cis*-photoisomerization, φ_{tc}, decrease. Conversely, the quantum yield of the back *cis–trans*-photoisomerization, φ_{ct}, is independent of viscosity, so that this reaction in glassy solutions proceeds at liquid nitrogen temperatures and even below [6]. Due to steric hindrance, *cis*-isomer has a strained nonplanar structure, so that *cis–trans*-isomerization is accompanied by a decrease in molecular volume [7].

For films containing *cis*- and *trans*-2N2N, irradiation at 365 nm gives rise to reversible photoisomerization. Irradiation at 313 nm initiates phototransformations only in the films containing *cis*-isomer. The quantum yields of the photoreactions under consideration are given in Table 1. It may be seen that these strongly differ for the films containing *cis*- and *trans*-isomers. In other words, the quantum yield of photoreactions in epoxy films depends on the sample history, i.e., the system exhibits the effect of matrix memory. The effect is more pronounced for *trans–cis*-isomerization, that is

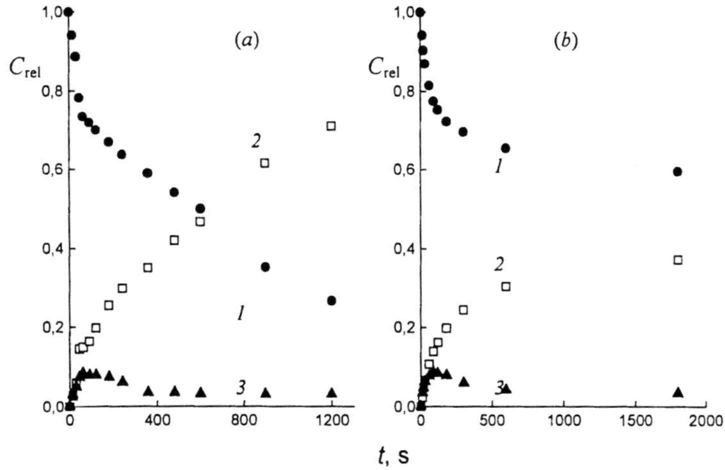

Figure 1. Evolution in relative concentrations of (*1*) *cis*-1N1N, (*2*) *trans*-1N1N, and (*3*) DHP during photolysis of *cis*-1N1N in (*a*) epoxy film and (*b*) epoxy film with an additive (PBUA): λ_{ex} = 313 nm.

consistent with the concept of larger volume occupied by a *cis*-molecule in a solid matrix. The dependence of φ on λ_{ex} can be explained by the presence of rotational conformers and/or structural inhomogeneity that leads to different reactivity of embedded molecules. Photolysis of the films containing *cis*-1N1N resulted in *trans–cis* conversion and formation of DHP (see Fig. 1*a*). The kinetic curves in Fig. 1*a* exhibit rapid initial stage (0–100 s) and slow stage (100–1200 s). [DHP] passes through a maximum. which is typical of all of the films initially containing *cis*-isomer.

For photolysis in an epoxy + PBUA matrix (Fig. 1*b*), the rapid stage is less pronounced. This may be related to the fact that, during phase separation, 1N1N turn out in the liquid PBUA. Annealing these films had no influence on the kinetic curves, which supports the above assumption (location of 1N1N in the liquid PBUA).

Upon photolysis of *trans*-1N1N at λ_{ex} = 365 or 313 nm, the absorption spectrum was found to remain unaffected (no change in relative concentrations) but the intensity of absorption diminished roughly proportional to a 40% decrease in amount of trans-1N1N. This was attributed to photodimerization yielding cyclobutane dimer in the dispersed phase where local concentration [*trans*-1N1N] in PBUA microinclusions may attain a value of 10^{-2}. Under these conditions, bimolecular dimerization becomes competitive with monomolecular *trans–cis*-isomerization.

The quantum yields for the films initially containing *cis*-1N1N as estimated from the initial portions of the kinetic curves in Fig. 1 are collected in Table 1. Note that the values of φ_{ct} change only slightly when changing

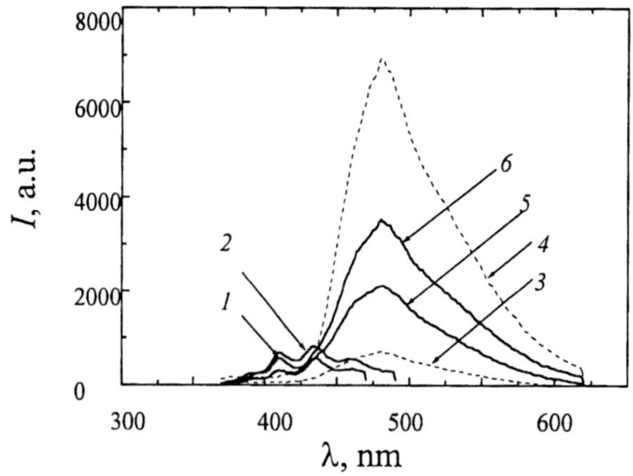

Figure 2. Fluorescence spectra of the DGEBA + 3% DMBA films containing (*1*) $2.8 \cdot 10^{-3}$ M An, (*2*) $3.02 \cdot 10^{-3}$ M An + 3.07% PBUA, (*3*) $1.03 \cdot 10^{-3}$ M AnE, (*4*) $1.04 \cdot 10^{-3}$ M AnE + 3.7% PBUA, (*5*) $3.1 \cdot 10^{-3}$ M An + $1.0 \cdot 10^{-3}$ M AnE, and (*6*) $0.9 \cdot 10^{-3}$ M An + $2.8 \cdot 10^{-3}$ M AnE + 3.3% PBUA.

from matrix to matrix or upon variation in λ_{ex}. Meanwhile, the value of φ_{DHP} increases almost by one order of magnitude while moving from $\lambda_{ex} = 313$ to 365 nm. This can be related to preferential excitation of coplanar conformations that are known to be more reactive relative to formation of DHP.

Luminescent Probes

The fluorescence spectra of cured epoxy films are presented in Fig. 2. It is easily seen that the spectra of the films containing both An and AnE (spectra *5*, *6*) exhibit the emission bands of both the donor and acceptor. For the samples containing no PBUA (spectrum *5*), contributions from An (donor) and AnE (acceptor) are additive. This can be explained by the absence of significant energy transfer. In the presence of PBUA (i.e., under the conditions of phase separation), a relative contribution from AnE increases (spectrum *6*). This implies that, in the process of phase separation, the efficiency of energy transfer from donor (An) to acceptor (AnE) increases.

However, note that the presence of PBUA markedly increases the fluorescence intensity. The effect is most pronounced for the samples containing AnE (Fig. 2, spectra *3*, *4*). This may be attributed to scattering of exciting light by finely dispersed PBUA. The absorption spectra of the samples are presented in Fig. 3.

It is clear that that added PBUA increases the optical density owing to light scattering. Since the starting optical density of luminophore is below 0.2, light scattering may markedly increase the amount of absorbed light. A similar effect was also observed for the samples containing An.

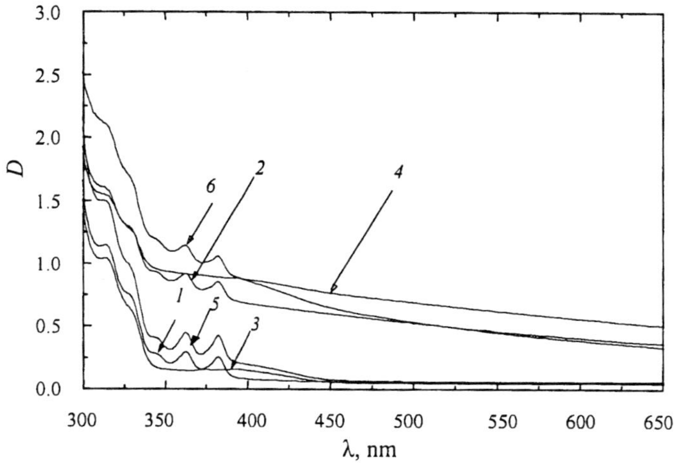

Figure 3. The absorption spectra for the samples characterized in Fig. 2.

Despite the fact that the efficiency of energy transfer from D to A increases, relative contribution from this effect is insignificant against the background of general increase due to an increase in the optical path length for exciting light caused by formation of finely dispersed PBUA.

We also measured the evolution of changes in the fluorescence intensity, light scattering, and absorption for the DGEBA + 3% DMBA + + 3.4% PBUA + $3 \cdot 10^{-3}$ M An layers during their hardening at 70°C (Fig. 4). The fluorescence intensity (curve *1*) increases virtually in parallel with light scattering (curve *3*) while absorption (curve *2*) behaves differently.

The complicated character of the curves indicates separation of the PBUA phase that is accompanied by variation in its structure. Initial growth in fluorescence intensity and light scattering can be attributed to separation and accumulation of finely dispersed PBUA. The subsequent decrease may be associated with the coarsening of fine PBUA particles upon their aggregation and coalescence, which leads to a decrease in total amount of light scattering particles. Upon cooling (from 70 to 20°C), a new buildup of fluorescence and light scattering (caused by increased amount of light-scattering centers) may be assigned to (1) additional separation of finely dispersed phase (dissolved at elevated temperature) and (2) partial diminution of large aggregates. Phase separation may be characterized by the cloud point. This is defined as the time corresponding to the onset of phase separation. This quantity can be determined as the intercept of the tangent of the light scattering curve (at the initial stage) with the abscissa axis (Fig. 4). In our case, this is about 50 min for fluorescence and light scattering and 110 min for absorption. Therefore, the optical density is less sensitive to phase separation and structural rearrangements in a precipitated phase.

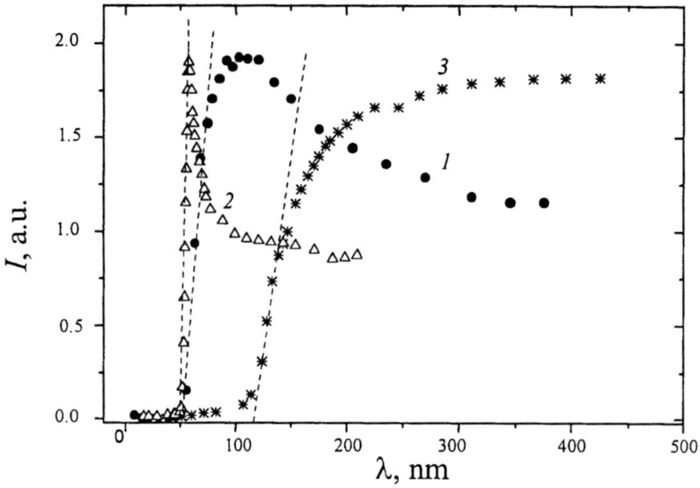

Figure 4. Evolution of (*1*) fluorescence intensity, (*2*) light scattering, and (*3*) absorption at 550 nm (*3*) for the DGEBA + 3% DMBA + 3.4% PBUA + 3×10^{-3} M An system during curing and microphase separation at 70°C.

CONCLUSIONS

Our data show that the environment of a photoactive molecule markedly affects its absorption/emission properties and photochemical behavior. The process of microphase separation is accompanied by a redistribution of probe molecules between a curing matrix and dispersed phases. All this changes the environment and favors light scattering that strongly affects the conditions for excitation and observation of luminescence. The technique of photoactive probes was found to adequately reflect the processes of microphase separation and structural rearrangement induced by cure reaction in thermosetting systems and suitable (in combination with other techniques) for investigating the microscopic characteristics of unordered solids.

Acknowledgments. This work was supported by the International Science and Technology Center (grant no. 358-96).

REFERENCES

1. Parker C., *Photoluminescence of Solutions*, Elsevier, Amsterdam, 1968.
2. Guillet J., Polymer Photophysics and Photochemistry: *An Introduction to the Study of Photoprocesses in Macromolecules*, Cambridge University Press, Cambridge (UK), 1985.

3. Vekshin N.L., *Fotonika biologicheskikh struktur* (Photonics of Biological Structures), Nauka, Pushchino, 1988.
4. Razumov V.F., Veretennikov A.V., Karpova T.P., Bogdanova L.M., Gur'eva L.L., and Rozenberg B.A., *Vysokomol. Soedin., Ser. A*, **40**, 748 (1998).
5. Agranovich V.M. and Galanin M.D., *Perenos energii elektronnogo vozbuzhdeniya v kondensirovannykh sredakh* (Transfer of Electronic Excitation in Condensed Media), Nauka, Moscow, 1978.
6. Filippov P.G., Razumov V.F., Rachinskii A.G., and Alfimov M.V., *Doкl. Akad. Nauk SSSR*, **295**, 434 (1987).
7. Gegiou D., Muszkat K.A., and Fischer E., *J. Am. Chem. Soc.*, **90**, 12 (1968).

Chapter 22

Investigation of CRIMPS by NMR

Viktor P. TARASOV*, Anatolii K. KHITRIN,
Lyudmila M. BOGDANOVA, and Boris A. ROZENBERG

Institute of Problems of Chemical Physics, Russian Academy of Sciences, Chernogolovka, Moscow Region, 142432 Russia

ABSTRACT

The kinetics of microphase separation induced by the cure reaction of diglycidyl ether of bis-phenol A (DGEBA) was studied by NMR under isothermal conditions and in the presence of modifying agents. Polymerization and polycondensation mechanisms induced by diamine and tertiary amine systems respectively have been considered. Method of spin diffusion was shown to be useful for studying the microphase separation in curing polymer systems. The onset of microphase separation, phase composition, and size of heterophase inclusions were detected and/or characterized.

---

*e-mail: vtarasov@icp.ac.ru

INTRODUCTION

ESR method was shown to provide additional information about phase separation compared to conventional experimental techniques [1]. In this work, the first effort was made to use NMR for the purpose.

NMR method allows investigating phase separation at the early stages of the process when inclusions of a new phase are very small (tens Å in size) and the sample is optically uniform. Existence of two phases manifests itself as the presence of two regions with strongly different molecular mobility, which gives rise to different line width of spectral lines[2]. Heterophase inclusions 10–100 Å in size can be detected by using the technique of spin diffusion [3, 4]. Its simplest option is the use of the Goldman–Shen pulse train (comprising 3 pulses) [4]. The first pulse produces a spatially nonuniform magnetization (different in different phases); relaxation of magnetization at the expense of spin diffusion occurs between the second and the third pulses; and after the third pulse, a signal of the free induction decay is recorded. A specific time of magnetization (polarization densities) relaxation is directly associated with the spin diffusion coefficients and inclusion size.

The data on polymer morphology [5–7] suggest that the spin diffusion method provides information about the shape and size of microheterogeneities. Here, spatially nonuniform spin polarization can arise not only at the expense of the time difference of free induction decay, but also by using more selective multipulse sequence, e.g., dipole filters [7], or multi-quantum filters [8].

EXPERIMENTAL

The following systems were studied:

- system **I**: diglycidyl ether of bisphenol A (DGEBA, $\overline{M}_n = 380$, $\rho = 1.16$ g/cm^3) + 3 wt % dimethylbenzylamine (DMBA, $\rho = 0.896$ g/cm^3, T_b = 100°C/65 mm Hg) + 20 wt % polypropyleneglycol-bis(toluylene-diurethanethyleneacrylate) (PBUA, $\overline{M}_n = 2540$, $\overline{M}_w = 3410$, $\rho = 1.10$ g/cm^3), cure temperature 70°C.
- system **II**: DGEBA + 4,4′-diaminodicyclohexylmethane (DADCM, $\rho = 0.972$ g/cm^3, T_b = 320°C) + 10 wt % butadiene and acrylonitrile copolymer with terminal amine groups (acrylonitrile content is 16% , η = 200 Pa·s,), cure temperature 50°C. The DGEBA to DADCM ratio was stoichiometric.

The polymerization kinetics for diphenylolpropane diglycidyl ether under the action of a tertiary amine and in the presence of additives was

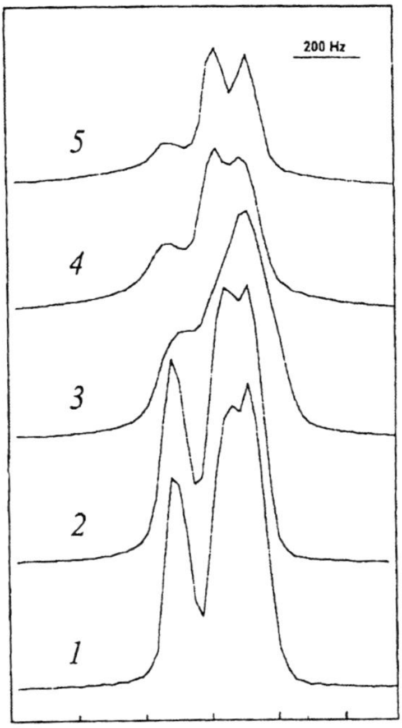

Figure 1. The shape of ^{1}H NMR line in the DGEBA–PBUA system (70°C) cured for t = (*1*) 0, (*2*) 15, (3) 30, (*4*) 45, and (*5*) 65 min.

taken from [9]. NMR spectra were taken with a pulsed NMR spectrometer (working frequency for protons 57 MHz).

RESULTS AND DISCUSSION

Figure 1 shows the change in the line shape for system **I** during isothermal cure. During first 20 min, no considerable change is observed. After 30 min (conversion α = 0.10), the line markedly broadens, thus indicating an overall decrease in molecular mobility. After 45 minute (α = 0.17), a broad spectral component (attributed to a low-mobility part of the system) emerges. Simultaneously, a resolved spectrum (similar to that of pure modifier) is observed. This implies that phase separation occurs between the 30th and 45th min. However, no optical heterogeneity is still observed.

Figure 2 shows the spectrum of the system after it has been cured for 65 min (α = 0.27). Here the narrow spectral component is comparable with the spectrum of pure modifier with a proton population of 25%. The number of protons in modifiers is higher than that in DGEBA by a factor of 1.28. Thus, the intensity shown in Fig. 2 is similar to a real concentration of modifier in the system. The intensity of the narrow component is somewhat higher than that of the spectrum of the pure modifier. Unfortunately, a strong overlapping of the lines prevents quantitative estimation of this difference. However, we may conclude that the mobile phase involves some amount of a low-molecular component of the epoxy oligomer, together with the modifier. This is also confirmed by the fact that the spectral lines of the system are narrower compared to those of pure modifier. If the mobile phase had consisted of pure modifier, a low-mobile environment would have inhibited the motion and broaden the line. Molecules of epoxy oligomer and modifier affect each other as plasticizers. The spectra in Fig. 3 confirm this: at the starting point, all lines of the blend are narrower than those of individual components (DGEBA and modifier).

Strong interaction between DGEBA and modifier makes the additive model inadequate (because the spectrum of the blend cannot be described as a sum of the DGEBA and modifier spectra). Therefore, more accurate quantitative analysis of the spectra is not possible.

As mentioned above, phase separation in the system occurs after curing for 30–45 min. Concentration of the more mobile phase is 25%, which slightly higher than the initial concentration of modifier.

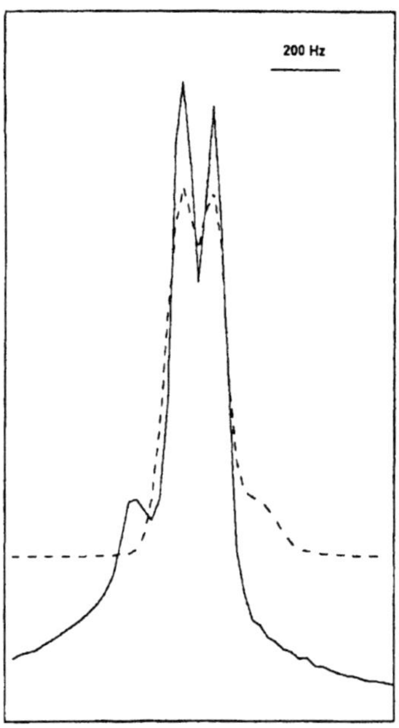

Figure 2. Normalized shape of the line at t = 65 min (solid line), and the line for the pure modifier with 25% intensity (dashed line).

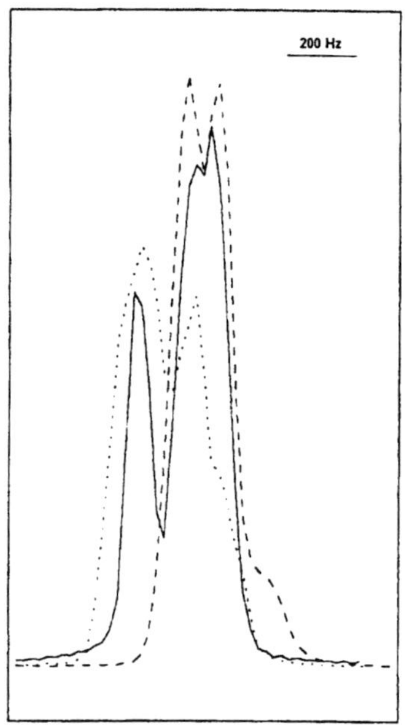

Figure 3. Normalized NMR lines of the blend at the starting point (solid line), DGEBA (dotted line) and pure modifier (dashed line).

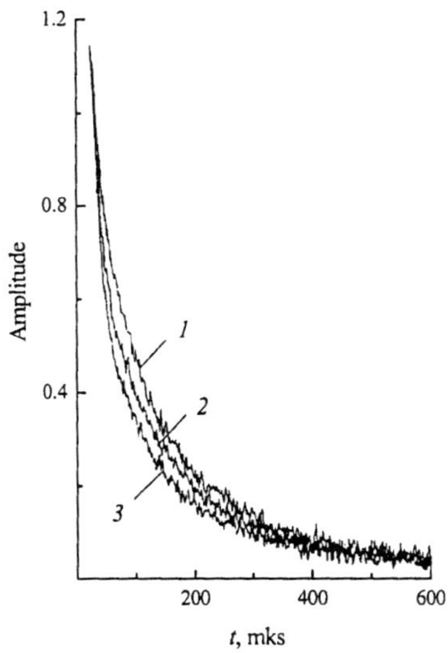

Figure 4. Free induction decay curves after the third pulse for different mixing time: t_m = (*1*) 1, (*2*) 3, (*3*) 6 ms.

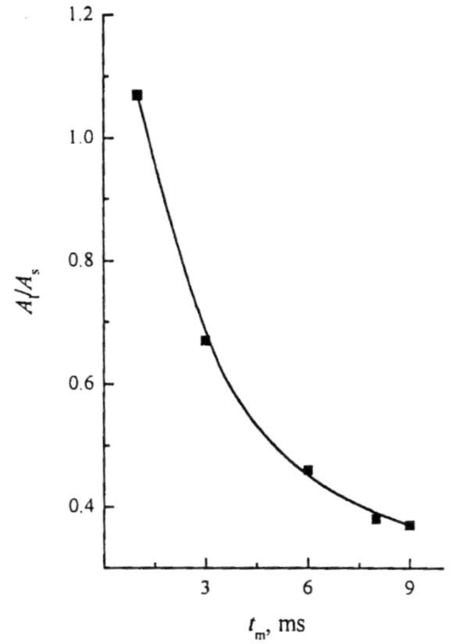

Figure 5. The ratio of the slow to rapid components vs. the mixing time t_m.

To estimate the size of the mobile phase inclusions at the beginning of phase separation, the sample was quenched in liquid nitrogen after curing for 40 min (α = 0.14). Then the temperature was allowed to increase until the motions in the more mobile phase were partially unfrozen. The temperature rise was stopped at −15°C when a distinct slow component appeared in the NMR signal of the free induction decay. Further experiments were performed at that temperature.

The Goldman–Shen three-pulse sequence was used [4]. After the first pulse, the free induction decay takes place. The time for the second pulse is determined by the moment when a rapid component has already decayed while the slow one not. The second pulse brings magnetization back to the direction of applied field, so that the spin diffusion equalizes the phase polarization. Then the third pulse is applied, and the amplitudes of the rapid and slow components are estimated from the free induction decay.

Figure 4 presents the curves of free induction decay after the third pulse for three values of t_m (1, 3, and 6 ms). With increasing t_m, the amplitude of the rapid component increases while that of the slow one, decreases. To calculate the amplitudes, the decay curves were treated as a sum of two exponential functions.

Figure 5 presents the ratio of the slow to rapid signal amplitudes A_l/A_s as a function of the mixing time t_m. The dotted line presents the following exponential function:

$$A_l/A_s = A_0 + A\exp(-t_m/t_s). \tag{1}$$

The characteristic time of mixing t_s as assessed from (1) is 2.60 ms. The maximum value of A_0 = 0.34 is close to the proton ratio of the mobile dispersed to solid matrix phases determined earlier.

To find a relationship between t_s and inclusion size, a simple model is considered where inclusions of a mobile phase are simulated by spherical particles with radius b uniformly distributed over a solid matrix. Upon unfreezing the molecular motions, dipole interactions responsible for mutual spin flip-flops begin to average. Simultaneously, a spatial transfer occurs, so that the overall diffusion coefficient increases. Therefore, we assume that spin diffusion in a solid matrix is a rate-controlling stage.

To estimate the coefficient of spin diffusion, we assume that protons in a solid matrix are located at the lattice sites of a cubic lattice with period a. Then, the coefficient of spin diffusion averaged over orientations is given by [3]:

$$D = a^2 M_2^{1/2}/30, \tag{2}$$

where M_2 is the second moment of an NMR absorption line. It is estimated from the short component of free induction decay assuming that the line is described by Gaussian curve; $M_2^{-1/2} = 9.5$ μs. For a three-dimensional case, the steady-state treatment can be applied to estimate the characteristic time of mixing. Taking into account that a volume fraction of the matrix phase is much larger than that of inclusions, this approximation yields:

$$t_c \approx b^2/3D. \tag{3}$$

Combining (2) and (3), we obtain:

$$t_c \approx 10\, b^2/(a^2 M_2^{1/2}). \tag{4}$$

For the measured values of t_s and M_2 (see above), we obtain: $b/a \approx 5$. As assessed from the proton density in the matrix phase, a = 2.9 Å. This implies that inclusions are about 30 Å in diameter.

From similar measurements for the system kept frozen after curing for 50 min ($\alpha = 0.19$), an average diameter of inclusions increases by a factor of 1.6 and attains a value of 50 Å. The sample with such a cure degree exhibits strong light scattering.

For system **II**, pulsed NMR yields the kinetic curve (Fig. 6) that is time-dependent. The kinetic curve levels off in 75–80 min of curing

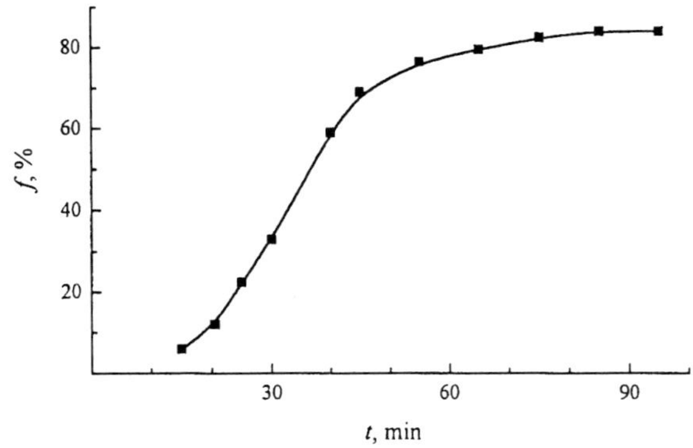

Figure 6. Kinetics of the solid phase formation in the presence of 10% additive (*f* is the volume fraction of the solid phase).

(α = 0.46–0.48). The amount of the liquid phase residue is slightly larger than that of the additive. This implies that a part of the components precipitates, together with the additive, into a separate phase.

CONCLUSIONS

The technique of nuclear magnetic resonance (NMR) was applied to investigate phase separation in curing systems. Phase separation in curing system **II** was studied by using pulsed NMR.

Method of spin diffusion was shown to be useful for studying the microphase separation in curing polymer systems. NMR method allows detection of the onset of microphase separation and the compositions of newly formed phases. Proper selection of system parameters (in particular, temperature) was found to minimize side effects (spin-lattice relaxation, multi-quantum coherence, and spin diffusion in both the phases). In addition, the size of heterophase inclusions can be determined by using the simplest pulse train (of 3 pulses).

Acknowledgments. This work was supported by the International Science and Technology Center (grant No. 358-96).

REFERENCES

1. Krisyuk B.E., Dzhavadyan E.A., Bogdanova L.M., and Rozenberg B.A. *Polym. Sci., Ser A*, **40**, 988 (1998).

2. Abraham A., *Principles of Nuclear Magnetism*, Clarendon Press, Oxford (UK), 1961.
3. Goldman M., *Spin Temperature and Nuclear Magnetic Resonance in Solids*, Clarendon Press, Oxford (UK), 1970.
4. Goldman M. and Shen L., *Phys. Rev.*, **144**, 321 (1960).
5. Assink R.A., *Macromolecules*, **11**, 1233 (1978).
6. Spiegel S., Schmidt-Rohr K., Boeffel C., and Spiess H.W., *Polymer*, **34**, 4566 (1993).
7. Demco D.E., Johanson A., and Tegenfeldt J., *Solid State Nucl. Magn. Reson.*, **4**, 13 (1995).
8. Yong Ba and Ripmester J.A., private communication.
9. Dzhavadyan E.A., Bogdanova L.M., Irzhak V.I., and Rozenberg B.A., *Polym. Sci., Ser. A*, **39**, 383 (1997).

Chapter 23

CRIMPS in an Epoxy–Amine System as Studied by ESR

Boris E. KRISYUK and Boris A. ROZENBERG*

Institute of Problems of Chemical Physics, Russian Academy of Sciences, Chernogolovka, Moscow Region, 142432 Russia

ABSTRACT

Curing and phase formation in an epoxy–amine system—diglycidyl ether of bisphenol A/cycloaliphatic diamine—were investigated by spin-probe and spin-label ESR in the presence of 0.0–17.0 vol % of oligomeric polypropyleneglycol-bis(toluylene-diurethanethyleneacrylate). This technique was found to be more sensitive to phase separation than light scattering. Formation of the secondary phase (just after gel formation) and kinetic inhomogeneity in starting epoxy–amine system are detected.

---

*e-mail: rozen@icp.ac.ru

INTRODUCTION

Synthesis of heterophase polymers from a starting mixture of homophase oligomers containing a polymer or oligomer modifier (that decomposes yielding two- or many-phase polymers) is a promising way for improving the properties of polymer materials [1–4]. In this case, the resultant material represents a matrix (α-phase) that contains small amount of dissolved modifying agent and inclusions (β-phase) enriched in modifier. Under some conditions, the β-phase may undergo secondary phase separation, thus forming a submatrix (polymer additive with a small amount of curing oligomer) and the γ-phase (cured oligomer containing low amount of dissolved polymer additive). The properties of the resultant heterophase polymers are defined largely by their morphology [1–5]. Understanding the processes taking place during phase separation and structure formation requires detailed information about (*i*) chemical reactions taking place in cured systems, (*ii*) processes of phase separation, (*iii*) gel formation and vitrification, (*iv*) structure of heterophase polymer, (*v*) amount and composition of new phases, (*vi*) particle size, etc. More or less sufficient information about these processes can be obtained only upon combined use of conventional experimental techniques (calorimetry, light scattering, DTA, electron microscopy, etc.). In past years, similar systems have been investigated at this Institute by ESR and NMR techniques [6–11].

In this work, the ESR method was used to investigate phase separation in the following system: diglycidyl ether of bisphenol A (**I**) + cycloaliphatic diamine (**II**) cured in the presence of poly(oxypropylene glycol) diacrylate (**III**). Although ESR is being used for investigating heterophase polymer blends for a long time [6–9], the process of curing-induced phase separation was studied by this method for the first time.

EXPERIMENTAL

In our experiments, we used diglycidyl ether of bisphenol A (**I**) (Epon-828 grade), 4,4'-diaminodicyclohexylmethane (**II**) as a curing agent, and polypropyleneglycol-bis(toluylene-diurethanethyleneacrylate) (**III**) as an additive. These compounds are characterized in Table 1.

The MMD for **I** was determined by GPC with a Milikhrom-1 chromatograph [4×200 mm column filled with a spherical silanized filler (Silasorb 600)]. Spectrally pure dioxane was used as a solvent. Analysis was carried out at 25°C at an elution rate of 200 μl/min.

An aliquot amount of **II** was added to **I** (containing additive **III**) immediately before curing. The system was outgassed. The amount of added **III** was varied between 0.0 and 17 vol %. Curing in the bulk was performed at 40–60°C.

Table 1. Characterization of the compounds used in experiments

Compound	$\overline{M}_n$, g mol^{-1}	$\overline{M}_w$, g mol^{-1}	d^{20}, g mol^{-1}
I	380	—	1.160
II	210	—	0.972
III	2040	5520	1.050

The kinetics and rheokinetics of curing and phase separation were measured as described elsewhere [12].

For ESR studies, outgassed reactive mixture was sampled by suction and sealed in glass tubes 1.5 mm in diameter. The stable nitroxyl radical 2,2,6,6-tetramethyl-4-benzoylhydroxypiperidine-1-oxyl was used as a spin probe. The spin probe was dissolved with **I** in acetone, after which acetone was removed in vacuum. The probe concentration was below 10^{-3} M. We used two types of spin labels: (*a*) nitroxyl radical attached to the epoxy group of **I** and (*b*) nitroxyl radical attached to the double bond of **III**. The spin-labels were introduced into **I** and **III** by reactions of the amino groups of radicals with the epoxy group or double bond. These reactions (shown in the scheme) were studied in detail earlier [13, 14].

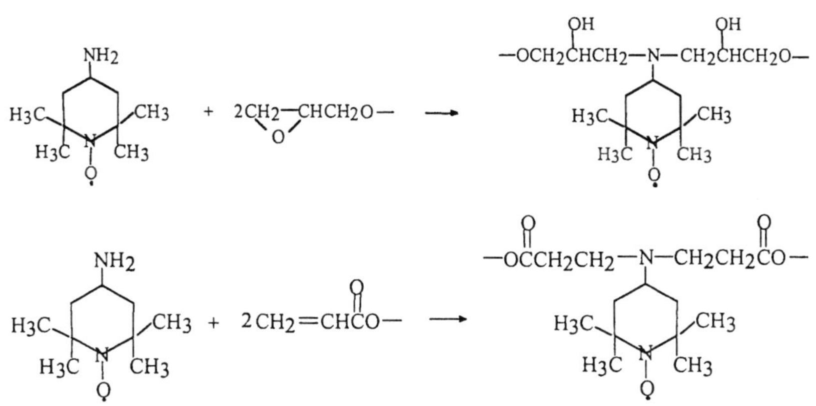

The curing kinetics was monitored (at 50°C) in the resonator of a Radiopan SE/X ESR spectrometer.

RESULTS AND DISCUSSION

Systems Containing Spin Probes

Evolution of ESR signals from spin probes during hardening the **I–II–III** system containing different amount of **III** is presented in Fig. 1. Similar

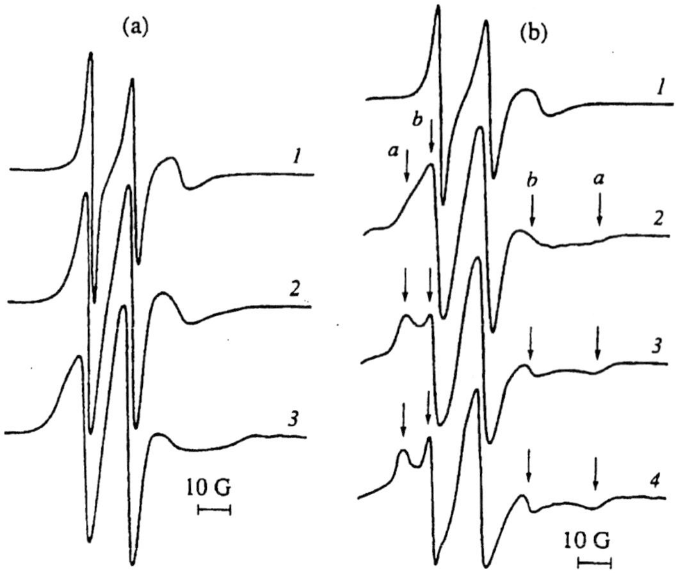

Figure 1. Evolution of ESR signals from spin probes during hardening the **I–II–III** system containing (a) 0.0 and (b) 10.9 vol % **III**: reaction time t = (*1*) 15, (*2*) 30, and (*3*) 45 min. Arrows *a* and *b* indicate the broad and narrow signal components.

behavior was observed for other concentrations of **III**. Initially, the spin probe behaves similarly, irrespective of additive concentration.

The systems are characterized by an identical correlation time $\tau_c = 2.5 \cdot 10^{-10}$ s. As could be expected, this value gradually increases in the course of curing. However, beginning with t = 30 min, the spectra of the systems containing 5.8 and 10.9 vol % **III** exhibit clear manifestations of a two-component structure. The spectra appear as a superposition of signals from the probes that are rotating 'rapidly' ($\tau_c = 10^{-8}$–10^{-10} s) and 'slowly' ($\tau_c = 10^{-7}$–10^{-8} s). Note that all this is preceded by phase separation as detected by light scattering. Such a behavior of the ESR spectra in the course of phase separation caused by curing seems natural for systems with limited compatibility of the components. Indeed, rapid rotation may be assigned to the spin probes in the β-phase enriched in **III**, whereas the slow component may be attributed to the spin probe embedded in the cured α-phase. This follows from variation in the probe mobility in two competitive processes: reactions of **I** and **III** with **II** (Fig. 1). As follows from Fig. 1*a*, the correlation time $\tau_c > 10^{-8}$ s at the end of curing, which is indicative of vitrification. Meanwhile, during reaction of **III** with **II**, the ESR spectrum exhibits only the narrow line, and the probe mobility remains relatively high until completion of the cure reaction ($\tau_c \ll 10^{-9}$ s). Note that heating the **II**–

III system has no effect on the ESR spectra: both at the beginning and end of cure reaction, the spectrum comprises a triplet of narrow lines corresponding to identical values of $\tau_c = 1.4 \cdot 10^{-10}$ s.

The ESR spectra of the samples that contain 2.4 vol % **III** exhibit no narrow line. This can be regarded as an evidence for the absence of phase separation in the system. However, this contradicts the data on light scattering, according to which the cloud point is attained in 41 min of curing [13].

In order to resolve this apparent discrepancy, we performed computer simulation of the EPR spectra for heterogeneous systems as suggested by Antsiferova *et al.* [15]. Upon variation in the τ_c values for each motion, blend composition, and the rate of spin probe exchange between the kinetically inhomogeneous phases, we tried to fit the measured spectra to the simulated. Similar analysis was also performed for the cured system containing no additive. Even in the absence of phase separation, the shape and character of the ESR spectra of the cured system were found by no means simple. As indicated above, initially the curing process is characterized by a unique correlation time $\tau_c = 2.5 \cdot 10^{-10}$ s. But in 30 min of curing, the observed spectra can be identified as those corresponding to a kinetically inhomogeneous spin system. Even in case of a single-phase system with a topological structure complicated in the course of curing, this can be explained by localization of the probe molecules within the areas of different kinetic complexity (hence, of different kinetic mobility). By that time, the amount of low-mobility spin probes is about 95%. Accordingly, their correlation time (equal to the lifetime of this state) is $1.4 \cdot 10^{-8}$ s, compared to $2.5 \cdot 10^{-9}$ s for the high-mobility probes. Note that the ESR spectrum retains the narrow lines. As the curing process goes on, the correlation times grow in both the kinetic phases, and the blend composition remains unchanged within the accuracy of modeling. After 60–80 min of curing, the correlation times grow to such an extent that the spectra of spin probes become insensitive to rotation. Therefore, we can take as reliably established only the very fact of the kinetic inhomogeneity arising in the course of curing but not the quantitative characteristics of the process.

In contrast to the systems with a higher content of additive, the absence of the narrow component in the spectra of the samples containing 2.4 vol % **III** can be explained by the onset of phase separation. At low [**III**], this happens at sufficiently high degree of conversion, when the probe motions in both phases are restrained. In our system, the additive also reacts with one of the components (diamine). Up to the moment of the phase dispersion, the conversion degree (with respect to double bonds) is known [14] to be relatively high (as well as the microviscosity of this phase). Note that the rate constant for the interaction between the amino and acrylate groups is twice as high as that between the amino and epoxy groups. Meanwhile, the concentration of the epoxy groups for an additive content of 2.4 vol % is about 30 times higher than that of the acrylate groups [14]. We must also bear in mind that phase separation is accompanied by fractionation of the components by their molecular weight. Consequently, the β-phase (rich in the additive) becomes diluted with the most low-molecular components of the matrix phase [4].

Therefore, up to the moment of phase separation, the spin probe in systems with a high amount modifying additive turns out in a medium with a markedly lower microviscosity compared to that of the matrix phase. As the additive is consumed, a conversion degree corresponding to the onset of phase separation increases, and the difference between the spin-probe mobility of the two phases levels off. The spin-probe signals from both the phases merge and look like the spectrum of a homogeneous system.

Secondary Phase Separation

Let us consider in more detail the behavior of a narrow component in the spectra of the β-phase (for relatively high [**III**]). In these systems, phase separation is started at a low conversion degree (η) and is readily detected by onset of the kinetic inhomogeneity in probe motions and formation of a phase characterized by rapid probe rotation. Figure 2 shows the line width ΔH_+ for a low-field narrow component in the ESR spectrum vs. the curing time t for the systems containing 5.8 and 10.9 vol % **III**. The lower the line width, the higher is the frequency of probe rotation. Kinetic curves *1* and *2* are seen to be similar: the line width initially increases, attains its maximum (in 50–60 min), and then sharply drops below the initial value. For the spin probe dissolved in pure additive, ΔH_+ = 2.4 G, which is close to that measured for long t.

The two systems under consideration were found to exhibit the kinetic inhomogeneity after a 30-min curing. Note that ΔH_+ keeps growing for some time, even after the onset of phase separation. This implies that curing proceeds in both the phases. No signal with ΔH_+ = 2.4 G (additive phase) is observed. Then the line width ceases to grow and begins to drop. This is indicative of the onset of secondary phase separation: the γ-phase precipitates from the β-phase in the form of isolated microscopic droplets. The phase of additive appears only after 150 min of curing, that is, when the process is virtually completed (according to the kinetic data, η ≈ 65%). Note that the secondary phase separation proceeds in a disperse phase, since the matrix phase becomes immobile (owing to gelation) at η = 58%.

Systems Containing Spin-Labeled Molecules

Despite the fact that the above considerations are self-consistent, it should be born in mind that the question of probe localization is always open to

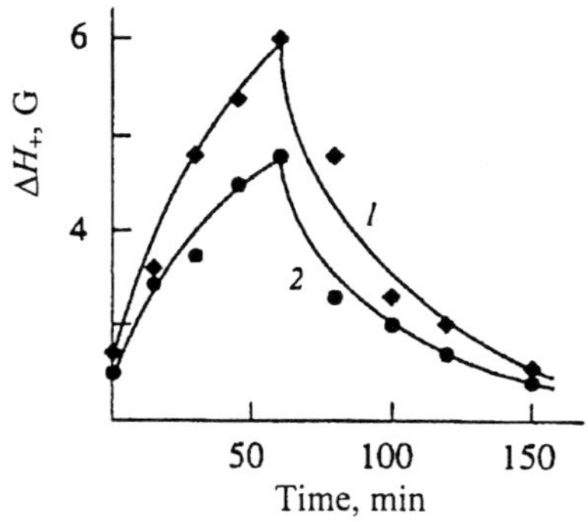

Figure 2. The line width ΔH_+ for a low-field narrow component in the ESR spectrum vs. curing time t (at 50°C) for the systems containing (*1*) 5.8 and (*2*) 10.9 vol % **III**.

discussion. Although there is no reason to expect localization of probes at some special areas, we performed control measurements with spin-labeled **I** and **III**. The spin labels chemically bound to one of system components provide information only about the state of given component.

Figure 3*a* shows the evolution of the ESR spectra taken for the cured system containing spin-labeled **I** and 2.4 vol % **III**. To save space, we omitted other spectra, because they were completely alike, irrespective of [**III**]. Note that the initial spectrum has a shape typical of a slow motion that is getting slower in the course of curing. This implies that the spin-label is strongly bound to **I** and cannot rotate freely, even in the starting system. Therefore, the spin-label is located within the area of slow motions from the very beginning, so that no phase separation in the system can be detected.

A more interesting result was obtained for spin-labeled additive **III**. Figure 3*b* exemplifies the ESR spectra for a system containing 10.9 vol % spin-labeled **III**. This series, similar to that observed for the spin-probe, exhibits a transition from situation with a unique correlation time to that taking place in a heterogeneous system. The time of the onset of heterogeneity agrees with that assessed from spin-probe and light-scattering measurements. As follows from Fig. 3*b*, a significant amount of spin labels after a 45-min curing (up to 90%) undergoes slow rotation. This implies that the labels are located in the α-phase (that is eventually vitrified). Saying glass transition (vitrification), we mean only transition to the mode of slow motion. However, for freely moving molecules (in contrast to the labels attached to the epoxy group), usually there exist some correlation between transition to slow motion and structural glass transition [16].

We should also like to emphasize a principal difference between the mobility of spin-labeled molecules **I** and **III** (similar from the chemical viewpoint). The presence of hydroxyl groups in the cured fragments of the

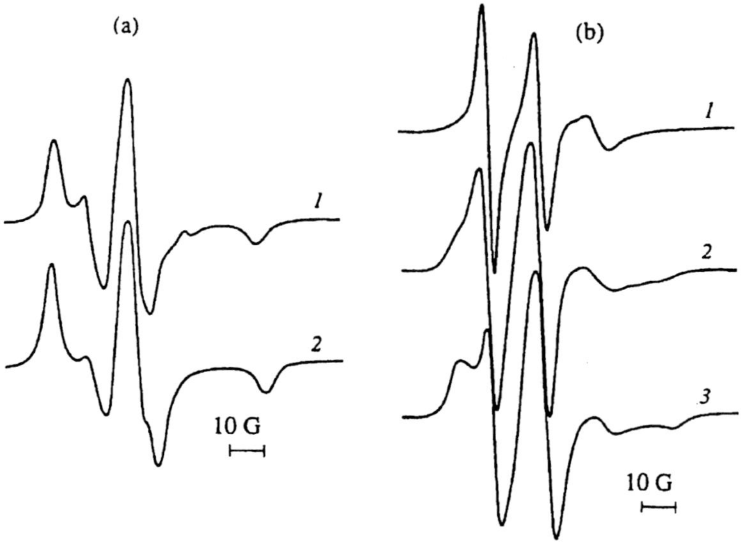

Figure 3. The ESR spectra of equifunctional **I**/**II** mixtures containing (a) 2.4 and (b) 10.9 vol % **III** at t = (*1*) 3, (*2*) 30, and (*3*) 80 min (at 50°C): (a) spin-labeled **I**, (b) spin-labeled **III**.

former compound leads to a strong intermolecular association of these groups by hydrogen bonding. Consequently, the spin-label (being strongly attached) loses its mobility because of self-association, even in a mobile environment.

CONCLUSIONS

ESR spectroscopy turns out to be useful for investigating phase formation in cured polymer systems. Analyzing the line shape in the spectra simulated with account of transitions of spin probes and spin labels between different rotational states, we can detect the effect of microphase separation that is induced by the curing process. ESR data reveal the kinetic heterogeneity in the behavior of spin probes in a homophase curing system. The time dependence of the line width of a narrow component may be indicative of the secondary phase separation in the β-phase. This event cannot be detected by other conventional techniques (light scattering or turbidimetry) used in the kinetics of phase separation. The very fact of secondary phase formation is normally determined only by analyzing a completely cured heterophase polymer by electron microscopy or small-angle x-ray diffraction [17]. Our preliminary qualitative results suggest that, upon appropriate selection of spin labels and spin probes, ESR may turn out useful for determining the composition and amount of newly formed phases.

Acknowledgments. This work was supported by the International Science and Technology Center (grant no. 358-96).

REFERENCES

1. Bucknall C.B., *Toughened Plastics*, Applied Science Publishers, London, 1977.
2. *Polymer Blends*, Paul D.R. and Newman S., Eds., Academic Press, New York, 1978.
3. *Toughened Plastics 1: Science and Engineering, Adv. Chem. Ser.*, **233,** Riew C.K. and Kinloch A.J., Eds., Amer. Chem. Soc., Washington DC, 1993.
4. Williams R.J.J., Rozenberg B.A., and Pascault J.-P., *Adv. Polym. Sci.*, **128**, 96 (1997).
5. Kozii V.V. and Rozenberg. B.A., *Vysokomol. Soedin., Ser. A*, **34**, 3 (1992).
6. Krisyuk B.E., Marikhin V.A., Myasnikova L.P., and Zaalishvili N.L., *Int. J. Polym. Mater.*, **22**, 161 (1993).
7. Krisyuk B.E. and Sandakov G.I., *Vysokomol. Soedin., Ser. A*, **37**, 615 (1995).
8. Krisyuk B.E., Popov A.A., and Kotov V.V., *Vysokomol. Soedin., Ser. A*, **37**, 94 (1995).
9. Volkova N.N., Sosikov A.I., Berezin M.P., *et al.*, *Vysokomol. Soedin., Ser. A*, **30**, 2133 (1988).
10. Bogdanova L.M., Volkova N.N., Lankin A.V., and Smirnov L.P., *Vysokomol. Soedin., Ser. A*, **40**, 397 (1998).
11. Tarasov V.P., Khitrin A.K., and Rozenberg B.A., this book, Ch. 22.
12. Krisyuk B.A., Dzhavadyan E.A., Bogdanova L.M., and Rozenberg B.A., *Polym. Sci., Ser. A*, **40**, 988 (1998).
13. Rozenberg B.A., *Adv. Polym. Sci.*, **75**, 113 (1985).
14. Komarov B.A., Dzhavadyan E.A., Irzhak V.I., and Rozenberg B.A., *Polym. Sci., Ser. A*, **39**, 153 (1997).
15. Antsiferova L.I. and Valova E.V., *Vysokomol. Soedin., Ser. A*, **38**, 1851 (1996).
16. Vasserman A.M. and Kovarskii A.L., *Spinovye metki i zondy v fizikokhimiii polimerov* (Spin Probes and Labels in the Physical Chemistry of Polymers). Nauka, Moscow, 1986.
17. Chen D., Pascault J.-P., Sautereau H., and Vigier G., *Polym. Int.*, **32**, 369 (1993).

Chapter 24

Gelation and Reaction-Induced Microphase Separation during Cure of Model Epoxy-Amine Systems as Studied by Dielectroscopy

Gennadii F. NOVIKOV*, Tatyana L. ELIZAROVA, Emma A. DZHAVADYAN, Lyudmila M. BOGDANOVA, and Boris A. ROZENBERG

Institute of Problems of Chemical Physics, Russian Academy of Sciences, Chernogolovka, Moscow Region, 142432 Russia

*e-mail: novikov@icp.ac.ru

ABSTRACT

Three different model polymer blends (based on diglycidyl ether of bisphenol-A) were investigated by dielectric spectroscopy in comparison with the relevant kinetic and rheokinetic data. These systems are known to set by polycondensation, polycondensation accompanied by microphase separation, and anionic polymerization. It is demonstrated that formation of microgel (and accompanying its microphase separation) and macrogel (of physical and chemical origin) taking place during cure in the systems under study can be reliably detected by dielectrometry.

INTRODUCTION

In the past decade, interest in the dielectric spectroscopy (micro-dielectrometry) has been increasingly growing [1]. This is a promising tool for independent (requiring no calibration) detection and identification of relaxation-related transitions (gelation, vitrification, phase separation) taking place during polymerization [2–6]. Initially, it was disputable whether the particular features (maxima, break points, etc.) of the measured kinetic curves for the active component of conductivity $\sigma(t)$, the real, $\varepsilon'(t)$, and imaginary, $\varepsilon''(t)$, parts of dielectric constant ε^*, and $\tan(\delta(t))$ indicate the 'moments of gelation, vitrification, or phase separation during setting of different oligomeric blends. This point of view was supported in [7–13]. Other workers [2, 14] believed that the point of chemical gelation could give no response in dielectric measurements altogether. Johari and co-workers (e.g., see [16, 17]) contributed greatly to resolution of the dilemma [18] and focused attention on the physical features of curing.

The relationship between the ionic and dipole conductivity is known to change during curing. These changes may be so significant that polymer blends undergo transition from the conducting state $[\sigma/(\omega\varepsilon'\varepsilon_0)] > 1$ [19, 20], where $\varepsilon_0 = 8.85\ 10^{-14}$ F/cm the permeability of vacuum, and ω is the angular frequency of electric field) to the dielectric state $(\sigma/(\omega\varepsilon'\varepsilon_0) < 1)$. In these cases, the relaxation maxima may overlap the throughout conductivity. This affects the time dependence of electric properties [9, 21–26] during curing polymer blends. Therefore, the presence or absence of dielectric response directly depends on experimental conditions. At the same time, a chemical process 'does not know' a probing frequency. In view of this, it is important to distinctly separate the ranges of the throughout ionic conductivity (independent of ω) and dipole relaxation (dependent on ω). In [16, 17] and then in [27–31], this was achieved by using the complex electric modulus $M^* = 1/\varepsilon^*$.

In this work, we investigated the dielectric response from three different polymer blends that are cured by polycondensation, polycondensation accompanied by microphase separation, and anionic polymerization mechanisms.

EXPERIMENTAL

The following model polymer blends (PBs) based on diglycidyl ester of bisphenol-A (DGEBA) were investigated. **PB1** was prepared by isothermal (50°C) curing of DGEBA with 4,4'-diaminodicyclohexyl-methane (DA) and cyclohexylamine (A) (in stoichiometric ratio of reacting functional groups). **PB2** was the same as **PB1** but doped with different amounts of butadiene-acrylonitrile rubber end-capped by amino groups (Hycar-16). **PB3** was prepared by curing DGEBA with dimethylbenzylamine (DMBA, 1–6 wt %) at 70°C. Before measurements, the samples were degassed at 40°C for 5 min. The conversion during samples preparation was below 0.05.

The measurements were performed in the frequency range 50 Hz–10 kHz. This range is most attractive because of low electrode effects. Experimental details are described elsewhere [28–30]. Physicochemical characteristics as well as the kinetic and rheokinetic parameters were taken from [32, 33].

RESULTS AND DISCUSSION

Polycondensation

The dependence of $M'' = \mathrm{Im}(1/\varepsilon^*)$ on $M' = \mathrm{Re}(1/\varepsilon^*)$ for **PB1** is presented in Fig. 1. Data points *1* and *2* refer to $f = 1$ and 10 kHz, respectively. Curve *1* can be approximated to a regular semi-circle in case of ionic conductivity and to a distorted semi-circle for relaxation of dipoles. The semi-circle (irrespective of f) is indicative of the Debye relaxation [16, 17]. Within this range, the conductivity in alternating field is identical to that in static field (i.e. to DC-conductivity). At the maximum, $\omega\tau = 1$ (τ is the relaxation time), so that $\tan\delta = M''/M' = \varepsilon''/\varepsilon' \approx 1$. The effective competition of dipole relaxation begins between the semi-circles and the skewed arcs.

Purely ionic conductivity can be analyzed in terms of existing theoretical models. Up to the moment of gelation, the mobility of ions μ_i is assumed [34] to drop sharply ($\mu_i \to 0$). In this case, the behavior of ions at shorter times can be used for *in situ* forecasting (by using ionic labels) and identification of gelation. In the most simple case, the ionic conductivity is only controlled by μ_i rather than by the charge of ions z_i or their concentration n_i (this holds true for numerous PBs [17]). Extrapolation to

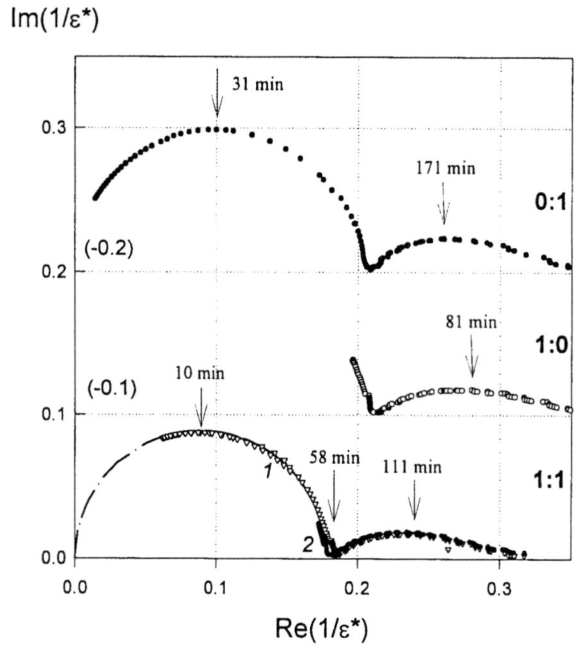

Figure 1. Im(1/ε*) vs. Re(1/ε*) for **PB1** at DA : A = 0 : 1, 1 : 0, and 1 : 1. The data for 0 : 1 и 1 : 0 are shifted along the ordinate axis (figures in parentheses indicate a necessary correction). Data points *1* and *2* refer to $f = 1$ and 10 kHz, respectively. Dash-dotted line for the 1 : 1 data shows the approximating function (semi-circle). Figures at arrows give the reaction time.

zero can be made by using the following power [34] or exponential [35] functions:

$$\sigma(T,t) = \sigma_0(T,t \to 0)\left[\frac{t_g(T) - t(T)}{t_g(T)}\right]^{x(T)} \quad (1)$$

$$\sigma(T,t) = \sigma_0(T,t \to 0)\exp\{-B(T)/[t_0(T) - t]\} \quad (2)$$

where σ_0 is the conductivity at $t \to 0$; T is the cure temperature; t_g, t_0 are the critical points, $x(T)$ is the index of power, and B is a temperature-dependent coefficient.

Functions (1) and (2) ensured good agreement with the experimental data obtained for **PB1** (irrespective of ω). The extrapolated values of t_g and t_0 as a function of amine concentration turned out to be close to those estimated from the viscosity data [32]. This indicates similar nature and frequency characteristics of both the sharp increase in viscosity and the arrested mobility of ions (gelation). However, the conversion corresponding to t_g (and close values of t_0) turned out to be lower than that determined by sol–gel analysis (for details, see [28, 29, 32]). This implies that the t_g value for **PB1** is associated with the 'liquid–physical gel' transition.

Polycondensation Accompanied by Microphase Separation

Addition of special agents to a blend stipulating microphase separation during cure is a well-known technique for modifying polymer properties. These systems are still studied by light scattering, a technique that is applicable only to relatively transparent polymers. The dielectric spectroscopy puts no strict requirements to polymer systems [36]. Since phase separation (crystallization) starts at low conversion, it can be expected that (just as in the case of **PB1**) the ionic conductivity is sensitive to the process of phase separation.

However when phase separation is initiated by a chemical reaction, the assumption of z_i = const and n_i = const may turn out to be invalid: ions are nonuniformly distributed between two phases and their μ_i and n_i values depend on t. In a polymer matrix, $\mu_i \to 0$ at the moment of gelation. But in a

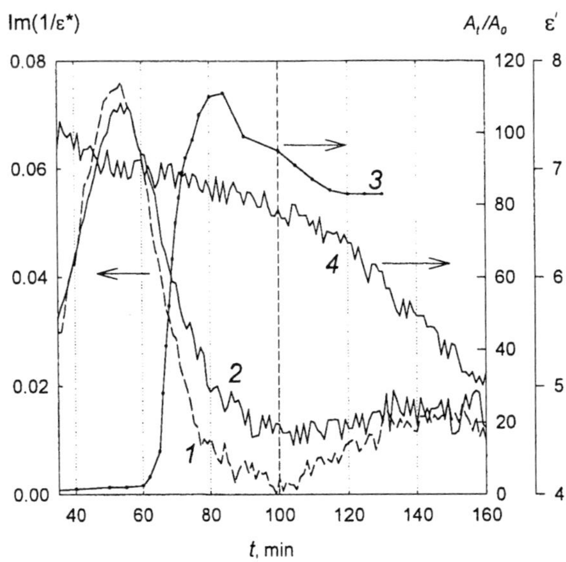

Figure 2. Im(1/ε*) for (*1*) starting blend and (*2*) a blend containing 15.4% Hycar-16; (*3*) parameter of light scattering A_t/A_0, and (*4*) ε' = Re(ε*); f = 50 Hz.

new phase (as a rule, with lower viscosity), variation in μ_i is weaker and governed by other factors.

In general, the contribution of ions from different phases to the overall conductivity is defined by such factors as particle size, phase composition, etc. During curing, the relative significance of these factors changes. This implies that we have to analyze ionic conductivity of a two-phase finely dispersed system, instead of applying the conventional additivity principle. The time dependence of ionic conductivity is difficult to predict because of the lack of detailed theory for field-induced infiltration of ions through interface between fine grains and matrix. However, some simplifying assumptions may sometimes be readily adopted. In particular, we can expect that the contribution from polymer matrix will obey expressions (1) or (2) while that from new phase will change only slightly. In this case, the dielectric response to phase separation (in the form of additional electric loss) can be expected prior to pronounced dipole relaxation [30].

As follows from Fig. 2, sharp increase in light scattering (at $t \approx 65$ min) correlates with an increase in electric loss for the samples containing rubber additive (compare curves *1* and *2*). Similar behavior was observed for the systems with other concentrations of Hycar-16. Since light scattering is caused by phase separation, an increase in electric loss can also be related to this process. This observation confirms the validity of the starting

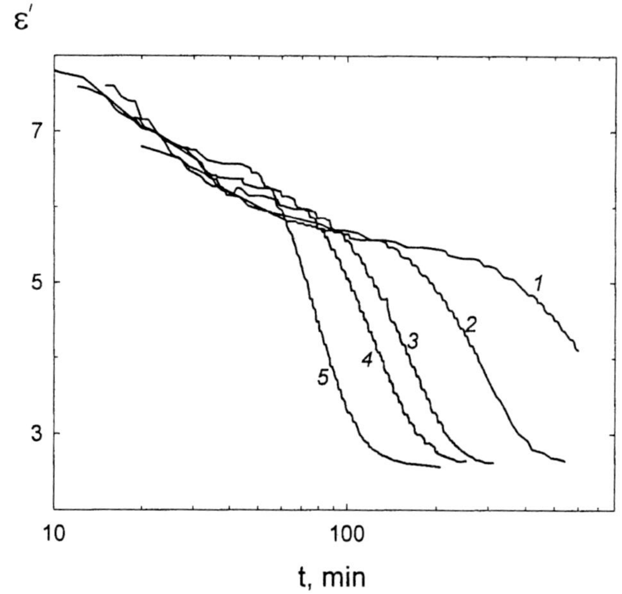

Figure 3. ε' for DGEBA + DMBA system vs. time (figures indicate DMBA concentrations, $f = 1$ kHz): *1*) 1%, *2*) 1.63%, *3*) 2.3%, *4*) 3%, *5*) 5.7%.

assumption of a direct relation between phase separation and electric loss. This result also agrees with [37]. Note (Fig. 2) that phase separation in **PB1** gives rise to a 4-fold increase in M'' at around $t \approx 100$ min (upon approaching t_g), ε' remaining virtually unchanged for $t = 70–100$ min (Fig. 2, curve *4*). In contrast to standard $\tan\delta(t)$ or $\varepsilon'(t)$ measurements, the variation in M'' is reliably detected.

Anionic Polymerization

The behavior of ε' for **PB3** is presented in Fig. 3. It follows that, with increasing DMBA concentration, the position of shoulder shifts toward shorter t (700, 250, 140, 120, and 80 min, respectively).

For DGEBA containing no DMBA, the conductivity is markedly (5-fold) lower. In the presence of DMBA, it attains a value of $2 \cdot 10^{-5}\ \Omega^{-1}\text{m}^{-1}$ (Fig. 4). For all DMBA concentrations, these curves also (just as with $\varepsilon'(t)$) exhibit shoulders whose position is concentration-dependent. Before the shoulder, σ is independent of ω and ω-dependent after the shoulder. This indicates formation of new growing reactive centers, anions [38, 39].

This is also supported by the shape of the $M''(M')$ plots (Fig. 5). It follows that the curves consist of semi-circles (the data can be readily approximated by semi-circles for $\text{Re}(1/\varepsilon^*) < 0{,}15$; in the range of ionic

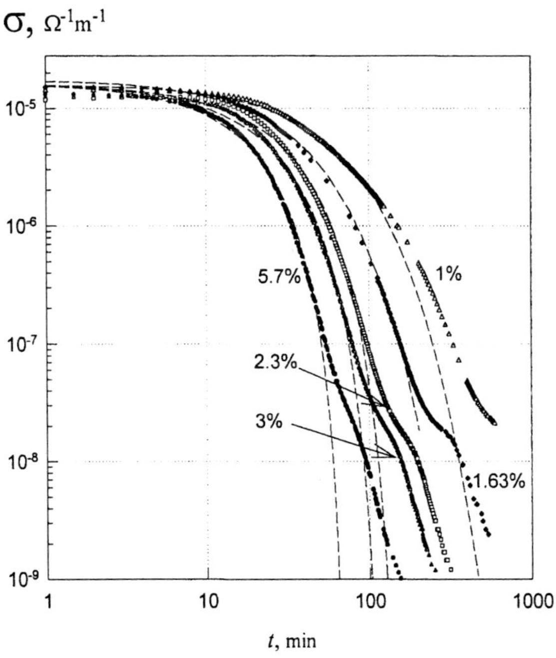

Figure 4. Conductivity of **PB3** vs. time (figures indicate DMBA concentration, f = 50 Hz). Dropped curves are as calculated from power law (1).

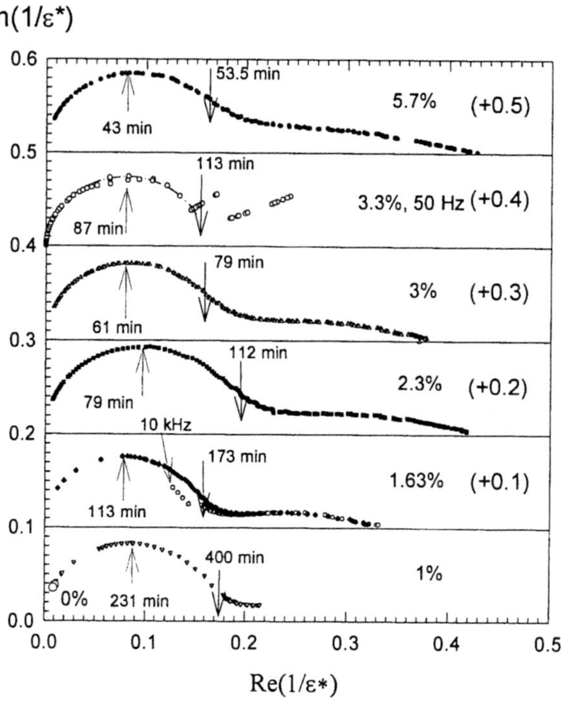

Figure 5. Im(1/ε*) vs. Re(1/ε*) for different DMBA concentration in PB3. The data for 0 : 1 and 1 : 0 Figures in parentheses indicate a shift along the vertical axis; $f = 1$ kHz (except for 1.63 and 3.3 wt%). Arrows indicate maxima, semi-circle diameter, and curing time. Dash-dotted line (in the 3.3% plot) indicates approximating semi-circle (no evacuation of the blend, 70°C).

conductivity) and by skew arcs (in the range of dipole relaxation). The $M''(M')$ plots in Fig. 5 differ from those in Fig. 1 by marked overlapping at the junction, so that the relevant M" values markedly differ from zero. This implies that, in contrast to **PB1**, contribution from the dipole relaxation becomes significant when the ionic conductivity still does not attain zero. In **PB3**, the dipole relaxation manifests itself much earlier. This 'advance' (observed for all DMBA concentration) that these events take place in different sample areas. This 'microheterogeneity' is similar to the known process of microgelation that is known [39] to take place at initial stages (calculated conversion below 0.1). At the same time, the range of 'overlapping is too wide and vaguely pronounced in order to immediately recognize this compliance.

The calculated function $\sigma(t)$ for different **PB3** is presented in Fig. 4. To fit the decline (for details, see [28, 40]), we used the power function (1) (dashed lines) and singular-function (2) (curves not shown). The approximation was performed for $t > 5$ min; the upper values, t^*, were limited by the boundary of ionic conductivity. In calculations, the range $t < t^*$ was gradually compressed until the extrapolated data (t_g and t_0) was independent of t^*. Both the functions gave good agreement with experiment. Within the accuracy of measurement and calculation, t_g and t_0 are independent of ω. Just as for **PB1** and **PB2**, calculated t_g and t_0 turned out to be different: $t_g < t_0$.

The inverse values of t_g as a function of DMBA concentration are presented in Fig. 6 [$t_{cal}^{-1}(t)$ curve]. It follows that $t_{cal}^{-1}(t)$ does not pass through the origin. Probably, this is due to the fact that reaction does not proceed at low DMBA concentration.

Despite good agreement with experiment attained in the case of the power law, let us note the following. Extrapolation turned out to give good results within markedly narrower ranges $t < t^*$. These ranges still further shrink with decreasing DMBA concentration (Fig. 4): compare t values at which calculated curves begin to deviate from those measured. This implies that the competition of the dipole relaxation at low DMBA concentration is less significant than that at high DMBA concentration. This also means that dipoles are formed already at low conversion degree but they *poorly manifest themselves* in dielectric measurements at high DMBA concentration. This conclusion is consistent with the above assumption about the microheterogeneity of the system.

In Fig. 6, the data on t_g (t_{cal}^{-1} curve) are given along with the time required for attaining the critical viscosity t_η (t_η^{-1} curve) and the conversion degree (α_{cal}, и α_η*) as estimated from the data for t_g and t_η. Detailed information about the kinetic and viscosimetric measurements can be found in [31]. If follows that the $t_{cal}^{-1}(t)$ and $t_\eta^{-1}(t)$ curves are close (within measurement accuracy). In other words, calculated t_g correlate with the time of relaxation onset as estimated from viscosimetric data. For [DMBA] = 2.9–5.7%, the extent of conversion (about 0.6) is independent of DMBA concentration. For [DMBA] ≤ 2.3%, the extent of conversion is hard to determine (markedly below 0.4). At [DMBA] = 1%, extrapolation gives a value of 0.1. All this agrees with estimates reported in [39].

For low DMBA concentration, the α values are too low (about 0.4, Fig. 6) to explain formation of chemical macrogel[1]. Interpretation of this transient range (between formation of microgel and chemical macrogel) based on dielectric and viscosimetric data imposes certain difficulties.

Formation of a microgel is accompanied by competition between (1) dipole relaxation and (2) immobilization of growing ions in microgel. These processes overlap in time, the degree of overlapping growing with the amount of microgel. The greater the amount of dipole relaxation oscillators,

[1]During cure, chain transfer to monomer yields hydroxyl groups that may lead to formation of 'physical' network rather than 'chemical' gel [38].

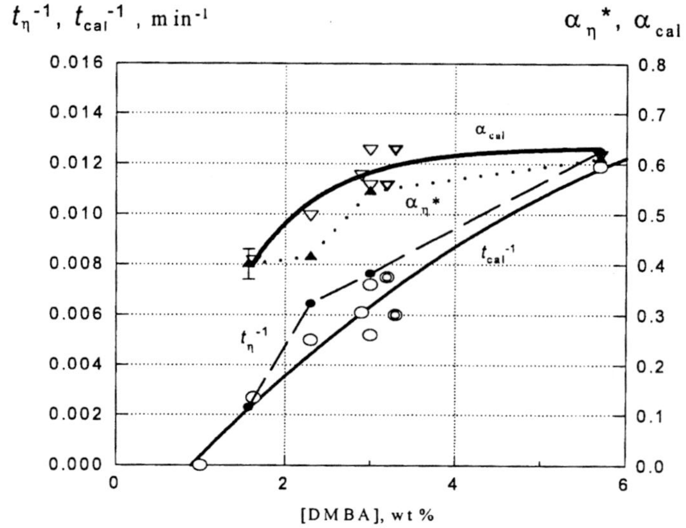

Figure 6. Extrapolated values for: $t_{cal}^{-1}(t)$ (open circles refer to second-order linear regression while double circles, to f = 50 Hz); $t_\eta^{-1}(t)$ viscosimetric data; α_{cal} give the conversion as estimated from t_{cal} (open triangles refer to nonlinear regression of ascending exponential function while double triangles, to f = 50 Hz); and α_η^* is the critical conversion degree as estimated from viscosimetric data—as a function of DMBA concentration.

the larger the amount of growing ions is immobilized in microgel particles. The overlapping gets stronger with increasing concentration of initiator that also increases the rate of both the processes. In view of this, high ionic conductivity can be associated with high mobility of ions in the sol microphase while relaxation of electric dipoles, with microheterogeneous structures in macrogel [41–44].

CONCLUSIONS

It was shown that dielectric measurements can be used for reliable detection of microgel formation, microphase separation, and formation of macrogel (of both physical and chemical origin) during curing the epoxyamine systems by polycondensation or ionic polymerization.

Acknowledgments. This work was supported by the International Science and Technology Center (grant no. 358-96) and the Russian Foundation for Basic Research (project no. 99-03-32397).

REFERENCES

1. Lushcheikin G.A., *Metody issledovaniya elktricheskikh svoistv polimerov* (Methods for Characterization the Electric Properties of Polymers), Khimiya, Moscow, 1988, p. 160.
2. Senturia S.D. and Sheppard N.F., *Adv. Polym. Sci.*, **80**, 1 (1986).
3. Senturia S.D. and Garverick L., US Patent 4 423 371, 1983.
4. Simpson J.O. and Bidsrap S.A., *Proc. ACS of Polymeric Materials Science and Engineering*, New York, 1991, vol. 65, p. 359.
5. Kranbuehl D., Delos S., Yi E., Mayer. J., Hou T., and Winfree W., *XXX Nat. SAMPE Symp.*, 1985, p. 638.
6. Lin C.R. and Hsief, P.Y., *XXXV Int. SAMPE Symp.*, 1990, p. 1233.
7. Zukas W.X., *Macromolecules*, **26**, 2390 (1993).
8. Parthun M.G. and Johari G.P., *Macromolecules*, **26**, 2392 (1993).
9. Boiteux G., Dublineau P., Feve M., Mathieu C., Seytre G., and Ulanski J., *Polymer Bulletin*, **30**, 441 (1993).
10. Mathieu C., Boiteux G., Seytre G., Villain R., and Dublineau P., *J. Non-Cryst. Solids*, **172–174**, 1012 (1994).
11. Bellucci F., Valentino M., Monetta T., Nicodemo L., Kenny J., Nicolais L., and Mijovic J., *J. Polym. Sci., Part B: Polym. Phys.*, **32**, 2519 (1994).
12. Stephan F., Seytre G., Boiteux G., and Ulanski J., *J. Non-Cryst. Solids*, **172–174**, 1001 (1994).
13. Delides C.G., Hayward D., Petrick R.A., and Vatalis A.S., *J. Appl. Polym. Sci.*, **47**, 2037 (1993).
14. Olyphant M., Proc. VI IEEE Electrical Insulation Conf., 1965, Suppl. P12, p. 1.
15. Almdal K., Dyre J., Hvidt S., and Kramer O., *Polym. Gels Networks*, **1**, 5 (1993).
16. Johari G.P., in: *Disorder Effects in Relaxation Processes*, Richert R. and Blumen A., Eds. Springer, Berlin, 1994, p. 627.
17. Johari G.P., in *Chemistry and Technology of Epoxy Resins*, Ellis B., Ed., Blackie and Sons, London, 1993, p. 175.
18. Novikov G.F., Elizarova T.L., Chukalin A.V., Bogdanova L.M., Dzhavadyan E.A., and Rosenberg B.A., *Proc. VI All-Russia Conf. on Structure and Dynamics of Molecular Systems (Yal'chik-99)*, Kazan'–Moscow–Yoshkar-Ola, 1999, p. 206.
19. Kittel C., *Solid State Physics*, Wiley, New York, 1971, pp. 190, 397.
20. Nikol'skii V.V., *Elektrodinamika i rasprostranenie radiovoln* (Electrodynamics and Propagation of Radiowaves), Nauka, Moscow, 1978.
21. Parthun M.G. and Johari G.P., *Macromolecules*, **26**, 2392 (1993).
22. Mathieu C., Boiteux G., Seytre G., Villain R., and Dublineau P., *J. Non-Cryst. Solids*, **172–174**, 1012 (1994).
23. Bellucci F., Valentino M., Monetta T., Nicodemo L., Kenny J., Nicolais L., and Mijovic J., *J. Polym. Sci.; Part B: Polym. Phys.*, **32**, 2519 (1994).
24. Cuve L., Pascault J.P., Boiteux G., and Seytre G., *Polymer*, **32**, 343 (1991).
25. Cuve L., Pascault J.P., and Boiteux G., *Polymer*, **32**, 641 (1991).
26. Boiteux G., Ho-Hoang A., Fache F., Lemaire M., Glowaski I., and Ulanski J., *Synthetic Metals*, **69**, 487 (1995).

27. Alili L, van Turnout J., and te Nijenhuis K., in *Chemical and Physical Networks: Formation and Control of Properties*, te Nijenhuis K. and Mijs W.J., Eds. Wiley, Chichester, 1998, p. 255.
28. Novikov G.F. and Chukalin A.V., *Zh. Fiz. Khim.*, **73**, 1707 (1999).
29. Novikov G.F., Chukalin A.V., Bogdanova L.M., Elizarova T.L., Dzhavadyan E.A., and Rozenberg B.A., *Polymer Science*, in press.
30. Novikov G.F., Elizarova T.L., and Rozenberg B.A., *Zh. Fiz. Khim.*, in press.
31. Novikov G.F., Chukalin A.V., Bogdanova L.M., Elizarova T.L., Dzhavadyan E.A., and Rozenberg B.A., *Polymer Science*, in press.
32. Dzhavadyan E.A., Irzhak V.I., and Rozenberg B.A., *Vysokomol. Soedin., Ser A*, **41**, 624 (1999).
33. Rozenberg B.A., Dzhavadyan E.A., and Irzhak V.I., *Wiley Polymer Networks Group Review*, 1999, vol. 2, in press.
34. Mangion M.B.M. and Johari G.P., *J. Polym. Sci. B: Polym. Phys.*, **28**, 1621 (1990).
35. Parthun M.G. and Johari G.P., *J. Polym. Sci. B: Polym. Phys.*, **30**, 655 (1992).
36. Maistros G.M., Block H., Bucknall C.B., and Partridge, I.K., *Polymer*, **23**, 4470 (1992).
37. Delides C.G., Hayward D., Petrick R.A., and Vatalis A.S., *J. Appl. Polym. Sci.*, **47**, 2037 (1993).
38. Rozenberg B.A., *Adv. Polym. Sci.*, **75**, 113 (1985).
39. Dzhavadyan E.A., Bogdanova L.M., Irzhak V.I., and Rozenberg B.A., *Vysokomol. Soedin., Ser. A*, **39**, 591 (1997) [Engl. Transl. *Polymer Science A*, **39**, 383 (1997)].
40. Novikov G.F., Chukalin A.V., Bogdanova L.M., Elizarova T.L., Dzhavadyan E.A., and Rozenberg B.A., *Polymer Science*, in press.
41. Rozenberg B.A. and Irzhak V.I., in: *Synthesis, Characterization, and Theory of Polymeric Networks and Gels*, Aharoni S., Ed., Plenum Press, New York, 1992, p. 147.
42. Irzhak V.I., Rozenberg B.A., and Enikolopyan N.S., *Setchatye polymery: Sintez, struktura, svoistva* (Network Polymers: Synthesis, Structure, and Properties). Nauka, Moscow, 1979.
43. Rozenberg B.A. and Irzhak V.I., *Macromol. Symp.*, **93**, 227 (1995).
44. Korolev G.V., Mogilevich M.M., and Golikov I.V., *Setchatye poliakrilaty: Mikrogeterogennye struktury, fizicheskie setki, deformatsionno-prochnostnye svoistva* (Network Polyacrylates: Microheterogeneous Structures, Physical Networks, Deformability, and Strength), Khimiya, Moscow, 1995.

Chapter 25

Pulsed NQR for Measuring Internal Stresses in Polymers and Composites

Viktor P. TARASOV, Leonid N. EROFEEV*, Emma A. DZHAVADYAN, Yurii N. SMIRNOV, and Boris A. ROZENBERG

Institute of Problems of Chemical Physics, Russian Academy of Sciences, Chernogolovka, Moscow Region, 142432 Russia

*e-mail: erofeev@icp.ac.ru

ABSTRACT

A new NQR technique for measuring internal stresses in polymer and composites is suggested. Copper protoxide was used as a stress detector (SD). The technique was tested on epoxy resins. Experimental conditions for reliable detection of internal stresses in epoxy polymers and composites are substantiated.

INTRODUCTION

Diagnostics of residual and internal stresses in solids is of current importance for detecting the wear resistance and strength of structural polymer materials. The existing techniques—optical detection by photoelasticity [1], acoustic tensometry [2], dielectric measurements [3], and x-ray diffraction [4]—are applicable largely to uniform and transparent materials. NQR first used in [5] for measuring the shrinkage stress has not however found its practical application. Development of pulsed NQR [6] in our Institute opens new horizons for monitoring internal stresses in polymers and composites. Epoxy composites are being used for manufacturing structural materials and construction units. These polymers are usually cured at elevated temperatures (100–200°C), which results in internal stresses that affect the strength of the resultant material.

In this work, we suggest a pulsed NQR technique for measuring internal stresses.

EXPERIMENTAL

For characterizing the thermoelastic properties of polymer composites, we designed a special NQR accessory for the multipulse NMR spectrometer developed previously [7]. The NQR accessory makes possible operation of the NMR spectrometer in the range 22–30 MHz that is suitable for NQR on the ^{63}Cu nuclei.

Characterization of Apparatus

Peak power	3 kW
Sensitivity of detector (at frequency range 20 kHz)	2 μV
Recovery time	3 μs
Marking frequency	1MHz
Probing pulse duration	5 μs

Measurement Principle

The polymer to be studied is doped with a crystalline stress detector (SD) that contains quadrupole nuclei giving a well-detectable NQR signal. Under the action of internal stresses in a matrix, the dopant crystals are subject to a pressure that shifts the frequency of the NQR signal. If the internal pressure is uniform, the NQR line is shifted without broadening [8]. The shift of NQR line is proportional to the mean value of internal pressure in the material (homogeneous component of pressure). The extent of broadening NQR line is a measure of deviation of the local pressure from its mean value (inhomogeneous component of pressure).

Preparation of Samples

The following two epoxy systems were used for preparation of polymer samples doped with Cu_2O: (**1**) tetraglycidyl ether of 3,3′-dichloro-4,4′-diaminodiphenylmethane cured by isomethyltetrahydropthalic anhydride at 100°C for 3 h, at 130°C for 4 h, and at 160°C for 3 h and (**2**) diglycidyl ether of bisphenol A (DGEBA) cured by polyethylene polyamine at 20°C for 24h and at 100°C for 2 h.

Samples of type A were made from standard prepregs [9] while type B, from binary prepregs upon separate admission of components. The epoxy resin based on DGEBA cured by Polyamine T was used as a binder. Prepregs were prepared by impregnating a glass cloth with mixtures or individual components from solutions in ethanol–acetone mixtures. All samples were cured at 180°C and 7.5 MPa for 2 h.

Choice of Stress Detector

A substance that is to be used as a stress detector should meet the following requirements:

- It must give a well-detectable NQR signal even when added in amounts below several percent.
- It is desirable that the position (ν) of NQR signal be proportional to applied pressure [linearity of $(\partial\nu/\partial p)_T$].
- SD must be chemically inert with respect to epoxy matrix.
- The coefficient of thermal expansion (α) for SD must be close to that of the filler added to epoxy resins.

The above requirements are best met by copper(I) oxide Cu_2O (^{63}Cu nuclei). At room temperature, the NQR spectrum of this compound represents a single line around 26 MHz with a natural line width of 11 kHz. Copper(I) oxide is chemically inert to a polymer matrix and stable up to 1200°C. For Cu_2O, $\alpha = 6\cdot10^{-6}$ K^{-1}, which is close to that of typical fillers (graphite, fiber glass) of epoxy resins. For this compound, $(\partial\nu/\partial p)_T$ is linear and constant over a wide temperature range: $(\partial\nu/\partial p)_T$ = = 3.1 kHz/(kg/mm^2) [1].

RESULTS AND DISCUSSION

Investigation of the Polymer–SD System

We measured the pressure that arise around SD particles (Cu_2O grains 3–15 μm in size) in an epoxy matrix upon its thermal compression in the temperature range between –20°C and +100°C.

As is already known [8], the effect of temperature on the shift of NQR signal (Δν) is much stronger than the effect of pressure. For this reason, we compared the NQR signals from a sample under investigation and from a reference sample (the same Cu_2O but not embedded into epoxy resin) measured at identical temperature. In this way, we determined the line shift and line broadening caused by polymer matrix. As is already known [10], the line shape of NQR signal from a sample can be represented as a superposition of two functions:

$$F(\omega) = G(\omega) \times H(\omega), \tag{1}$$

where $G(\omega)$ is the natural line shape for Cu_2O and $H(\omega)$ is the broadening caused by external influences (in our case, applied pressure). The second momentum of $H(\omega)$ has the form [11]:

$$M_2[H(\omega)] = \int \omega^2 H(\omega) d\omega \tag{2}$$

This expression defines the rms value of internal pressure in the sample.

Functions $G(\omega)$ и $H(\omega)$ are well approximated by the Gaussian curves with close line widths. For this reason, line broadening can be assessed from the expression:

$$M_2[H(\omega)] = M_2[F(\omega)] - M_2[G(\omega)] \tag{3}$$

As follows from Fig. 1, the internal uniform pressure is virtually independent (for $C < 20$ vol %) of the amount of added SD.

Irrespective of C, the nonuniform component of internal pressure grows (Fig. 2). This may be associated with the following factors:

(i) Low size of Cu_2O particles favors formation of the surface developed at the interface with epoxy matrix, which may affect the cure kinetics .

(ii) Cu_2O particles are nonspherical, so that resultant internal stresses in a particle may turn out nonuniform.

(iii) Local deviation from the stoichiometric epoxy/curing agent ratio may give rise to local inhomogeneity in the elasticity of matrix and, hence, to local inhomogeneity of internal stresses in a sample.

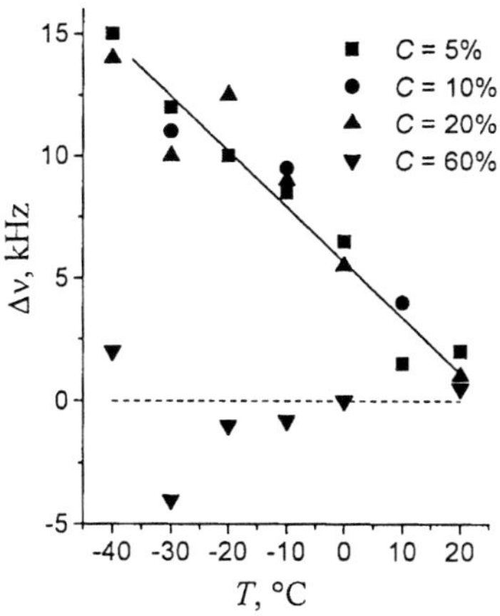

Figure 1. Shift of the NQR signal from cured polymer matrix **1** (uniform internal pressure) doped with Cu_2O powder vs. temperature: $(\partial\nu/\partial p)_T$ = 3.1 Hz/(kg/mm^2), C is given in vol % Cu_2O.

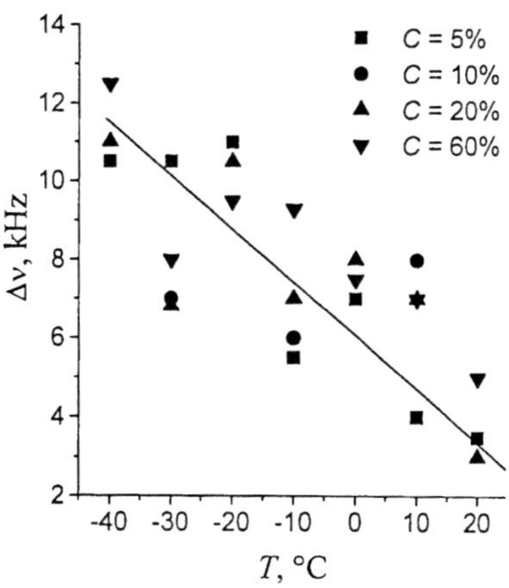

Figure 2. Line width of the NQR signal from polymer matrix **1** (uniform internal pressure) doped with Cu_2O vs. T: $(\partial\nu/\partial p)_T$ = 3.1 kHz/(kg/mm^2), C is given in vol % Cu_2O.

(iv) Cu_2O particles can also carry some amount of air. Since the volume of the carried air is markedly lower than the particle volume, the amount of impurity (particles + air bubbles) in a resin may be expected to be independent of C (for $C < 20$ vol %).

Thorough stirring of mixture prior to addition of Cu_2O can eliminate factor (iii). According to equation (1), internal pressure must be independent of particle size. In order to check this assumption, we performed control experiments with the epoxy resin doped with Cu_2O of different particle sizes. The particles were classified by the time of their deposition from a water suspension and subsequent viewing through a microscope.

The data obtained are given in Fig. 3. The fact the curves are different is indicative of line broadening by a mechanism involving the surface effects at the interface. This is also supported by some difference between the temperature of transition to the viscoelastic state. With decreasing temperature, nonuniform internal pressure diminishes to 3.8 and 4.6 kg/(kg/mm^2) in case (a) and (b), respectively. This can be related to particle size distribution within a given fraction. The difference between uniform and nonuniform internal pressure is lower in case of the narrower fraction of Cu_2O particles.

Another mechanism is operative in the polymer matrix 2 (Fig. 4).

The temperature dependence of uniform pressure (curve a) exhibits three characteristic portions. Above 50°C, p is independent of temperature. This implies that the matrix is in its rubbery state, its Young's modulus is

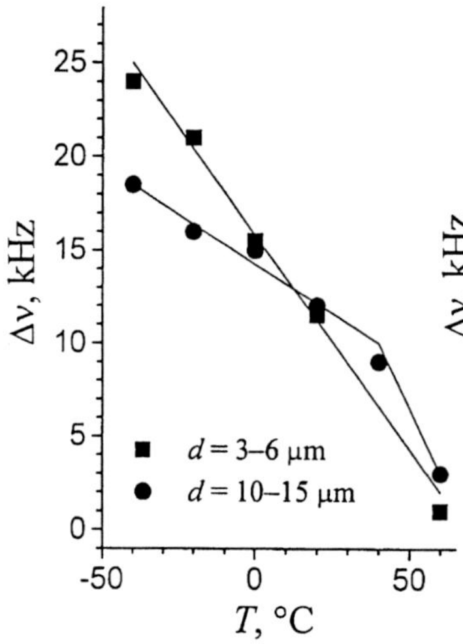

Figure 3. Temperature dependence of internal uniform pressure (shift of NQR line) in polymer matrix **1** filled with Cu_2O of different particle size *d*: $(\partial\nu/\partial p)_T = 3.1$ kHz/(kg/mm^2).

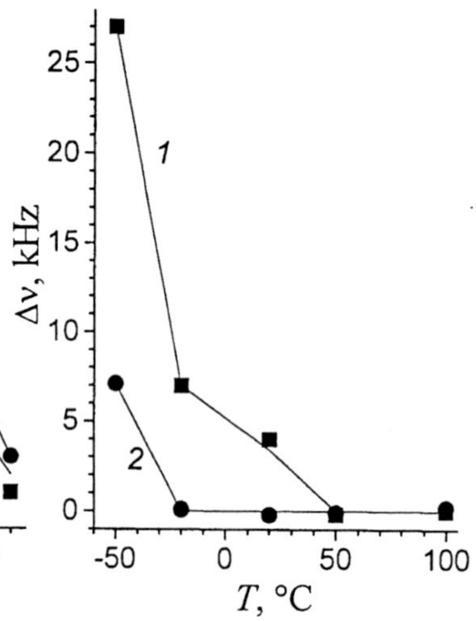

Figure 4. Temperature dependence of uniform and nonuniform pressure in polymer matrix **2** as measured in terms of (*1*) line shift and (*2*) line broadening: $(\partial\nu/\partial p)_T = 3.1$ kHz/(kg/mm^2).

too low to produce noticeable change in p upon cooling-related shrinkage. In the range between +50°C and –20°C, p increases with decreasing T, the nonuniform component remaining virtually unchanged. (Fig. 4, curve *a*). This means that internal stresses are relaxed within the range of particle size. Below –20°C the matrix becomes rigid, so that the NQR line undergoes sharp broadening.

Internal Stresses in Composites

In preliminary experiments, formation (in contrast to type A composite) of highly and rarely crosslinked areas in the type B composite was detected by spin-locking NMR [12], dilatometry, and thermomechanical analysis [13]. Being embedded into a polymer matrix with a highly crosslinked microstructure, rarely crosslinked areas may improve the physico-mechanical properties of type B composites by relaxing internal stresses arising in the polymer matrix.

To check this possibility, we measured (by NQR) the internal stresses arising because of the different thermal expansions of glass and of polymer ($\alpha = 40\cdot10^{-6}$ K^{-1}). In the range between –20 and +100°C, we determined the uniform and nonuniform components of p. Just as for the polymer–powder system [14], the nonuniform component of p was calculated from expression (4).

In contrast to the polymer–Cu_2O system, the uniform portion of p is independent of temperature within measurement error while the nonuniform component exhibits a well-pronounced temperature dependence (Fig. 5). In type A composite (curve *1*), internal pressure is higher than that in type B composite (curve *2*) over the entire temperature range studied. The polymer matrix undergoes the glass transition at 100°C for type A composite and at 60°C for type B composite. Figure 5 also presents the values of p measured at 40° and 60°C after a 5-min annealing at 150°C. The effect of particle size was investigated for the samples of type B composite cured at 130°C and then annealed at 150°C for 5 min.

The temperature dependence of nonuniform stresses in type B composite is illustrated in Fig. 6. Within the measurement accuracy, these

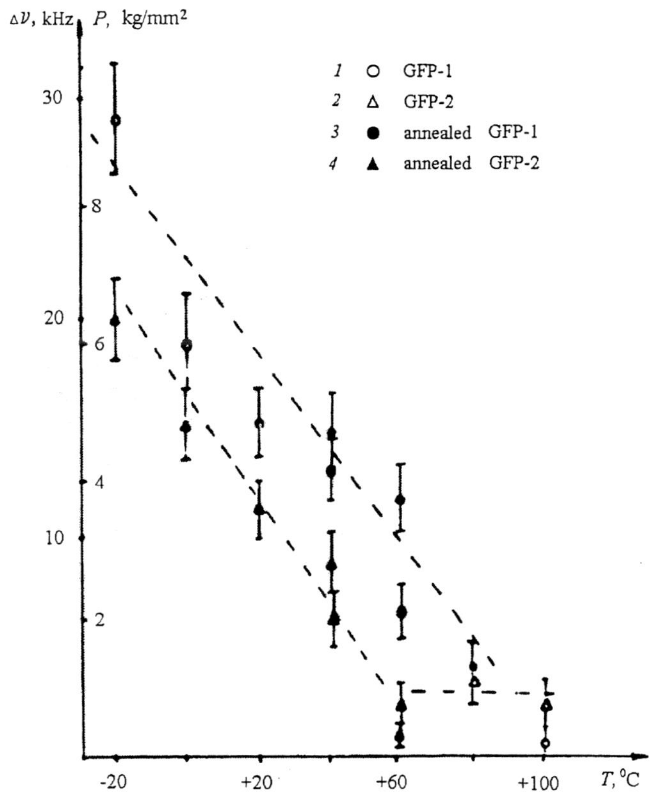

Figure 5. Temperature dependence of the NQR linewidth (nonuniform pressure) for Cu_2O-doped composites: (*1*) type A composite, (*2*) type B composite, (*3*) annealed type A composite, and (*4*) annealed type B composite. $(\partial\nu/\partial p)_T$ = 3.1 kHz/(kg/mm²), T = 130°C.

functions coincide and are below curve *2* in Fig. 5. Comparing the data in Figs. 5 and 6, we conclude that the observed line broadening reflects distribution of local internal stresses in polymer matrix is not affected by inhomogeneity of stresses in a given Cu_2O particle. Using the expression

$$H(\omega) = F(t)/G(t), \tag{5}$$

where $F(t)$ and $G(t)$ are the Fourier transforms of $F(\omega)$ and $G(\omega)$, we can obtain the distribution of internal stresses that coincides with the function $H(\omega)$. In this case, the value of p is defined by the difference between the α values for Cu_2O and polymer matrix and a pressure produced due to the adhesion between matrix and fiber. It follows from [15], that these components are nearly identical in their magnitude but act in opposite directions.

The distributions of nonuniform internal stresses in type B composite (cured at 130°C) for three different temperatures are illustrated in Fig. 7. In

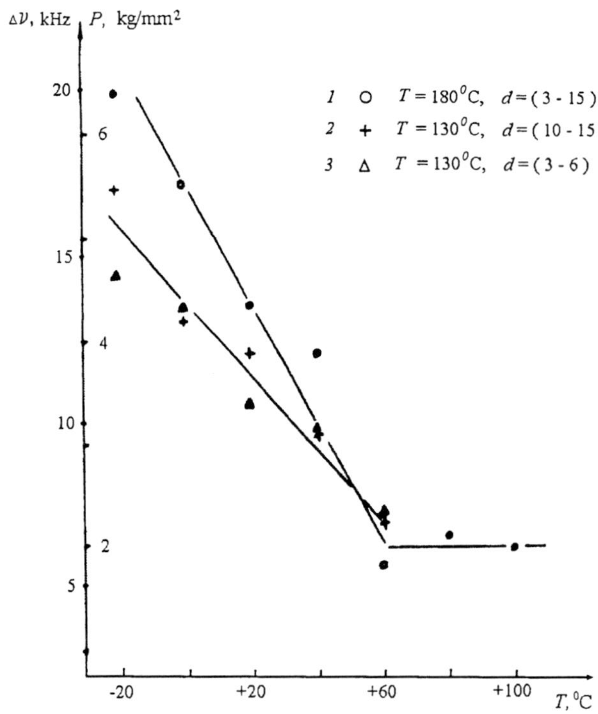

Figure 6. Temperature dependence of nonuniform pressure in type B composite doped with Cu_2O: (*1*) cure temperature $T_c = 180°C$, Cu_2O particle size d = 3–15 μm; (*2*) $T_c = 130°C$, d = 10–15 μm; and (*3*) $T_c = 130°C$, d = 3–6 μm; $(\partial\nu/\partial p)_T = 3.1$ kHz/(kg/mm²).

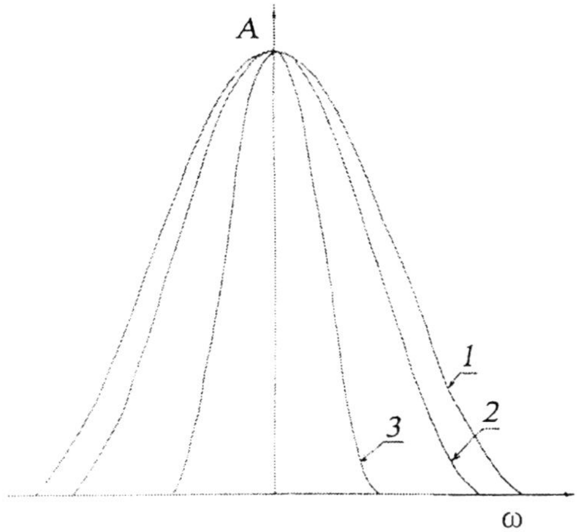

Figure 7. Distribution of internal stresses in type B composite, measured at the following temperatures: (*1*) – 0°C; (*2*) – 20°C;. (*3*) – 60°C.

this case, NQR is unable to determine the uniform component of internal stresses, that is, the shift of the lines shown in Fig. 7 with respect to the origin. Using a stress detector whose α is identical to that of polymer can solve this problem. Nevertheless, comparative analysis of internal stresses in materials with different polymer matrix can be performed by using only the nonuniform component of internal stresses.

CONCLUSIONS

Internal thermal stresses in structural epoxy-based materials were investigated by NQR. For the 'polymer–SD' system, the amplitude of internal stresses was measured as a function of SD particle size and SD content. The mean pressure arising around SD particles in polymer matrixes as well as the distribution of pressure in a sample also were determined.

Copper oxide powder ($d \leq 15$ μm) added to polymer in amounts up to 20% was found to be a suitable stress indicator that does not affect the final result. The technique was used for monitoring internal stresses in conventional (type A) and specially prepared (type B) fiberglass plastics. Internal stresses in type B composites were found to be lower because of a lower-glass transition temperature. The distribution of internal stresses in a sample was determined experimentally. The thermoelastic properties of polymer matrix in type A and type B composites as well as the spatial distribution of internal stresses were found to depend on the temperature of curing.

Acknowledgments. This work was supported by the International Science and Technology Center (grant no. 358-96).

REFERENCES

1. Collacot R.A., Mechanical Fault Diagnosis and Condition Monitoring, Chapman and Hall, London, 1977.
2. Bobrenko V.M., Averbukh I.I., and Chugunov A.A., *Defektoskopiya*, No. 1, 72 (1974).
3. Guz' A.N., Garf M.E., and Malashenko S.V., *Nerazrushayushchii kontrol' materialov i elementov konstruktsii* (Nondestructive Defectoscopy of Materials and Construction Units), Naukova Dumka, Kiev, 1981.
4. Predecki P. and Barrett Ch.S., *J. Comp. Mater.*, **13**, 61 (1979).
5. Gutowsky H.S., Bernheim R.A., and Tarkow H., *J. Polym. Sci.*, **44**, 143 (1960).
6. Hitrin A.K., Karnauch G.E., and Provotorov B.N., *J. Mol. Struct.*, **83**, 269 (1982).
7. Erofeev L.N., Vetrov O.D., Isaev M.Sh., and Manelis G.B., *Prib. Tekh. Eksp.*, No. 2, 145 (1977).
8. Kushida T., Benedek G.B., and Blombergen N., *Phys. Rev.*, **104**, 1364 (1956).
9. Natrusov V.I., Shatskaya T.E., Lapitskii V.A., *et al.*, *Mekh. Comp. Mater.*, No. 2, 315 (1987).
10. Azheganov A. and Ainbinder N., *X Int. Symp. on Nuclear Quadrupole Resonance Spectroscopy*, Tokyo, 1989, Vol. 3, p. 37.
11. Andrew E.R., *Nuclear Magnetic Resonance*, Cambridge University Press, Cambridge (UK), 1955, Ch. 6.
12. Tarasov V.P., Smirnov Yu.N., Shatskaya, T.E., *et al.*, *Vysokomolek. Soedin., Ser. B*, **31**, 428 (1994).
13. Smirnov Yu.N., Magomedov, G.M., Shatskaya T.E., and Natrusov V.I., *Vysokomolek. Soedin., Ser. A*, **38**, 1717 (1996). Ermolaev K.V., Tarasov V.P., Erofeev L.N., *et al.*, *Vysokomolek. Soedin., Ser. B*, **40**, 1715 (1998).
14. Landau L.D. and Lifshits E.M., *Teoriya uprugosti* (Theory of Elasticity), Nauka, Moscow, 1965.

SUBJECT INDEX